Java

8th Edition

by Barry Burd

Java® For Dummies®, 8th Edition

Published by: **John Wiley & Sons, Inc.**, 111 River Street, Hoboken, NJ 07030-5774, www.wiley.com

For general information on our other products and services, please contact our Customer Care Department within the U.S. at 877-762-2974, outside the U.S. at 317-572-3993, or fax 317-572-4002. For technical support, please visit https://hub.wiley.com/community/support/dummies.

Wiley publishes in a variety of print and electronic formats and by print-on-demand. Some material included with standard print versions of this book may not be included in e-books or in print-on-demand. If this book refers to media such as a CD or DVD that is not included in the version you purchased, you may download this material at http://booksupport.wiley.com. For more information about Wiley products, visit www.wiley.com.

Library of Congress Control Number: 2022932287

ISBN: 978-1-119-86164-5; 978-1-119-86165-2 (ebk); 978-1-119-86168-3 (ebk)

SKY10032668_022422

Contents at a Glance

Table of Contents

Introduction

What's all the fuss about Java? To help answer that question, I offer a few facts:

- More than a third of the world's programmers use Java as one of their primary programming languages. That's at least 5.2 million programmers. And Java powers more than 52 percent of the world's back-end web services.*
- Ninety percent of all Fortune 500 companies use Java.**
- Websites that use Java include Google, YouTube, LinkedIn, Amazon, and eBay.***
- In 2021, Glassdoor, Inc., ranked jobs based on earnings potential, job satisfaction, and number of available job openings. Among the company's "50 Best Jobs in America for 2021," a career as a Java developer ranked number one.****

Sounds good. Right?

Please, read on.

About This Book

This book isn't the usual dry techie guide. It's written for normal human beings — people with little or no programming experience. In this book, I divide Java into manageable chunks. Each chunk is (more or less) a chapter on its own. I explain concepts in plain language using complete code examples that you can download

---

* https://blog.jetbrains.com/idea/2020/09/a-picture-of-java-in-2020

** https://blogs.oracle.com/oracleuniversity/post/why-does-java-remain-so-popular

*** www.frgconsulting.com/blog/why-is-java-so-popular-developers

**** www.glassdoor.com/List/Best-Jobs-in-America-LST_KQ0,20.htm

and run. I keep each code example focused on a few key concepts. I resist the urge to use fancy tricks that impress professional programmers. I expand on concepts that may be difficult for newcomers. I add diagrams to help you visualize important ideas. I provide exercises with each chapter along with solutions to the exercises on the book's website.

Finally, and most importantly — and without question the most significant of all this book's features — I throw in some jokes. I've written some good jokes and lots of bad jokes. (I should say "lots and lots" of bad jokes.) I've hidden Easter eggs in the text. I've added anecdotes about all kinds of topics. Some of the anecdotes are true, and many of them are . . . well, you figure it out.

Foolish Assumptions

In this book, I make a few assumptions about you, the reader. If one of these assumptions is incorrect, you're probably okay. If all these assumptions are incorrect, please buy the book anyway:

- **I assume that you have access to a computer.** Here's the good news: You can run most of the code in this book on almost any computer. The only computers you can't use to run this code are ancient boxes that are more than ten years old (give or take a few years).
- **I assume that you can navigate your computer's common menus and dialog boxes.** You don't have to be a Windows, Linux, or Macintosh power user, but you should be able to start a program, find a file, put a file into a certain directory — that sort of thing. Most of the time, when you follow instructions in this book, you're typing code on the keyboard, not pointing-and-clicking the mouse.
- **I assume that you can think logically.** That's all there is to programming in Java — thinking logically. If you can think logically, you have it made. If you don't believe that you can think logically, read on. You may be pleasantly surprised.
- **I make few assumptions about your computer programming experience (or your lack of such experience).** In writing this book, I've tried to do the impossible: Make the book interesting for experienced programmers yet accessible to people with little or no programming experience. So I assume no particular programming background on your part. If you've never created a loop or indexed an array, that's okay.

On the other hand, if you've done these things (maybe in Visual Basic, Python, or C++), you'll discover some interesting plot twists in Java. The developers of Java took the best ideas in object-oriented programming, streamlined them, reworked them, and reorganized them into a sleek, powerful way of thinking about problems. You'll find many new, thought-provoking features in Java. As you find out about these features, many of them will seem quite natural to you. One way or another, you'll feel good about using Java.

Icons Used in This Book

If you could watch me write this book, you'd see me sitting at my computer, talking to myself. I say each sentence in my head. Most of the sentences, I mutter several times. When I have an extra thought or a side comment that doesn't belong in the regular stream, I twist my head a little bit. That way, whoever's listening to me (usually, nobody) knows that I'm off on a momentary tangent.

Of course, in print, you can't see me twisting my head. I need some other way to set a side thought in a corner by itself. I do it with icons. When you see a Tip icon or a Remember icon, you know that I'm taking a quick detour.

Here's a list of icons that I use in this book:

TIP

A tip is an extra piece of information — a helpful tidbit that the other books may forget to tell you.

WARNING

Everyone makes mistakes. Heaven knows that I've made a few in my time. Anyway, when I think people are especially prone to make a mistake, I mark it with a Warning icon.

REMEMBER

Sometimes I want to hire a skywriting airplane crew. "Barry," says the white smoky cloud, "if you want to compare two numbers, use the double equal sign. Please don't forget to do this." Because I can't afford skywriting, I have to settle for a more modest option: I create a paragraph marked with the Remember icon.

CROSS REFERENCE

"If you don't remember what such-and-such means, see blah-blah-blah," or "For more information, read blahbity-blah-blah."

TRY IT OUT

Writing computer code is an activity, and the best way to learn an activity is to practice it. That's why I've created things for you to try in order to reinforce your knowledge. Many of these are confidence-builders, and some are more challenging. When you first start putting concepts into practice, you'll discover all kinds of issues, quandaries, and roadblocks that didn't occur to you when you started reading about the material. But that's a good thing. Keep at it! Don't become frustrated. Or, if you do become frustrated, visit this book's website (`http://javafordummies.allmycode.com`) for hints and solutions.

This icon calls attention to useful material that you can find online. Check it out!

TECHNICAL STUFF

Occasionally, I run across a technical tidbit. The tidbit may help you understand what the people behind the scenes (the people who developed Java) were thinking. You don't have to read it, but you may find it useful. You may also find the tidbit helpful if you plan to read other (geekier) books about Java.

Beyond the Book

In addition to what you're reading right now, this book comes with a free, access-anywhere Cheat Sheet containing code that you can copy and paste into your own Java program. To get this Cheat Sheet, simply go to `www.dummies.com` and type **Java For Dummies Cheat Sheet** in the Search box.

Where to Go from Here

If you've gotten this far, you're ready to start reading about Java application development. Think of me (the author) as your guide, your host, your personal assistant. I do everything I can to keep things interesting and, most importantly, to help you understand.

If you like what you read, send me a note. My email address, which I created just for comments and questions about this book, is `JavaForDummies@allmycode.com`. If email and chat aren't your favorites, you can reach me instead on Twitter (`@allmycode`) and on Facebook (`www.facebook.com/allmycode`). And don't forget — for the latest updates, visit this book's website. The site's address is `http://javafordummies.allmycode.com`.

1

Getting Started with Java

IN THIS PART . . .

Install the software you need for developing Java programs.

Find out how Java fits into today's technology scene.

Run your first complete Java program.

» What Java is

» Where Java came from

» Why Java is so cool

» How to orient yourself to object-oriented programming

Chapter 1
All about Java

Say what you want about computers. As far as I'm concerned, computers are good for just two simple reasons:

- **When computers do work, they feel no resistance, no stress, no boredom, and no fatigue.** Your computer can work 24/7 making calculations for `www.climateprediction.net` — a distributed computing project to model the world's climate change. Or, have your computer crunch numbers for `Rosetta@home` — a site that models proteins to help cure major illnesses. Will you feel sorry for my computer because it's working so hard? Will the computer complain? No.

  You can make demands, give the computer its orders, and crack the whip. Will you (or should you) feel the least bit guilty? Not at all.

- **Computers move ideas, not paper.** Not long ago, whenever you wanted to send a message to someone, you hired a messenger. The messenger mounted a horse and delivered your message personally. The message was recorded on paper or parchment or a clay tablet or whatever other physical medium was available at the time.

  This whole process seems wasteful now, but that's only because you and I are sitting comfortably in the electronic age. Messages are ideas, and physical objects like ink, paper, and horses have little or nothing to do with real ideas; they're just temporary carriers for ideas (even though people used them for several centuries to carry ideas). Nevertheless, the ideas themselves are paperless, horseless, and messengerless.

> The neat thing about computers is that they carry ideas efficiently. They carry nothing but the ideas, a couple of photons, and some electrical power. They do this with no muss, no fuss, and no extra physical baggage.

When you start dealing efficiently with ideas, something very nice happens: Suddenly, all overhead is gone. Instead of pushing paper and trees, you're pushing numbers and concepts. Without the overhead, you can do things much faster and do things that are far more complex than ever.

What You Can Do with Java

It would be nice if all this complexity were free, but, unfortunately, it isn't. Someone has to think hard and decide exactly what to ask the computer to do. After that thinking takes place, someone has to write a set of instructions for the computer to follow.

Given the current state of affairs, you can't write these instructions in English or any other language that people speak. Science fiction is filled with stories about people who make simple requests of robots and get back disastrous, unexpected results. English and other such languages are unsuitable for communication with computers, for several reasons:

- **An English sentence can be misinterpreted.** "Chew one tablet three times a day until finished."
- **It's difficult to weave a complicated command in English.** "Join flange A to protuberance B, making sure to connect only the outermost lip of flange A to the larger end of the protuberance B while joining the middle and inner lips of flange A to grommet C."
- **An English sentence has lots of extra baggage.** "Sentence has unneeded words."
- **English can be difficult to interpret.** "As part of this Publishing Agreement between John Wiley & Sons, Inc. ('Wiley') and the Author ('Barry Burd'), Wiley shall pay the sum of one-thousand-two-hundred-fifty-seven dollars and sixty-three cents ($1,257.63) to the Author upon submittal of *Java For Dummies,* 8th Edition ('the Work') either in whole or in part as determined by Clause 9 in Section 16 of this agreement or its subsequent amendments under the laws of the State of Indiana."

To tell a computer what to do, you have to use a special language to write terse, unambiguous instructions. A special language of this kind is called a *computer programming language*. A set of instructions written in such a language is called a *program*. When looked at as a big blob, these instructions are called *software* or *code*. Here's what code looks like when it's written in Java:

```
public class PayBarry {

    public static void main(String args[]) {
        double checkAmount = 1257.63;
        System.out.print("Pay to the order of ");
        System.out.print("Dr. Barry Burd ");
        System.out.print("$");
        System.out.println(checkAmount);
    }
}
```

Why You Should Use Java

It's time to celebrate! You've just picked up a copy of *Java For Dummies*, 8th Edition, and you're reading Chapter 1. At this rate, you'll be an expert Java programmer* in no time at all, so rejoice in your eventual success by throwing a big party.

To prepare for the party, I'll bake a cake. I'm lazy, so I'll use a ready-to-bake cake mix. Let me see: Add water to the mix and then add butter and eggs — hey, wait! I just looked at the list of ingredients. What's MSG? And what about propylene glycol? That's used in antifreeze, isn't it?

I'll change plans and make the cake from scratch. Sure, it's a little harder, but that way, I get exactly what I want.

Computer programs work the same way: You can use somebody else's program or write your own. If you use somebody else's program, you use whatever you get. When you write your own program, you can tailor the program especially for your needs.

---

* In professional circles, a developer's responsibilities are usually broader than those of a programmer. But, in this book, I use the terms *programmer* and *developer* almost interchangeably.

Writing computer code is a big, worldwide industry. Companies do it, freelance professionals do it, hobbyists do it — all kinds of people do it. A typical big company has teams, departments, and divisions that write programs for the company. But you can write programs for yourself or for someone else, for a living or for fun. In a recent estimate, the number of lines of code written each day by programmers in the world exceeds the number of methane molecules on the planet Jupiter.** Take almost anything that can be done with a computer — with the right amount of time, you can write your own program to do it. (Of course, the "right amount of time" may be quite long, but that's not the point. Many interesting and useful programs can be written in hours or even minutes.)

Gaining Perspective: Where Java Fits In

Here's a brief history:

- **1954–1957: FORTRAN is developed.**

  FORTRAN was the first modern computer programming language. For scientific programming, FORTRAN is a real racehorse. Year after year, FORTRAN is a leading language among computer programmers throughout the world.

- **1959: Grace Hopper at Remington Rand develops the COBOL programming language.**

  The letter *B* in COBOL stands for *Business,* and business is just what COBOL is all about. The language's primary feature is the processing of one record after another, one customer after another, or one employee after another.

  Within a few years after its initial development, COBOL became the most widely used language for business data processing.

- **1972: Dennis Ritchie at AT&T Bell Labs develops the C programming language.**

  The "look and feel" that you see in this book's examples comes from the C programming language. Code written in C uses curly braces, `if` statements, `for` statements, and other elements.

  In terms of power, you can use C to solve the same problems that you can solve by using FORTRAN or Java or any other modern programming language. (You can write a scientific calculator program in COBOL but doing that sort of thing would feel quite strange.) The difference between one programming

** I made up this fact all by myself.

language and another isn't power — the difference is ease and appropriateness of use. That's where the Java language excels.

- **1986: Bjarne Stroustrup (also at AT&T Bell Labs) develops C++.**

  Unlike its C language ancestor, the language C++ supports object-oriented programming. This support represents a huge step forward. (See the next section in this chapter.)

- **May 23, 1995: Sun Microsystems releases its first official version of the Java programming language.**

  Java improves upon the concepts in C++. Java not only supports object-oriented programming but also *enforces the use of* object-oriented programming.

  Additionally, Java is a great general-purpose programming language. A program written in Java runs seamlessly on all major platforms, including Windows, Macintosh, and Linux. With Java, you can write windowed applications, build and explore databases, control handheld devices, and more. Within five short years, the Java programming language has 2.5 million developers worldwide. (I know — I have a commemorative T-shirt to prove it.)

- **November 2000: Java goes to school.**

  In the US, the College Board announces that, starting in the year 2003, the Computer Science Advanced Placement exams will be based on Java.

- **2004: Java is the top language on the world-famous TIOBE Index, and stays on top for the next 15 years.**

- **Also in 2004: Java goes into space!**

  A robotic rover, named *Spirit,* runs Java code to explore Mars.

- **January 2010: Oracle Corporation purchases Sun Microsystems, bringing Java technology into the Oracle family of products.**

- **August 2017: Oracle announces its plan to release new versions of Java every six months.**

  Until then, new Java versions became available once every few years. But the release of Java 9 in September 2017 is followed by the rollout of Java 10 in March 2018. Up next is Java 11 in September 2018.

  In September 2021, Java 17 is a *l*ong-*t*erm *s*upport (LTS) release. This means that Oracle promises to keep Java running smoothly until at least September 2026. These LTS releases come every two years, so the next rock-solid, take-no-prisoners version of Java is Java 21 in September 2023.

  The new release cycle has injected energy into the evolution of the Java programming language.

- **May 2020: Java celebrates its 25th birthday.**

Java technology powers applications of companies like Netflix, Alibaba, Tinder, Uber, PayPal, the *New York Times*, Pinterest, Slack, Shopify, Twitter, Target, and Wells Fargo.* The job search site Monster.com says:

> "Java is one of the most popular programming languages in use, so it's no surprise it came in as the No. 1 skill tech companies were looking for. According to Oracle, 3 billion mobile phones run Java, along with 125 million TV devices and 89% of desktop computers in the U.S. Java is everywhere and the demand for strong developers is high." **

Well, I'm impressed.

Object-Oriented Programming (OOP)

It's three in the morning. I'm dreaming about the history course I failed in high school. The teacher is yelling at me, "You have two days to study for the final exam, but you won't remember to study. You'll forget and feel guilty, guilty, guilty."

Suddenly, the phone rings. I'm awakened abruptly from my deep sleep. (Sure, I disliked dreaming about the history course, but I like being awakened even less.) At first, I drop the telephone on the floor. After fumbling to pick it up, I issue a grumpy, "Hello, who's this?" A voice answers, "I'm a reporter from the Reuters news agency. I'm writing an article about Java, and I need to know all about the programming language in five words or less. Can you explain it?"

My mind is too hazy. I can't think. So I say the first thing that comes to my mind and then go back to sleep.

Come morning, I hardly remember the conversation with the reporter. In fact, I don't remember how I answered the question. Did I utter a few obscenities and then go back to sleep?

---

* Sources:

www.softwaretestinghelp.com/real-world-applications-of-java, https://newrelic.com/blog/nerd-life/what-you-can-do-with-java,https://vaadin.com/blog/the-state-of-java, https://discovery.hgdata.com/product/spring-boot

** www.monster.com/career-advice/article/programming-languages-you-should-know

I put on my robe and rush out to my driveway. As I pick up the morning paper, I glance at the front page and see this 2-inch headline:

> **Burd Calls Java "A Great Object-Oriented Language"**

Object-oriented languages

Java is object-oriented. What does that mean? Unlike languages, such as FORTRAN, that focus on giving the computer imperative "Do this/Do that" commands, object-oriented languages focus on data. Of course, object-oriented programs still tell the computer what to do. They start, however, by organizing the data, and the commands come later.

Object-oriented languages are better than "Do this/Do that" languages because they organize data in a way that helps people do all kinds of things with it. To modify the data, you can build on what you already have rather than scrap everything you've done and start over each time you need to do something new. Although computer programmers are generally smart people, they took a while to figure this out. For the full history lesson, see the nearby sidebar, "The winding road from FORTRAN to Java" (but I won't make you feel guilty if you don't read it).

THE WINDING ROAD FROM FORTRAN TO JAVA

In the mid-1950s, a team of people created a programming language named FORTRAN. It was a good language, but it was based on the idea that you should issue direct, imperative commands to the computer. "Do this, computer. Then do that, computer." (Of course, the commands in a real FORTRAN program were much more precise than "Do this" or "Do that.")

In the years that followed, teams developed many new computer languages, and many of the languages copied the FORTRAN "Do this/Do that" model. One of the more popular "Do this/Do that" languages went by the 1-letter name *C*. Of course, the "Do this/Do that" camp had some renegades. In languages named SIMULA and Smalltalk, programmers moved the imperative "Do this" commands into the background and concentrated on descriptions of data. In these languages, you didn't come right out and say, "Print a list of delinquent accounts." Instead, you began by saying, "This is what it means to be an account. An account has a name and a balance." Then you said, "This is how you ask an account whether it's delinquent." Suddenly, the data became king. An account was a thing that had a name, a balance, and a way of telling you whether it was delinquent.

(continued)

(continued)

Languages that focus first on the data are called *object-oriented* programming languages. These object-oriented languages make excellent programming tools. Here's why:

- Thinking first about the data makes you a good computer programmer.
- You can extend and reuse the descriptions of data over and over again. When you try to teach old FORTRAN programs new tricks, however, the old programs show how brittle they are. They break.

In the 1970s, object-oriented languages, such as SIMULA and Smalltalk, were buried in the computer hobbyist magazine articles. In the meantime, languages based on the old FORTRAN model were multiplying like rabbits.

So, in 1986 a fellow named Bjarne Stroustrup created a language named C++. The C++ language became popular because it mixed the old C language terminology with the improved object-oriented structure. Many companies turned their backs on the old FORTRAN/C programming style and adopted C++ as their standard.

But C++ had a flaw. Using C++, you could bypass all the object-oriented features and write a program by using the old FORTRAN/C programming style. When you started writing a C++ accounting program, you could take either fork in the road:

- Start by issuing direct "Do this" commands to the computer, saying the mathematical equivalent of "Print a list of delinquent accounts, and make it snappy."
- Choose the object-oriented approach and begin by describing what it means to be an account.

Some people said that C++ offered the best of both worlds, but others argued that the first world (the world of FORTRAN and C) shouldn't be part of modern programming. If you gave a programmer an opportunity to write code either way, that person would too often choose to write code the wrong way.

So, in 1995 James Gosling of Sun Microsystems created the language named Java. In creating Java, Gosling borrowed the look and feel of C++. But Gosling took most of the old "Do this/Do that" features of C++ and threw them in the trash. Then he added features that made the development of objects smoother and easier. All in all, Gosling created a language whose object-oriented philosophy is pure and clean. When you program in Java, you have no choice but to work with objects. That's the way it should be.

Objects and their classes

In an object-oriented language, you use objects *and* classes to organize your data.

Imagine that you're writing a computer program to keep track of the houses in a new condominium development (still under construction). The houses differ only slightly from one another. Each house has a distinctive siding color, an indoor paint color, a kitchen cabinet style, and so on. In your object-oriented computer program, each house is an object.

But objects aren't the whole story. Although the houses differ slightly from one another, all the houses share the same list of characteristics. For instance, each house has a characteristic known as *siding color*. Each house has another characteristic known as *kitchen cabinet style*. In your object-oriented program, you need a master list containing all characteristics that a house object can possess. This master list of characteristics is called a *class*.

So there you have it. Object-oriented programming is misnamed. It should be called "programming with classes and objects."

Now notice that I put the word *classes* first. How dare I do this! Well, maybe I'm not so crazy. Think again about a housing development that's under construction. Somewhere on the lot, in a rickety trailer parked on bare dirt, is a master list of characteristics known as a *blueprint*. An architect's blueprint is like an object-oriented programmer's class. A blueprint is a list of characteristics that each house will have. The blueprint says "siding." The actual house object has gray siding. The blueprint says "kitchen cabinet." The actual house object has Louis XIV kitchen cabinets.

The analogy doesn't end with lists of characteristics. Another important parallel exists between blueprints and classes. A year after you create the blueprint, you use it to build ten houses. It's the same with classes and objects. First, the programmer writes code to describe a class. Then when the program runs, the computer creates objects from the (blueprint) class.

So that's the real relationship between classes and objects. The programmer defines a class, and from the class definition, the computer makes individual objects.

What's so great about an object-oriented language?

Based on the preceding section's story about home building, imagine that you've already written a computer program to keep track of the building instructions for houses in a new development. Then the big boss decides on a modified plan — a plan in which half the houses have three bedrooms and the other half have four.

If you use the old FORTRAN/C style of computer programming, your instructions look like this:

```
Dig a ditch for the basement.
Lay concrete around the sides of the ditch.
Place two-by-fours along the sides for the basement's frame.
...
```

This would be similar to an architect creating a long list of instructions instead of a blueprint. To modify the plan, you have to sort through the list to find the instructions for building bedrooms. To make matters worse, the instructions might be scattered among pages xvii, 234, 394–410, 739, 10, and 2. If the builder had to decipher other peoples' complicated instructions, the task would be ten times harder.

Starting with a class, however, is like starting with a blueprint. If you decide to have both three- and four-bedroom houses, you can start with a blueprint called the *house blueprint* — it has a ground floor and a second floor, but has no indoor walls drawn on the second floor. Then you make two more second-floor blueprints — one for the three-bedroom house and another for the four-bedroom house. (You name these new blueprints the *three-bedroom house* blueprint and the *four-bedroom house* blueprint.)

Your builder colleagues are amazed at your sense of logic and organization, but they have concerns. They pose a question. "You called one of the blueprints the 'three-bedroom house' blueprint. How can you do this if it's a blueprint for a second floor and not for a whole house?"

You smile knowingly and answer, "The three-bedroom house blueprint can say, 'For info about the lower floors, see the original house blueprint.' That way, the three-bedroom house blueprint describes a whole house. The four-bedroom house blueprint can say the same thing. With this setup, we can take advantage of all the work we already did to create the original house blueprint and save lots of money."

In the language of object-oriented programming, the three- and four-bedroom house classes are *inheriting* the features of the original house class. You can also say that the three- and four-bedroom house classes are *extending* the original house class. (See Figure 1-1.)

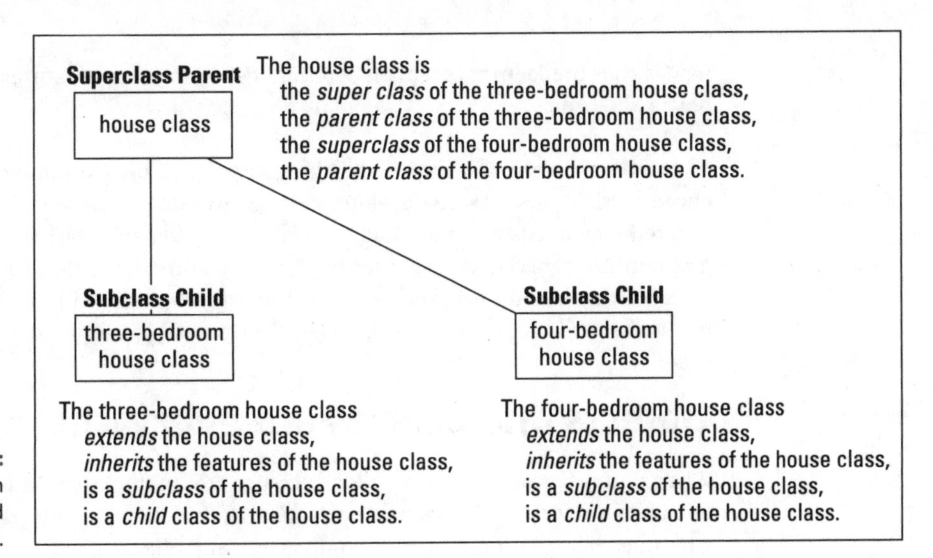
Superclass Parent
house class
The house class is
the super class of the three-bedroom house class,
the parent class of the three-bedroom house class,
the superclass of the four-bedroom house class,
the parent class of the four-bedroom house class.
Subclass Child
three-bedroom house class
Subclass Child
four-bedroom house class
The three-bedroom house class
extends the house class,
inherits the features of the house class,
is a subclass of the house class,
is a child class of the house class.
The four-bedroom house class
extends the house class,
inherits the features of the house class,
is a subclass of the house class,
is a child class of the house class.

FIGURE 1-1: Terminology in object-oriented programming.

The original house class is called the *superclass* of the three- and four-bedroom house classes. In that vein, the three- and four-bedroom house classes are *subclasses* of the original house class. Put another way, the original house class is called the *parent class* of three- and four-bedroom house classes. The three- and four-bedroom house classes are *child classes* of the original house class. (Refer to Figure 1-1.)

Needless to say, your homebuilder colleagues are jealous. A crowd of homebuilders is mobbing around you to hear about your great ideas. So, at that moment, you drop one more bombshell: "By creating a class with subclasses, we can reuse the blueprint in the future. If someone comes along and wants a five-bedroom house, we can extend our original house blueprint by making a five-bedroom house blueprint. We'll never have to spend money for an original house blueprint again."

"But," says a colleague in the back row, "what happens if someone wants a different first-floor design? Do we trash the original house blueprint or start scribbling all over the original blueprint? That'll cost big bucks, won't it?"

In a confident tone, you reply, "We don't have to mess with the original house blueprint. If someone wants a Jacuzzi in their living room, we can make a new, small blueprint describing only the new living room and call it the Jacuzzi-in-living-room house blueprint. Then this new blueprint can refer to the original house blueprint for info on the rest of the house (the part that's not in the living room)." In the language of object-oriented programming, the Jacuzzi-in-living-room house blueprint still *extends* the original house blueprint. The Jacuzzi blueprint is still a subclass of the original house blueprint. In fact, all the terminology about superclass, parent class, and child class still applies. The only thing that's new is that the Jacuzzi blueprint *overrides* the living room features in the original house blueprint.

In the days before object-oriented languages, the programming world experienced a crisis in software development. Programmers wrote code and then discovered new needs and then had to trash their code and start from scratch. This problem happened over and over again because the code that the programmers were writing couldn't be reused. Object-oriented programming changed all this for the better (and, as Burd said, Java is "A Great Object-Oriented Language").

Objects and classes are everywhere

When you program in Java, you work constantly with classes and objects. These two ideas are vitally important. That's why, in this chapter, I hit you over the head with one analogy after another about classes and objects.

Close your eyes for a minute and think about what it means for an item to be a chair:

A chair has a seat, a back, and legs. Each seat has a shape, a color, a degree of softness, and other characteristics. These are the properties a chair possesses. What I describe is *chairness* — the notion of an item being a chair. In object-oriented terminology, I'm describing the `Chair` class.

Now peek over the edge of this book's margin and take a minute to look around the room. (If you're not sitting in a room right now, fake it.)

Several chairs are in the room, and each chair is an object. Each of these objects is an example of that ethereal thing called the `Chair` class. So that's how it works — the class is the idea of *chairness,* and each individual chair is an object.

REMEMBER

A class isn't quite a collection of things. Instead, a class is the idea behind a certain kind of thing. When I talk about the class of chairs in your room, I'm talking about the fact that each chair has legs, a seat, a color, and so on. The colors may be different for different chairs in the room, but that doesn't matter. When you talk about a class of things, you're focusing on the properties that each of the things possesses.

It makes sense to think of an object as being a concrete instance of a class. In fact, the official terminology is consistent with this line of thinking. If you write a Java program in which you define a `Chair` class, each actual chair (the chair you're sitting on, the empty chair next to you, and so on) is called an *instance* of the `Chair` class.

Here's another way to think about a class. Imagine a table displaying all three of your bank accounts (see Table 1-1).

TABLE 1-1 **A Table of Accounts**

Account Number	Type	Balance
16-13154-22864-7	Checking	174.87
1011 1234 2122 0000	Credit	–471.03
16-17238-13344-7	Savings	247.38

Think of the table's column headings as a class and think of each row of the table as an object. The table's column headings describe the `Account` class.

According to the table's column headings, each account has an account number, a type, and a balance. Rephrased in the terminology of object-oriented programming, each object in the `Account` class (that is, each instance of the `Account` class) has an account number, a type, and a balance. So the bottom row of the table is an object with account number *16-17238-13344-7*. This same object has type *Savings* and a balance of *247.38*. If you opened a new account, you would have another object and the table would grow an additional row. The new object would be an instance of the same `Account` class.

What's Next?

This chapter is filled with general descriptions of things. A general description is good when you're just getting started, but you don't really understand things until you get to know some specific info, as laid out in the next several chapters.

So please, turn the page. The next chapter can't wait for you to read it.

» **Peeking inside the tools**

» **Writing and running your own Java programs**

Chapter 2 All about Software

The best way to get to know Java is to do Java. When you're doing Java, you're writing, testing, and running your own Java programs. This chapter describes the kind of software you use for all those Java-related tasks. The chapter has *general* instructions to help you set up your computer, but it has no detailed instructions. If you want detailed instructions, visit this book's website: `http://JavaForDummies.allmycode.com`.

Get Ready for Java

If you're a seasoned veteran of computers and computing (whatever that means) and you're too jumpy to follow the detailed instructions on this book's website, you can try installing the required software by reading these general instructions:

1. **Install a Java Development Kit (JDK).**

   A Java Development Kit is a bunch of software that makes all Java programs work.

   But wait! What does it mean to make "all Java programs work"? I answer that question later in this chapter, in the section "The Inside Scoop."

   To install a Java Development Kit, visit `https://adoptium.net` and follow that website's instructions.

The `https://adoptium.net` site has several JDKs, written by several different companies, and almost any of these kits will work with this book's examples. If you dislike the adoptium.net alternatives and you prefer instead to get the "official" JDK, you can download it from `www.oracle.com/java/technologies/downloads`. The problem with Oracle's official version is" that it comes with a long, somewhat confusing list of legal requirements. It's probably okay to run this book's examples with Oracle's JDK, but I'm not a lawyer and I've never even played one on TV. So I always recommend taking the safest possible route — get Java from `https://adoptium.net`.

2. **Install an integrated development environment.**

   An *integrated development environment (IDE)* is a program to help you compose and test new software. It's like a glorified version of Microsoft Word for writing computer code. For this book's examples, you can use almost any IDE that supports Java.

   Here's a list of IDEs that are most popular among professional developers:

   - **Eclipse** (`www.eclipse.org/downloads`)
   - **IntelliJ IDEA** (`www.jetbrains.com/idea`)
   - **NetBeans** (`https://netbeans.apache.org`)
   - **Visual Studio Code, also known as VS Code** (`https://code.visualstudio.com`)

   Some IDEs are made especially for students, educators, and other specialized communities. These include BlueJ, DrJava, Greenfoot, JCreator, jGrasp, and several others.

If you like roughing it, you can write and run Java programs without using an IDE: Just type your Java program in a plain text editor (such as Windows Notepad) and run the program on your operating system's command line (Windows MS-DOS, macOS Terminal, or whatever). It's not fun to develop software this way, but it makes you feel like a big shot.

This book's website has detailed instructions for installing and using the most commonly used IDEs.

3. **Test your installed software.**

   What you do in this step depends on which IDE you choose in Step 2. Anyway, here are some general instructions:

   a. Launch your IDE (Eclipse, IntelliJ IDEA, NetBeans, or whatever).

   b. In the IDE, create a new Java project.

   c. Within the Java project, create a new Java class named `Main`. (Choosing File ⇨ New ⇨ Java Class works in most IDEs.)

   d. Edit the new `Main.java` file by typing the following lines of code:

```
public class Main {

    public static void main(String[] args) {
        System.out.print(12345);
    }
}
```

For most IDEs, you add the code into a big (mostly blank) editor pane. Try to type the code exactly as you see it here. If you see an uppercase letter, type an uppercase letter. Do the same with all lowercase letters.

What? You say you don't want to type a bunch of code from the book? Well, all right then! Visit this book's website (`http://JavaForDummies.allmycode.com`) to download all the code examples and copy them into the IDE of your choice.

e. Run `Main.java` and make sure that the run's output reads `12345`.

TIP

You may find variations on the picture that I paint in the preceding steps. For example, some IDEs come with options for you to install a JDK. In those cases, you can skip Step 1 and march straight to Step 2. Nevertheless, the picture that I paint with the preceding steps is useful and reliable. When you follow my instructions, you might end up with two copies of the JVM, or two IDEs, but that's okay. You never know when you'll need a spare.

That's it! But remember: Not everyone (computer geek or not) can follow these skeletal instructions flawlessly. So, if you want more details, visit `http://JavaForDummies.allmycode.com`.

The Inside Scoop

One of my acquaintances is a tool-and-die maker. She uses tools to make tools (and dies). I once asked, "Who makes the tools that you use to make tools?" After ignoring her smart-aleck answer, I guessed that a tool-and-die-toolmaker makes tools for tool-and-die makers so that tool-and-die makers can make tools.

A computer programmer does the same kind of thing: A programmer uses existing programs as tools to create new programs. The existing programs and new programs might perform very different kinds of tasks. For example, a Java program (a program you create) might keep track of a business's customers. To create that customer-tracking program, you use several programs belonging to a Java Development Kit. With a JDK's programs, you can create many other useful programs — customer-tracking programs, weather-predicting programs, gaming programs, or programs that run on your mobile phone.

A JDK contains at least 30 different programs — many more, if you count compressed archives and other such items. This section deals with two of the most important JDK components:

- A *compiler* is a program that takes your Java code and turns that code into a bunch of instructions called *bytecode*.

  Humans can't readily compose or decipher bytecode instructions. But certain software that you run on your computer can interpret and carry out bytecode instructions.

- A *Java virtual machine (JVM)* is a program that interprets and carries out bytecode instructions.

The rest of this section describes compilers and Java virtual machines.

What is a compiler?

A compiler is a program that takes your Java code turns that code into a bunch of instructions called bytecode.

—Barry Burd, Java For Dummies, 8th Edition

You're a human being. (Sure, every rule has exceptions. But if you're reading this book, you're probably human.) Anyway, humans can write and comprehend the code in Listing 2-1.

LISTING 2-1: Looking for a Vacant Room

```
// This is part of a Java program.
// It's not a complete Java program.
roomNum = 1;
while (roomNum < 100) {
    if (guests[roomNum] == 0) {
        out.println("Room " + roomNum + " is available.");
        exit(0);
    } else {
        roomNum++;
    }
}
out.println("No vacancy");
```

The Java code in Listing 2-1 checks for vacancies in a small hotel (a hotel with room numbers from 1 to 99). You can't run the code in Listing 2-1 without adding several lines. But here in Chapter 2, those additional lines aren't important. What's important is that, by staring at the code, squinting a bit, and looking past all the code's strange punctuation, you can see what the code is trying to do:

```
Set the room number to 1.
As long as the room number is less than 100,
    Check the number of guests in the room.
    If the number of guests in the room is 0, then
        report that the room is available,
        and stop.
    Otherwise,
        prepare to check the next room by
        adding 1 to the room number.
If you get to the nonexistent room number 100, then
    report that there are no vacancies.
```

If you see no similarities between Listing 2-1 and its English equivalent, don't worry: You're reading *Java For Dummies*, 8th Edition, and, like most human beings, you can learn to read and write the code in Listing 2-1. The code in Listing 2-1 is called *Java source code.*

Here's the catch: Computers aren't human beings. Computers don't normally follow instructions like the instructions in Listing 2-1. That is, computers don't follow Java source code instructions. Instead, computers follow cryptic instructions like the ones in Listing 2-2.

LISTING 2-2: **A Translation of Listing 2-1 into Java Bytecode**

```
aload_0
iconst_1
putfield Hotel/roomNum I
goto 32
aload_0
getfield Hotel/guests [I
aload_0
getfield Hotel/roomNum I
iaload
ifne 26
getstatic java/lang/System/out Ljava/io/PrintStream;
new java/lang/StringBuilder
```

(continued)

LISTING 2-2: (continued)

```
dup
ldc "Room "
invokespecial java/lang/StringBuilder/<init>(Ljava/lang/String;)V
aload_0
getfield Hotel/roomNum I
invokevirtual java/lang/StringBuilder/append(I)Ljava/lang/StringBuilder;
ldc " is available."
invokevirtual
  java/lang/StringBuilder/append(Ljava/lang/String;)Ljava/lang/StringBuilder;
invokevirtual java/lang/StringBuilder/toString()Ljava/lang/String;
invokevirtual java/io/PrintStream/println(Ljava/lang/String;)V
iconst_0
invokestatic java/lang/System/exit(I)V
goto 32
aload_0
dup
getfield Hotel/roomNum I
iconst_1
iadd
putfield Hotel/roomNum I
aload_0
getfield Hotel/roomNum I
bipush 100
if_icmplt 5
getstatic java/lang/System/out Ljava/io/PrintStream;
ldc "No vacancy"
invokevirtual java/io/PrintStream/println(Ljava/lang/String;)V
return
```

The instructions in Listing 2-2 aren't Java source code instructions. They're *Java bytecode* instructions. When you write a Java program, you write source code instructions (like the instructions in Listing 2-1). After you finish writing the source code, your IDE runs a program (a translation tool) on your source code. The program is a *compiler*. The compiler translates your source code instructions into Java bytecode instructions. In other words, the compiler takes code that you can write and understand (like the code in Listing 2-1) and translates it into code that a computer has a fighting chance of carrying out (like the code in Listing 2-2).

TECHNICAL STUFF

You might put your source code in a file named `Hotel.java`. If so, the compiler probably puts the Java bytecode in another file named `Hotel.class`. Normally, you don't bother looking at the bytecode in the `Hotel.class` file. In fact, the compiler doesn't encode the `Hotel.class` file as ordinary text, so you can't examine the bytecode with an ordinary editor. If you try to open `Hotel.class` with

Notepad, TextEdit, KWrite, or even Microsoft Word, you see nothing but dots, squiggles, and other gobbledygook. To create Listing 2-2, I had to apply yet another tool to my `Hotel.class` file. That tool displays a text-like version of a Java bytecode file. I used Ando Saabas's Java Bytecode Editor (`https://set.ee/jbe`).

No one (except for a few crazy programmers in some isolated labs in faraway places) writes Java bytecode. You run software (a compiler) to create Java bytecode. The only reason to look at Listing 2-2 is to understand what a hard worker your computer is.

What is a Java virtual machine?

A Java virtual machine (JVM) is a program that interprets and carries out bytecode instructions.

—Barry Burd, Java For Dummies, 8th Edition

In the preceding "What is a compiler?" section, I make a big fuss about computers following instructions like the ones in Listing 2-2. As fusses go, it's a very nice fuss. But if you don't read every fussy word, you may be misguided. The exact wording is ". . . computers follow cryptic instructions *like* the ones in Listing 2-2." The instructions in Listing 2-2 are a lot like instructions that a computer can execute, but generally, computers don't execute Java bytecode instructions. Instead, each kind of computer processor has its own set of executable instructions, and each computer operating system uses the processor's instructions in a slightly different way.

Here's a hypothetical situation: The year is 1992 (a few years before Java was made public) and you run the Linux operating system on a computer that has an old Pentium processor. Your friend runs Linux on a computer with a different kind of processor — a PowerPC processor. (In the 1990s, Intel Corporation made Pentium processors, and IBM made PowerPC processors.)

Listing 2-3 contains a set of instructions to display `Hello world!` on the computer screen.* The instructions work on a Pentium processor running the Linux operating system.

---

* I paraphrase these Intel instructions from Konstantin Boldyshev's Linux Assembly HOWTO document (`http://tldp.org/HOWTO/Assembly-HOWTO/hello.html`).

LISTING 2-3: A Simple Program for a Pentium Processor

```
.data
msg:
        .ascii  "Hello, world!\n"
        len = . - msg
.text
    .global _start
_start:
        movl    $len,%edx
        movl    $msg,%ecx
        movl    $1,%ebx
        movl    $4,%eax
        int     $0x80

        movl    $0,%ebx
        movl    $1,%eax
        int     $0x80
```

Listing 2-4 contains another set of instructions to display `Hello world!` on the screen.** The instructions in Listing 2-4 work on a PowerPC processor running Linux.

LISTING 2-4: A Simple Program for a PowerPC Processor

```
.data
msg:
        .string "Hello, world!\n"
        len = . - msg
.text
        .global _start
_start:
        li      0,4
        li      3,1
        lis     4,msg@ha
        addi    4,4,msg@l
        li      5,len
        sc

        li      0,1
        li      3,1
        sc
```

** I paraphrase the PowerPC code from Hollis Blanchard's PowerPC Assembly (`www.ibm.com/developerworks/library/l-ppc`). Hollis also reviewed and critiqued this "What is a Java virtual machine?" section for me. Thank you, Hollis.

The instructions in Listing 2-3 run smoothly on a Pentium processor. But these instructions mean nothing to a PowerPC processor. Likewise, the instructions in Listing 2-4 run nicely on a PowerPC, but these same instructions are complete gibberish to a computer with a Pentium processor. So your friend's PowerPC software might not be available on your computer. And your Intel computer's software might not run at all on your friend's computer.

Now go to your cousin's house. Your cousin's computer has a Pentium processor (just like yours), but your cousin's computer runs Windows instead of Linux. What does your cousin's computer do when you feed it the Pentium code in Listing 2-3? It screams, "Not a valid Win32 application" or "Windows can't open this file." What a mess!

Java bytecode creates order from all this chaos. Unlike the code in Listings 2-3 and 2-4, Java bytecode isn't specific to one kind of processor or to a single operating system. Instead, any kind of computer can have a Java virtual machine, and Java bytecode instructions run on any computer's Java virtual machine. The JVM that runs on a Pentium with Linux translates Java bytecode instructions into the kind of code you see in Listing 2-3. And the JVM that runs on a PowerPC with Linux translates Java bytecode instructions into the kind of code you see in Listing 2-4. The same kind of translation takes place for modern processors, like the Intel i9 and the ARM M1.

If you write a Java program and compile that Java program into bytecode, then the JVM on your computer can run the bytecode, the JVM on your friend's computer can run the bytecode, and the JVM on your grandmother's supercomputer can run the bytecode.

CROSS REFERENCE

For a look at some Java bytecode, see Listing 2-2. ***Remember:*** You never have to write or decipher Java bytecode. Writing bytecode is the compiler's job. Deciphering bytecode is the Java virtual machine's job.

With Java, you can take a bytecode file that you created with a Windows computer, copy the bytecode to who-knows-what kind of computer, and then run the bytecode with no trouble. That's one of the many reasons Java has become popular so quickly. This outstanding feature, which gives you the ability to run code on many different kinds of computers, is called *portability.*

What makes Java bytecode so versatile? This fantastic universality enjoyed by Java bytecode programs comes from the Java virtual machine. The Java virtual machine is one of those three tools that you *must* have on your computer.

Imagine that you're the Windows representative to the United Nations Security Council. (See Figure 2-1.) The Macintosh representative is seated to your right, and the Linux representative is to your left. (Naturally, you don't get along with either of these people. You're always cordial to one another, but you're never sincere. What do you expect? It's politics!) The distinguished representative from Java is at the podium. The Java representative is speaking in bytecode, and neither you nor your fellow ambassadors (Mac and Linux) understand a word of Java bytecode.

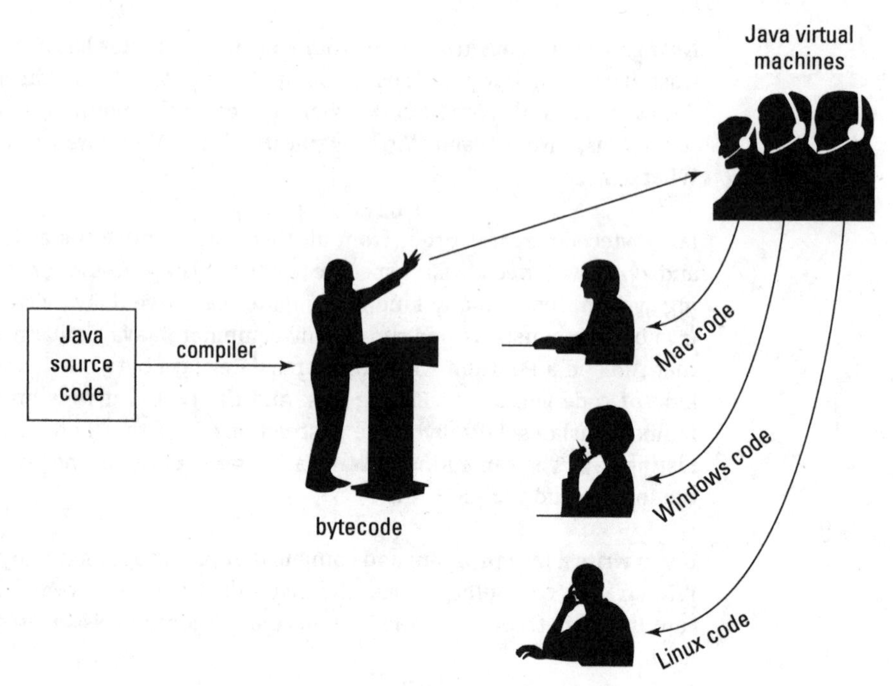

FIGURE 2-1: An imaginary meeting of the UN Security Council.

But each of you has an interpreter. Your interpreter translates from bytecode to Windows while the Java representative speaks. Another interpreter translates from bytecode to Macintosh-ese. And a third interpreter translates bytecode into Linux-speak.

Think of your interpreter as a virtual ambassador. The interpreter doesn't really represent your country, but the interpreter performs one of the important tasks that a real ambassador performs. The interpreter listens to bytecode on your behalf. The interpreter does what you would do if your native language were Java bytecode. The interpreter pretends to be the Windows ambassador and sits through the boring bytecode speech, taking in every word and processing each word in some way or another.

You have an interpreter — a virtual ambassador. In the same way, a Windows computer runs its own bytecode-interpreting software. That software is the Java virtual machine.

A Java virtual machine is a proxy, an errand boy, a go-between. The JVM serves as an interpreter between Java's run-anywhere bytecode and your computer's own system. While it runs, the JVM walks your computer through the execution of bytecode instructions. The JVM examines your bytecode, bit by bit, and carries out the instructions described in the bytecode. The JVM interprets bytecode for your Windows system, your Mac, or your Linux box, or for whatever kind of computer you're using. That's a good thing. It's what makes Java programs more portable than programs in any other language.

Developing Software

All this has happened before, and it will all happen again.

—Peter Pan (J. M. Barrie) and Battlestar Galactica (2003–2009, NBC Universal)

When you create a Java program, you repeat the same steps over and over again. Figure 2-2 illustrates the cycle.

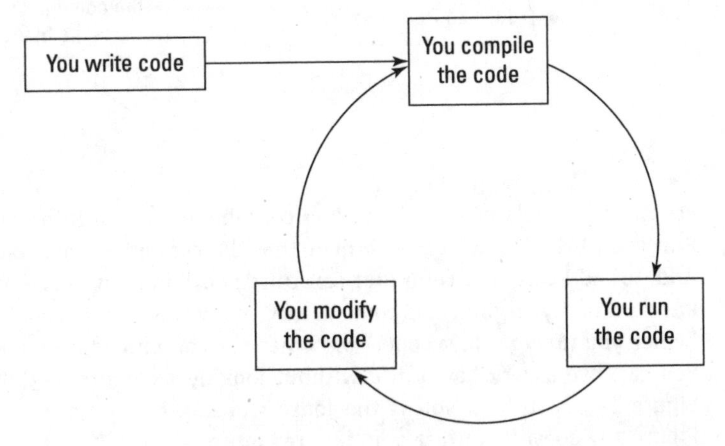

FIGURE 2-2: Developing a Java program.

First, you write a program. After writing the first draft, you repeatedly compile, run, and modify the program. With a little experience, the compile and run steps become easy to carry out. In many cases, one mouse-click starts the compilation or the run.

However, writing the first draft and modifying the code are not one-click tasks. Developing code requires time and concentration.

REMEMBER

Never be discouraged when the first draft of your code doesn't work. For that matter, never be discouraged when the 25th draft of your code doesn't work. Rewriting code is one of the most important things you can do (aside from ensuring world peace).

When people talk about writing programs, they use the wording in Figure 2-2. They say, "You compile the code" and "You run the code." But the "you" isn't always accurate, and the "code" differs slightly from one part of the cycle to the next. Figure 2-3 describes the cycle from Figure 2-2 in a bit more detail.

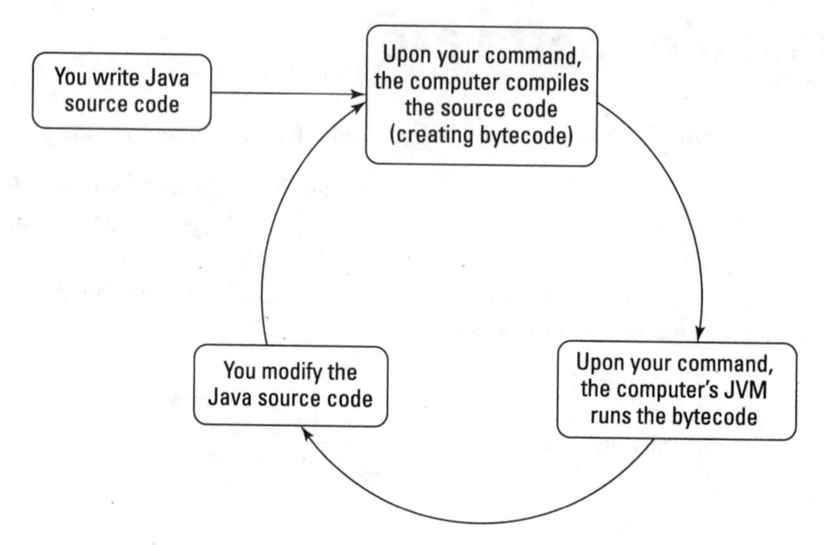

FIGURE 2-3: Who does what with which code?

TIP

For most people's needs, Figure 2-3 contains too much information. If I click a Run icon, I don't have to remember that the computer runs code on my behalf. And, for all I care, the computer can run my original Java code or some bytecode knockoff of my original Java code. In fact, many times in this book, I casually write "when you run your Java code," or "when the computer runs your Java program." You can live a very happy life without looking at Figure 2-3. The only use for Figure 2-3 is to help you if the loose wording in Figure 2-2 confuses you. If Figure 2-2 doesn't confuse you, ignore Figure 2-3.

Spoiler Alert!

In the next chapter, you start running Java code. That's what you really want from this book. Isn't it?

» Taking a first glance at Java code

» Understanding the parts of a simple program

» Documenting your code

Chapter 3 Using the Basic Building Blocks

"Все мысли, которые имеют огромные последствия всегда просты. (All great ideas are simple.)"

—LEO TOLSTOY

The quotation applies to all kinds of things — things like life, love, and computer programming. That's why this chapter takes a multilayered approach. In this chapter, you get your first details about Java programming. And in discovering details, you'll see the simplicities.

Speaking the Java Language

If you try to picture in your mind the entire English language, what do you see? Maybe you see words, words, words. (That's what Hamlet saw.) Looking at the language under a microscope, you see one word after another. The bunch-of-words image is fine, but if you step back a bit, you may see two other things:

» The language's grammar

» Thousands of expressions, sayings, idioms, and historical names

The first category (the grammar) includes rules like this: "The verb agrees with the noun in number and person." The second category (expressions, sayings, and stuff) includes knowledge like this: "Julius Caesar was a famous Roman emperor, so don't name your son Julius Caesar, unless you want him to get beaten up every day after school."

The Java programming language has all the aspects of a spoken language like English. Java has words, grammar, commonly used names, stylistic idioms, and other such elements.

The grammar and the common names

The people at Sun Microsystems who created Java thought of Java as having separate parts. Just as English has its grammar and commonly used names, the Java programming language has its specification (its grammar) and its application programming interface (its commonly used names). Along with these parts come two important documents:

- **The Java Language Specification:** This documentation includes rules like this: "Always put an open parenthesis after the word *for*" and "Use an asterisk to multiply two numbers."
- **The API Specification:** Java's *application programming interface (API)* contains thousands of names that were added to Java after the language's grammar was defined. These names range from the commonplace to the exotic. For example, one name — the name *JFrame* — represents a window on your computer's screen. A more razzle-dazzle name — *pow* — helps you raise 5 to the tenth power or raise whatever to the whatever-else power. Other names help you listen for the user's button clicks, query databases, and do all kinds of useful things.

You can download the language specification by poking around at `http://docs.oracle.com/javase/specs`, but I don't recommend doing it. With the language spec, you can settle subtle arguments about edge cases in the behavior of Java programs. But the spec is far too detailed for the day-to-day study of Java.

The second document — the API Specification — is the go-to document for most of your Java programming needs. The API contains thousands and thousands of names and keeps growing with each new Java language release. That may seem scary, but there's good news — you don't have to memorize anything in the API. Nothing. None of it. You can look up the stuff you need to use in the documentation and ignore the stuff you don't need. What you use often, you'll remember. What you don't use often, you'll forget (like any other programmer).

The API document for Java 17 lives online at `https://docs.oracle.com/en/java/javase/17/docs/api/index.html`.

No one knows all there is to know about the Java API. If you're a Java programmer who frequently writes programs that open new windows, you know how to use the API `JFrame` class. If you seldom write programs that open windows, the first few times you need to create a window, you can look up the `JFrame` class in the API documentation. My guess is that if you prevented a typical Java programmer from looking up anything in the API documentation, the programmer would be able to use less than 2 percent of all names in the Java API.

You may love the *For Dummies* style, but unfortunately, Java's official API documentation isn't written that way. The API documentation is both concise and precise. For some help in deciphering the API documentation's language and style, see this book's website (`http://JavaForDummies.allmycode.com`).

In a way, nothing about the Java API is special. Whenever you write a Java program — even the smallest, simplest Java program — you create a class that's on par with any of the classes defined in the official Java API. The API is just a set of classes and other names that were created by ordinary programmers who happen to participate in the official Java Community Process (JCP). Unlike the names you create, the names in the API are distributed with every version of Java. (I'm assuming that you, the reader, are not a participant in the Java Community Process. But, with a fine book like *Java For Dummies,* 8th Edition, one never knows.)

If you're interested in the JCP's activities, visit `www.jcp.org`.

The folks at the JCP don't keep the Java programs in the official Java API a secret. If you want, you can look at all these programs. When you install Java on your computer, the installation puts a file named `src.zip` on your hard drive. You can open `src.zip` with your favorite unzipping program. There, before your eyes, is all the Java API code.

The words in a Java program

A hard-core Javateer will say that the Java programming language has four kinds of words: keywords, restricted keywords, literals, and identifiers. This is true. But the bare truth, with no other explanation, isn't useful. So, I dress up the truth a bit by thinking in terms of three kinds of words: keywords, identifiers that ordinary programmers like you and I create, and identifiers from the API.

The differences among these three kinds of words are similar to the differences among words in the English language. In the sentence "Sam is a person," the word *person* is like a Java keyword. No matter who uses the word *person,* the word always means roughly the same thing. (Sure, you can think of bizarre exceptions in English usage, but please don't.)

The word *Sam* is like a Java identifier because Sam is a name for a particular person. Words like *Sam*, *Dinswald*, and *McGillimaroo* aren't prepacked with meaning in the English language. These words apply to different people, depending on the context, and become names when parents pick one for their newborn kid.

Now consider the sentence "Julius Caesar is a person." If you utter this sentence, you're probably talking about the fellow who ruled Rome until the Ides of March. Although the name Julius Caesar isn't hardwired into the English language, almost everyone uses the name to refer to the same person. If English were a programming language, the name Julius Caesar would be an API identifier.

Here's how I, in my mind, divide the words in a Java program into categories:

- **Keywords:** A *keyword* is a word that has its own special meaning in the Java programming language, and that meaning doesn't change from one program to another. Examples of keywords in Java are `if`, `else`, and `do`.

  The JCP committee members, who have the final say on what constitutes a Java program, have chosen all the Java keywords. If you think about the two parts of Java, which I discuss earlier, in the section "The grammar and the common names," the Java keywords belong solidly to the language specification.

- **Identifiers:** An *identifier* is a name for something. The identifier's meaning can change from one program to another, but some identifiers' meanings tend to change more:

  - *Identifiers created by you and me:* As a Java programmer (yes, even as a novice Java programmer), you create new names for classes and other items you describe in your programs. Of course, you may name something `Prime`, and your coworker writing code two cubicles down the hall can name something else `Prime`. That's okay because Java has no predetermined meaning for `Prime`. In your program, you can make `Prime` stand for the Federal Reserve's prime rate. And your friend down the hall can make `Prime` stand for the "bread, roll, preserves, and prime rib." No conflict arises, because you and your coworker are writing two different Java programs.

  - *Identifiers from the API:* The JCP members have created names for many things and thrown tens of thousands of these names into the Java API. The API comes with each version of Java, so these names are available to anyone who writes a Java program. Examples of such names are `String`, `Integer`, `JWindow`, `JButton`, `JTextField`, and `File`.

Strictly speaking, the meanings of the identifiers in the Java API aren't cast in stone. Although you can make up your own meanings for `JButton` or `JWindow`, this isn't a good idea. If you did, you would confuse the dickens out of other programmers, who are used to the standard API meanings for these familiar identifier names. But even worse, when your code assigns a new meaning to an identifier like `JButton`, you lose any computational power that was created for the identifier in the API code. The programmers at Sun Microsystems, Oracle, and the Java Community Process did all the work of writing Java code to handle buttons. If you assign your own meaning to `JButton`, you're turning your back on all the progress made in creating the API.

To see the list of Java keywords, go to `www.dummies.com` and enter *Beginning Programming with Java For Dummies cheat sheet* in the Search box.

Checking Out Java Code for the First Time

The first time you look at somebody else's Java program, you may tend to feel a bit queasy. The realization that you don't understand something (or many things) in the code can make you nervous. I've written hundreds (maybe thousands) of Java programs, but I still feel insecure when I start reading someone else's code.

The truth is that finding out about a Java program is a bootstrapping experience. First, you gawk in awe of the program. Then you run the program to see what it does. Then you stare at the program for a while or read someone's explanation of the program and its parts. Then you gawk a little more and run the program again. Eventually, you come to terms with the program. (Don't believe the wise guys who say they never go through these steps. Even the experienced programmers approach a new project slowly and carefully.)

In Listing 3-1, you get a blast of Java code. (Like all novice programmers, you're expected to gawk humbly at the code.) Hidden in the code, I've placed some important ideas, which I explain in detail in the next section. These ideas include the use of classes, methods, and Java statements.

LISTING 3-1: The Simplest Java Program

```
public class Displayer {

    public static void main(String[] args) {
        System.out.println("You'll love Java!");
    }
}
```

You don't have to type the code in Listing 3-1 (or in any of this book's listings). To download all the code in this book, visit the book's website (`http://JavaForDummies.allmycode.com`).

When you run the program from Listing 3-1, the computer displays `You'll love Java!`. (Figure 3-1 shows the output of the Displayer program.) Now, I admit that writing and running a Java program is a lot of work just to get `You'll love Java!` to appear on somebody's computer screen, but every endeavor has to start somewhere.

```
You'll love Java!
```

FIGURE 3-1: Running the program in Listing 3-1.

This book's website (`http://JavaForDummies.allmycode.com`) has instructions to help you run Java programs such as the code in Listing 3-1.

In the following section, you do more than just admire the program's output. After you read the following section, you actually understand what makes the program in Listing 3-1 work.

Understanding a Simple Java Program

This section presents, explains, analyzes, dissects, and otherwise demystifies the Java program shown previously in Listing 3-1.

The Java class

Because Java is an object-oriented programming language, your primary goal is to describe classes and objects. (If you're not convinced about this, read the sections on object-oriented programming in Chapter 1.)

On those special days when I'm feeling sentimental, I tell people that Java is more pure in its object-orientation than many other so-called object-oriented languages. I say this because, in Java, you can't do anything until you create a class of some kind. It's like being on *Jeopardy!* and hearing the host say, "Let's go to a commercial" and then interrupting that person by saying, "I'm sorry — you can't issue an instruction without putting your instruction inside a class."

The code in Listing 3-1 is a Java program, and that program describes a class. I wrote the program, so I get to make up a name for my new class. I chose the name `Displayer` because the program displays a line of text on the computer screen. That's why the first line in Listing 3-1 contains the words `class Displayer`. (See Figure 3-2.)

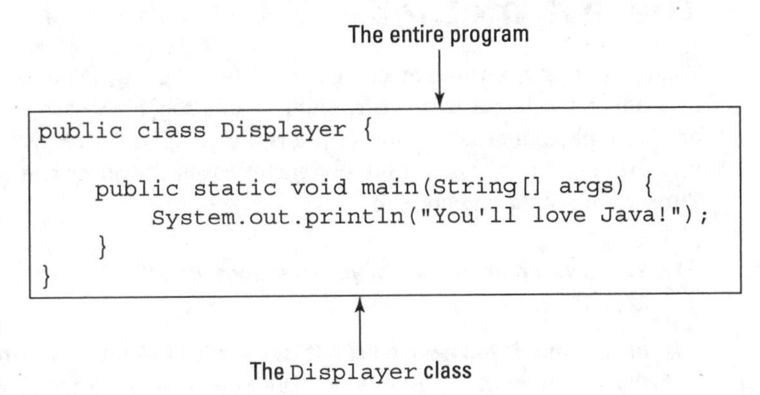

The entire program
public class Displayer {
public static void main(String[] args) {
System.out.println("You'll love Java!");
}
}
The Displayer class

FIGURE 3-2: A Java program is a class.

The first two words in Listing 3-1, `public` and `class`, are Java keywords. (See the section "The words in a Java program," earlier in this chapter.) No matter who writes a Java program, the words `public` and `class` are always used in the same way. On the other hand, `Displayer` in Listing 3-1 is an identifier. (I made up the word `Displayer` while I was writing this chapter.) `Displayer` is the name of a particular class — the class that I'm creating by writing this program.

CROSS REFERENCE

This book is filled with talk about classes, but for the best description of a Java class (the reason for using the word `class` in Listing 3-1), visit Chapter 7. The word `public` means that other Java classes (classes other than the `Displayer` class in Listing 3-1) can use the features declared in Listing 3-1. For more details about the meaning of `public` and the use of the word `public` in a Java program, see Chapters 7 and 14.

WARNING

tHE jAVA PROGRAMMING LANGUAGE IS cASe-sEnsITiVE. If you change a lowercase letter in a word to an UpperCase letter, you can change the word's meaning. cHANGING case can make the entire word go from being meaningful to being meaningless. In the first line of Listing 3-1, you can't replace `class` with `Class`. iF YOU DO, THE WHOLE PROGRAM STOPS WORKING. The same holds true, to some extent, for the name of a file containing a particular class. For example, the name of the class in Listing 3-1 is `Displayer`, starting with an uppercase letter `D`. So, it's a good idea to save the code of Listing 3-1 in a file named `Displayer.java`, starting with an uppercase letter `D`.

Normally, if you define a class named `DogAndPony`, the class's Java code is in a file named `DogAndPony.java`, spelled and capitalized exactly the same way that the class name is spelled and capitalized. In fact, this file naming convention is mandatory for most examples in this book.

The Java method

You're working as an auto mechanic in an upscale garage. Your boss, who's always in a hurry and has a habit of running words together, says, "fixTheAlternator on that junkyOldFord." Mentally, you run through a list of tasks: "Drive the car into the bay, lift the hood, find a wrench, loosen the alternator belt," and so on. Three things are going on here:

- **You have a name for what you're supposed to do.** The name is *fixTheAlternator.*
- **In your mind, you have a list of tasks associated with the name *fixTheAlternator.*** The list includes "Drive the car into the bay, lift the hood, find a wrench, loosen the alternator belt," and so on.
- **You have a grumpy boss who's telling you to do all this work.** Your boss gets you working by saying, "fixTheAlternator." In other words, your boss gets you working by saying the name of what you're supposed to do.

In this scenario, using the word *method* wouldn't be a big stretch. You have a method for doing something with an alternator. Your boss calls that method into action, and you respond by doing all the things in the list of instructions that you associate with the method.

If you believe all that (and I hope you do), you're ready to read about Java methods. In Java, a *method* is a list of things to do. Every method has a name, and you tell the computer to do the things in the list by using the method's name in your program.

I've never written a program to get a robot to fix an alternator. But, if I did, the program might include a `fixTheAlternator` method. The list of instructions in my `fixTheAlternator` method would look something like the text in Listing 3-2.

WARNING

Don't scrutinize Listings 3-2 and 3-3 too carefully — all the code in them is fake! I made up this code so that it looks a lot like real Java code, but it's not real. What's more important, the code in Listings 3-2 and 3-3 isn't meant to illustrate all the rules about Java. So, if you have a grain of salt handy, take it with Listings 3-2 and 3-3.

LISTING 3-2: A Method Declaration

```
void fixTheAlternator(onACertainCar) {
   driveInto(car, bay);
   lift(hood);
   get(wrench);
   loosen(alternatorBelt);
   ...
}
```

Somewhere else in my Java code (somewhere outside of Listing 3-2), I need an instruction to call my `fixTheAlternator` method into action. The instruction to call the `fixTheAlternator` method into action may look like the line in Listing 3-3.

LISTING 3-3: A Method Call

```
fixTheAlternator(junkyOldFord);
```

Now that you have a basic understanding of what a method is and how it works, you can dig a little deeper into some useful terminology:

- If I'm being lazy, I refer to the code in Listing 3-2 as a *method.* If I'm not being lazy, I refer to this code as a *method declaration.*
- The method declaration in Listing 3-2 has two parts. The first line (the part with `fixTheAlternator` in it, up to but not including the open curly brace) is a *method header.* The rest of Listing 3-2 (the part surrounded by curly braces) is a *method body.*
- The term *method declaration* distinguishes the list of instructions in Listing 3-2 from the instruction in Listing 3-3, which is known as a *method call.*

REMEMBER

A *method's declaration* tells the computer what happens if you call the method into action. A *method call* (a separate piece of code) tells the computer to actually call the method into action. A method's declaration and the method's call tend to be in different parts of the Java program.

The main method in a program

Figure 3-3 has a copy of the code from Listing 3-1. The bulk of the code contains the declaration of a method named `main`. (Just look for the word *main* in the code's method header.) For now, don't worry about the other words in the method header: `public`, `static`, `void`, `String`, and `args`. I explain these words in the next several chapters.

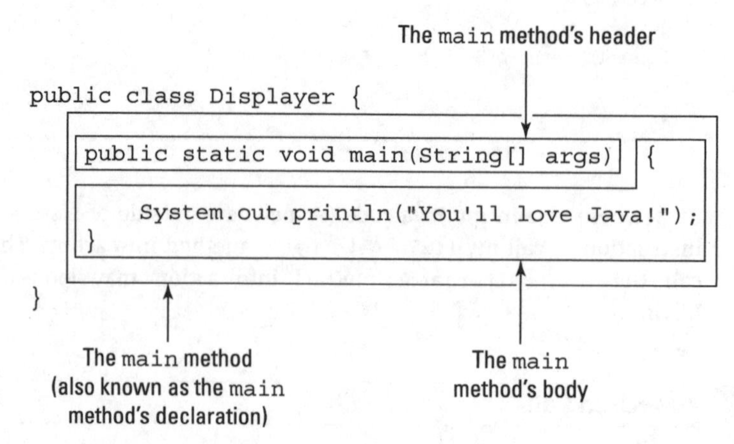
The main method's header
public class Displayer {
public static void main(String[] args)
{
System.out.println("You'll love Java!");
}
}
The main method (also known as the main method's declaration)
The main method's body

FIGURE 3-3: The main method.

Like any Java method, the `main` method is a recipe:

```
How to make biscuits:
    Heat the oven.
    Roll the dough.
    Bake the rolled dough.
```

or

```
How to follow the main instructions for a Displayer:
    Print "You'll love Java!" on the screen.
```

The word *main* plays a special role in Java. In particular, you never write code that explicitly calls a `main` method into action. The word *main* is the name of the method that is called into action automatically when the program begins running.

Look back at Figure 3-1. When the `Displayer` program runs, the computer automatically finds the program's `main` method and executes any instructions inside the method's body. In the `Displayer` program, the `main` method's body has only one instruction. That instruction tells the computer to print `You'll love Java!` on the screen. So, in Figure 3-1, `You'll love Java!` appears on the computer screen.

REMEMBER

The instructions in a method aren't executed until the method is called into action. But, if you give a method the name *main*, that method is called into action automatically.

TECHNICAL STUFF

Almost every computer programming language has something akin to Java's methods. If you've worked with other languages, you may remember terms like *functions, procedures, subprograms, subroutines,* and good ol' `PERFORM` statements. Whatever you call it in your favorite programming language, a method is a bunch of instructions collected and given a new name.

How you finally tell the computer to do something

Buried deep in the heart of Listing 3-1 is the single line that actually issues a direct instruction to the computer. The line, which is highlighted in Figure 3-4, tells the computer to display `You'll love Java!`. This line is a statement. In Java, a *statement* is a direct instruction that tells the computer to do something (for example, display this text, put 7 in that memory location, make a window appear).

```
public class Displayer {

    public static void main(String[] args)  {
        System.out.println("You'll love Java!");
    }
}
```

A statement
(a call to the `System.out.println` method)

FIGURE 3-4: A Java statement.

REMEMBER

In `System.out.println`, the next-to-last character is a lowercase letter *l*, not a digit *1*.

Of course, Java has different kinds of statements. A method call, which I introduce in the earlier section "The Java method," is one of the many kinds of Java statements. Listing 3-3 shows you what a method call looks like, and Figure 3-4 also contains a method call that looks like this:

```
System.out.println("You'll love Java!");
```

When the computer executes this statement, the computer calls into action a method named `System.out.println`. (Yes, in Java, a name can have dots in it. The dots mean something.)

WARNING

I said it already, but it's worth repeating: In `System.out.println`, the next-to-last character is a lowercase letter *l (as in the word* line), not a digit *1 (as in the number* one). If you use a digit *1*, your code won't work. Just think of `println` as a way of saying "print line" and you won't have any problem.

To learn the meaning behind the dots in Java names, see Chapters 7 and 14.

Figure 3-5 illustrates the `System.out.println` situation. Actually, two methods play active roles in the running of the `Displayer` program. Here's how they work:

- **There's a declaration for a `main` method.** I wrote the `main` method myself. This `main` method is called automatically whenever I run the `Displayer` program.
- **There's a call to the `System.out.println` method.** The method call for the `System.out.println` method is the only statement in the body of the main method. In other words, calling the `System.out.println` method is the only item on the `main` method's to-do list.

  The declaration for the `System.out.println` method is buried inside the official Java API. For a refresher on the Java API, see the sections "The grammar and the common names" and "The words in a Java program," earlier in this chapter.

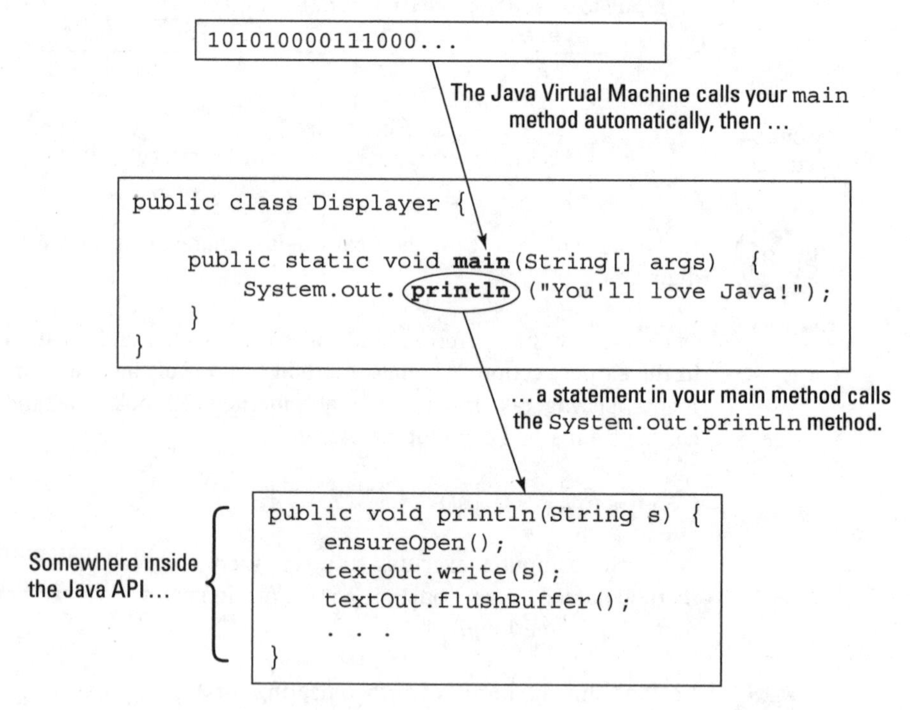

FIGURE 3-5: Calling the `System.out.println` method.

TECHNICAL STUFF

When I say things like, "`System.out.println` is buried inside the API," I'm not doing justice to the API. True, you can ignore all the nitty-gritty Java code inside the API. All you need to remember is that `System.out.println` is defined somewhere inside that code. But I'm not being fair when I make the API code sound like something magical. The API is just another bunch of Java code. The statements in the API that tell the computer what it means to carry out a call to `System.out.println` look a lot like the Java code in Listing 3-1.

In Java, each statement (like the boxed line in Figure 3-4) ends with a semicolon. Other lines in Figure 3-4 don't end with semicolons, because the other lines in Figure 3-4 aren't statements. For instance, the method header (the line with the word *main* in it) doesn't directly tell the computer to do anything. The method header announces, "Just in case you ever want to do `main`, the next few lines of code tell you how to do it."

REMEMBER

Every complete Java statement ends with a semicolon.

Brace yourself

Long ago, or maybe not so long ago, your schoolteachers told you how useful outlines are. With an outline, you can organize thoughts and ideas, help people see forests instead of trees, and generally show that you're a member of the Tidy Persons Club. Well, a Java program is like an outline. The program in Listing 3-1 starts with a header line that says, "Here comes a class named `Displayer`." After that header, a subheader announces, "Here comes a method named `main`."

Now, if a Java program is like an outline, why doesn't a program *look* like an outline? What takes the place of the Roman numerals, capital letters, and other items? The answer is twofold:

- In a Java program, curly braces enclose meaningful units of code.
- You, the programmer, can (and should) indent lines so that other programmers can see at a glance the outline form of your code.

In an outline, everything is subordinate to the item in Roman numeral *I*. In a Java program, everything is subordinate to the top line — the line with `class` in it. To indicate that everything else in the code is subordinate to this `class` line, you use curly braces. Everything else in the code goes inside these curly braces. (See Listing 3-4.)

LISTING 3-4: Curly Braces for a Java Class

```
public class Displayer {

    public static void main(String[] args) {
        System.out.println("You'll love Java!");
    }
}
```

In an outline, some stuff is subordinate to a capital letter *A* item. In a Java program, some lines are subordinate to the method header. To indicate that something is subordinate to a method header, you use curly braces. (See Listing 3-5.)

LISTING 3-5: Curly Braces for a Java Method

```
public class Displayer {

    public static void main(String[] args) {
        System.out.println("You'll love Java!");
    }
}
```

In an outline, some items are at the bottom of the food chain. In the `Displayer` class, the corresponding line is the line that begins with `System.out.println`. Accordingly, this `System.out.println` line goes inside all the other curly braces and is indented more than any other line.

REMEMBER

Never lose sight of the fact that a Java program is, first and foremost, an outline.

If you put curly braces in the wrong places or omit curly braces where the braces should be, your program probably won't work at all. If your program works, it'll probably work incorrectly.

If you don't indent lines of code in an informative manner, your program will still work correctly, but neither you nor any other programmer will be able to figure out what you were thinking when you wrote the code.

If you're a visual thinker, you can picture outlines of Java programs in your head. One friend of mine visualizes an actual numbered outline morphing into a Java program. (See Figure 3-6.) Another person, who shall remain nameless, uses more bizarre imagery. (See Figure 3-7.)

```
I. The Displayer class
 A. The main method
   1. Print "You'll love Java!"
```

```
I. public class Displayer
 A. public static void main (String[] args)
   1. System.out.println("You'll love Java!");
```

```
public class Displayer {

    public static void main(String[] args)  {
        System.out. println ("You'll love Java!");
    }
}
```

FIGURE 3-6: An outline turns into a Java program.

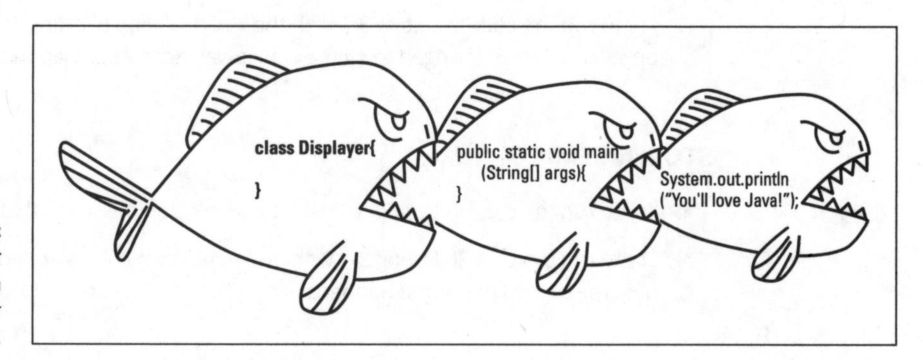
class Displayer{
}
public static void main
(String[] args){
}
System.out.println
("You'll love Java!");

FIGURE 3-7: A class is bigger than a method; a method is bigger than a statement.

TIP

I appreciate a good excuse as much as the next guy, but failing to indent your Java code is inexcusable. In fact, many IDEs have tools to indent your code automatically. In your favorite IDE, look for menus such as Source⇨Format or Code⇨Reformat Code.

TRY IT OUT

Here are some things for you to try to help you understand the material in this section. If trying these things builds your confidence, that's good. If trying these things makes you question what you've read, that's good too. If trying these things makes you nervous, don't be discouraged. You can find answers and other help at this book's website (`http://JavaForDummies.allmycode.com`). You can also email me with your questions (`JavaForDummies@allmycode.com`).

BEGIN THE BEGUINE

- » If you've downloaded the code from this book's website, import Listing 3-1 (from the downloaded `03-01` folder) into your IDE. If you don't plan to download the code, create a new project in your IDE. In the new project, create a class named `Displayer` with the code from Listing 3-1. With the downloaded project, or with your own, newly created project, run the program and look for the words `You'll love Java!` in the output.
- » Try running the code in Listing 3-1 with the text `"You'll love Java!"` changed to `"No more baked beans!"`. What happens?

CAPITAL PAINS

- » Try to run the code in Listing 3-1 with the word `public` (all lowercase) changed to `Public` (starting with an uppercase letter). What happens?
- » Try to run the code in Listing 3-1 with the word `main` (all lowercase) changed to `Main` (starting with an uppercase letter). What happens?
- » Try to run the code in Listing 3-1 with the word `System` (starting with an uppercase letter) changed to `system` (all lowercase). What happens?

STOLEN SEMICOLON

- » Try to run the code in Listing 3-1 with the semicolon missing. What happens?
- » Try to run the code in Listing 3-1 with additional semicolons added at the ends of some of the lines. What happens?

MISCELLANEOUS MISCHIEF

- » Try to run the code in Listing 3-1 with the indentation changed. For example, don't indent any lines. Also, for good measure, remove the line breaks between the first curly brace and the word `public` (so that the code reads `public class Displayer { public ...`). What happens?
- » Try to run the code in Listing 3-1 with the word `println` changed to `print1n` (with the digit `1` near the end). What happens?
- » Try to run the code in Listing 3-1 with the text `"You'll love Java!"` changed to `" Use a straight quote \", not a curly quote \u201D"`. What happens?

And Now, a Few Comments

People gather around campfires to hear the old legend about a programmer whose laziness got her into trouble. To maintain this programmer's anonymity, I call her Jane Pro. Jane worked many months to create the holy grail of computing: a program that thinks on its own. If completed, this program could work independently, learning new concepts without human intervention. Day after day, night after night, Jane Pro labored to give the program that spark of creative, independent thought.

One day, when she was almost finished with the project, she received a disturbing piece of paper mail from her health insurance company. No, the mail wasn't about a serious illness. It was about a routine office visit. The insurance company's claim form had a place for Jane's date of birth, as if her date of birth had changed since the last time she sent in a claim. She had absentmindedly scribbled *2016* as her year of birth, so the insurance company refused to pay the bill.

Jane dialed the insurance company's phone number. Within 20 minutes, she was talking to a live person. "I'm sorry," said the live person. "To resolve this issue, you must dial a different number." Well, you can guess what happened next. "I'm sorry. The other operator gave you the wrong number." And then, "I'm sorry. You must call back the original phone number."

Five months later, Jane's ear ached, but after 800 hours on the phone, she had finally gotten a tentative promise that the insurance company would eventually reprocess the claim. Elated as she was, she was anxious to get back to her programming project. Could she remember what all those lines of code were supposed to be doing?

No, she couldn't. Jane stared and stared at her own work and, like a dream that doesn't make sense the next morning, the code was completely meaningless to her. She had written a million lines of code, and not one line was accompanied by an informative explanatory comment. She had left no clues to help her understand what she'd been thinking, so in frustration she abandoned the whole project.

Adding comments to your code

Listing 3-6 holds an enhanced version of this chapter's sample program. In addition to all the keywords, identifiers, and punctuation, Listing 3-6 has text that's meant for human beings to read.

LISTING 3-6: Three Kinds of Comments

```
/*
 * Listing 3-6 in "Java For Dummies, 8th Edition"
 *
 * Copyright 2022 Wiley Publishing, Inc.
 * All rights reserved.
 */

/**
 * The Displayer class displays text
 * on the computer screen.
 *
 * @author  Barry Burd
 * @version 1.0 1/24/22
 * @see     java.lang.System
 */
public class Displayer {

    /**
     * The main method is where
     * execution of the code begins.
     *
     * @param  args   (See Chapter 11.)
     */
    public static void main(String[] args) {
        System.out.println("I love Java!");  //Replace "I" with "You"?
    }
}
```

A *comment* is a special section of text, inside a program, whose purpose is to help people understand the program. A comment is part of a good program's documentation.

The Java programming language has three kinds of comments:

- **Traditional comments:** The first five lines of Listing 3-6 form one *traditional* comment. The comment begins with `/*` and ends with `*/`. Everything between the opening `/*` and the closing `*/` is for human eyes only. No information about `"Java For Dummies, 8th Edition"` or `Wiley Publishing, Inc.` is translated by the compiler.

CROSS REFERENCE

To read about compilers, see Chapter 2.

The second, third, fourth, and fifth lines in Listing 3-6 have extra asterisks (`*`). I call them extra because these asterisks aren't required when you create a comment. They just make the comment look pretty. I include them in Listing 3-6 because, for some reason that I don't entirely understand, most Java programmers add these extra asterisks.

- **End-of-line comments:** The text `//Replace "I" with "You"?` in Listing 3-6 is an end-of-line comment. An *end-of-line* comment starts with two slashes and goes to the end of a line of type. Once again, the compiler doesn't translate the text inside the end-of-line comment.
- **Javadoc comments:** A *javadoc* comment begins with a slash and two asterisks (/**). Listing 3-6 has two javadoc comments: one with the text `The Displayer class ...` and another with the text `The main method is where ....`

  A *javadoc* comment, which is a special kind of traditional comment, is meant to be read by people who never even look at the Java code. But that doesn't make sense. How can you see the javadoc comments in Listing 3-6 if you never look at Listing 3-6?

  Well, a certain program called *javadoc* (what else?) can find all the javadoc comments in Listing 3-6 and turn these comments into a nice-looking web page. Figure 3-8 shows the page.

Javadoc comments are great. Here are several great things about them:

- The only person who has to look at a piece of Java code is the programmer who writes the code. Other people who use the code can find out what the code does by viewing the automatically generated web page.
- Because other people don't look at the Java code, other people don't make changes to the Java code. (In other words, other people don't introduce errors into the existing Java code.)
- Because other people don't look at the Java code, other people don't have to decipher the inner workings of the Java code. All these people need to know about the code is what they read on the code's web page.
- The programmer doesn't create two separate files — some Java code over here and some documentation about the code over there. Instead, the programmer creates one piece of Java code and embeds the documentation (in the form of javadoc comments) right inside the code.
- The generation of web pages from javadoc comments is automatic. So everyone's documentation has the same format. No matter whose Java code you use, you find out about that code by reading a page like the one in Figure 3-8. That's good because the format in Figure 3-8 is familiar to anyone who uses Java.

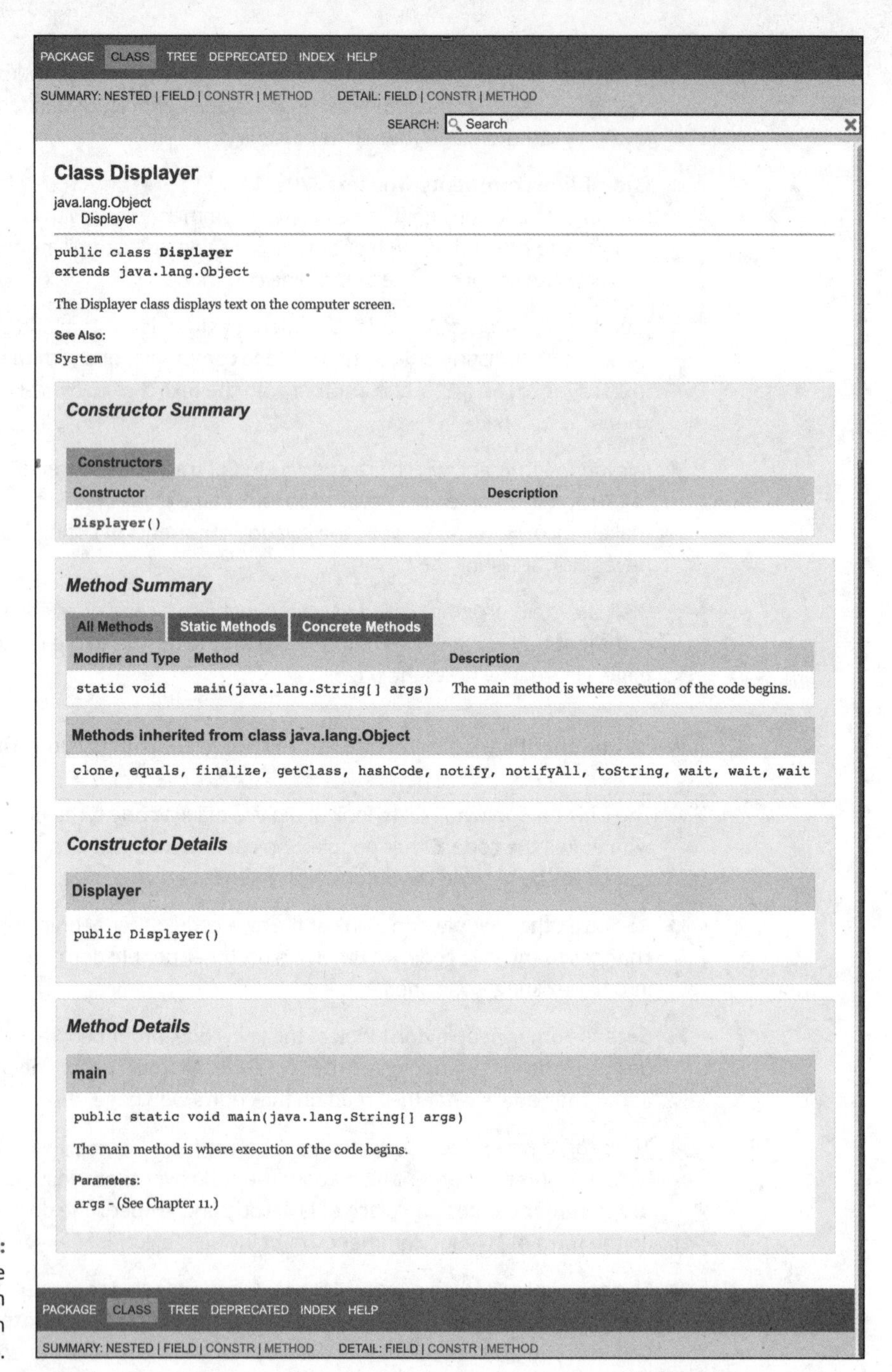

FIGURE 3-8: The javadoc page generated from the code in Listing 3-6.

» Best of all, Java's online API Specification, which describes all of Java's standard identifiers, comes entirely from runs of the *javadoc* program on comments in Java code. There's nothing arbitrary or haphazard about Java's API Specification. The Specification is always in sync with the actual Java code.

You can generate your own web pages from the javadoc comments that you put in your code. From your IDE's menu bar, select Tools or Project and look for a command labeled Generate JavaDoc.

What's Barry's excuse?

For years I've been telling my students to put comments in their code, and for years I've been creating sample code (like the code in Listing 3-1) with no comments in it. Why?

Three little words: *Know your audience.* When you write complicated, real-life code, your audience is other programmers, information technology managers, and people who need help deciphering what you've done. When I write simple samples of code for this book, my audience is you — the novice Java programmer. Rather than read my comments, your best strategy is to stare at my Java statements — the statements that Java's compiler deciphers. That's why I put so few comments in this book's listings.

Besides, I'm a little lazy.

Using comments to experiment with your code

You may hear programmers talk about *commenting out* certain parts of their code. When you're writing a program and something's not working correctly, it often helps to try removing some of the code. If nothing else, you find out what happens when that suspicious code is removed. Of course, you may not like what happens when the code is removed, so you don't want to delete the code completely. Instead, you turn your ordinary Java statements into comments. For instance, you turn the statement

```
System.out.println("I love Java!");
```

into the comment

```
// System.out.println("I love Java!");
```

This change keeps the Java compiler from seeing the code while you try to figure out what's wrong with your program.

Traditional comments aren't very useful for commenting out code. The big problem is that you can't put one traditional comment inside of another. Suppose that you want to comment out the following statements:

```
System.out.println("Parents,");
System.out.println("pick your");
/*
 * Intentionally displays on four separate lines
 */
System.out.println("battles");
System.out.println("carefully!");
```

If you try to turn this code into one traditional comment, you get the following mess:

```
/*
  System.out.println("Parents,");
  System.out.println("pick your");
  /*
   * Intentionally displays on four separate lines
   */
  System.out.println("battles");
  System.out.println("carefully!");
*/
```

The first `*/` (after `Intentionally displays`) ends the traditional comment prematurely. Then the `battles` and `carefully` statements aren't commented out, and the last `*/` chokes the compiler. You can't nest traditional comments inside one another. Because of this, I recommend end-of-line comments as tools for experimenting with your code.

Most IDEs can comment out sections of your code for you automatically. From your IDE's menu bar, select Code or Source and look for menu items pertaining to comments.

2 Writing Your Own Java Programs

IN THIS PART . . .

Create new values and modify existing values.

Add decision-making to your application's logic.

Use repetition as a tool in problem-solving.

» Making things store certain types of values

» Applying operators to get new values

Chapter 4

Making the Most of Variables and Their Values

The following conversation between Mr. Van Doren and Mr. Barasch never took place:

> *Charles:* A sea squirt eats its brain, turning itself from an animal into a plant.
>
> *Jack:* Is that your final answer, Charles?
>
> *Charles:* Yes, it is.
>
> *Jack:* How much money do you have in your account today, Charles?
>
> *Charles:* I have fifty dollars and twenty-two cents in my checking account.
>
> *Jack:* Well, you had better call the IRS, because your sea squirt answer is correct. You just won a million dollars to add to your checking account. What do you think of that, Charles?
>
> *Charles:* I owe it all to honesty, diligence, and hard work, Jack.

Some aspects of this dialogue can be represented in Java by a few lines of code.

Varying a Variable

No matter how you acquire your million dollars, you can use a variable to tally your wealth. Listing 4-1 shows the code.

LISTING 4-1: Using a Variable

```
amountInAccount = 50.22;
amountInAccount = amountInAccount + 1000000.00;
```

You don't have to type the code in Listing 4-1 (or in any of this book's listings). To download all the code in this book, visit the book's website (`http://JavaForDummies.allmycode.com`).

The code in Listing 4-1 makes use of the `amountInAccount` variable. A *variable* is a placeholder. You can stick a number like 50.22 into a variable. After you place a number in the variable, you can change your mind and put a different number into the variable. (That's what varies in a variable.) Of course, when you put a new number in a variable, the old number is no longer there. If you didn't save the old number somewhere else, the old number is gone.

Figure 4-1 gives a before-and-after picture of the code in Listing 4-1. After the first statement in Listing 4-1 is executed, the variable `amountInAccount` has the number 50.22 in it. Then, after the second statement of Listing 4-1 is executed, the `amountInAccount` variable suddenly has 1000050.22 in it. When you think about a variable, picture a place in the computer's memory where wires and transistors store 50.22, 1000050.22, or whatever. On the left side of Figure 4-1, imagine that the box with 50.22 in it is surrounded by millions of other such boxes.

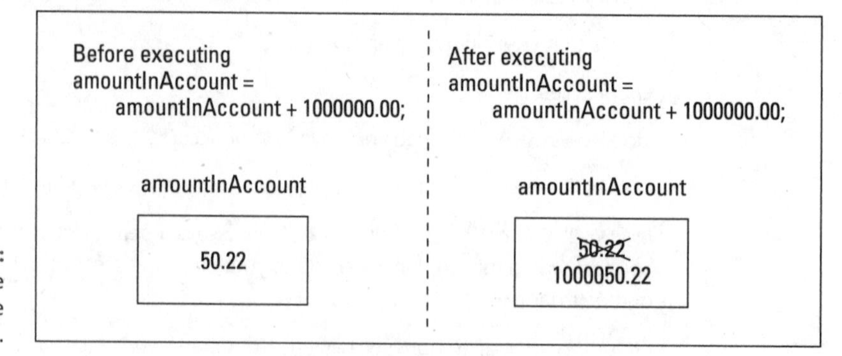

FIGURE 4-1: A variable (before and after).

Now you need some terminology. The thing stored in a variable is a *value*. A variable's value can change during the run of a program (when Jack gives you a million bucks, for instance). The value that's stored in a variable isn't necessarily a number. (For instance, you can create a variable that always stores a letter.) The kind of value that's stored in a variable is a variable's *type*.

You can read more about types in the section "The types of values that variables may have," later in this chapter.

TECHNICAL STUFF

A subtle, almost unnoticeable difference exists between a variable and a variable's *name*. Even in formal writing, I often use the word *variable* when I mean *variable name*. Strictly speaking, `amountInAccount` is a variable name, and all the memory storage associated with `amountInAccount` (including the type that `amountInAccount` has and whatever value `amountInAccount` currently represents) is the variable itself. If you think this distinction between *variable* and *variable name* is too subtle for you to worry about, join the club.

Every variable name is an identifier — a name that you can make up in your own code. In preparing Listing 4-1, I made up the name *amountInAccount*.

CROSS REFERENCE

For more information on the kinds of names in a Java program, see Chapter 3.

Before the sun sets on Listing 4-1, you need to notice one more part of the listing. The listing has `50.22` and `1000000.00` in it. Anybody in their right mind would call these things *numbers*, but in a Java program it helps to call these things *literals*.

And what's so literal about `50.22` and `1000000.00`? Well, think about the variable `amountInAccount` in Listing 4-1. The variable `amountInAccount` stands for 50.22 some of the time, but it stands for 1000050.22 the rest of the time. You could use the word *number* to talk about `amountInAccount`. But really, what `amountInAccount` stands for depends on the fashion of the moment. On the other hand, 50.22 literally stands for the value 50 22/100.

REMEMBER

A variable's value changes; a literal's value doesn't.

TIP

You can add underscores to numeric literals. Rather than use the plain old `1000000.00` in Listing 4-1, you can write `amountInAccount = amountInAccount + 1_000_000.00`. Unfortunately, you can't easily do what you're most tempted to do. You can't write `1,000,000.00` (as you would in the United States), nor can you write `1.000.000,00` (as you would in Germany). If you want to display a number such as `1,000,000.00` in the program's output, you have to use some fancy formatting tricks. For more information about formatting, check Chapters 10 and 11.

Assignment statements

Statements like the ones in Listing 4-1 are called assignment statements. In an *assignment statement*, you assign a value to something. In many cases, this something is a variable.

I recommend getting into the habit of reading assignment statements from right to left. Figure 4-2 illustrates the action of the first line in Listing 4-1.

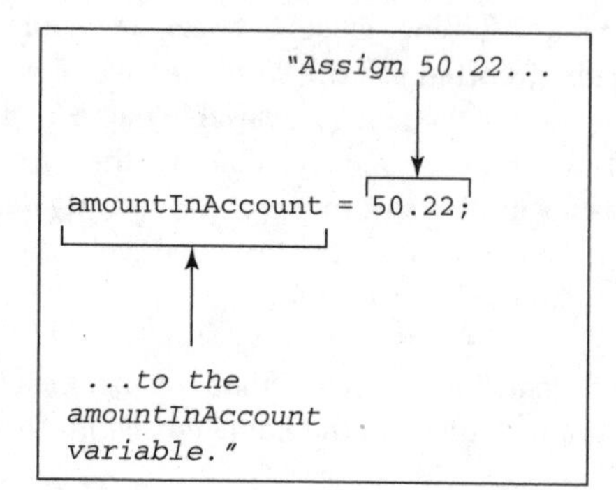

FIGURE 4-2: The action of the first line in Listing 4-1.

The second line in Listing 4-1 is just a bit more complicated. Figure 4-3 illustrates the action of the second line in Listing 4-1.

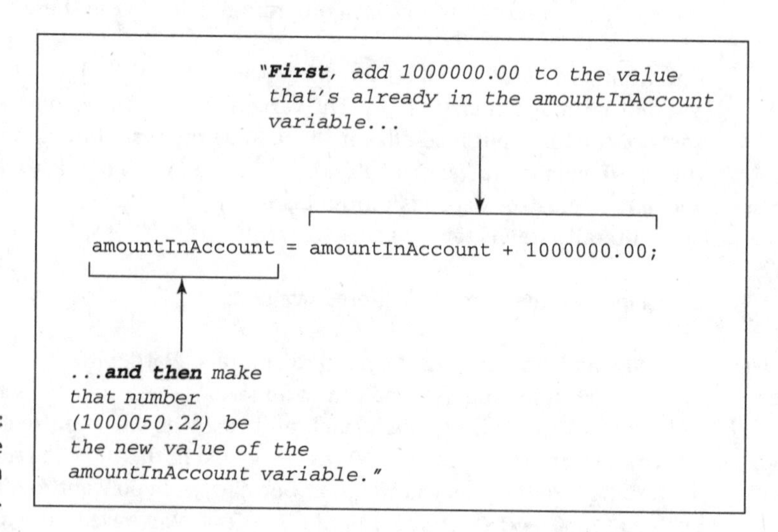

FIGURE 4-3: The action of the second line in Listing 4-1.

REMEMBER

In an assignment statement, the thing being assigned a value is always on the left side of the equal sign.

The types of values that variables may have

Have you seen the TV commercials that make you think you're flying among the circuits inside a computer? Pretty cool, eh? These commercials show 0s (zeros) and 1s (ones) sailing by because 0s and 1s are the only things that computers can deal with. When you think a computer is storing the letter *J*, the computer is really storing 01001010. Everything inside the computer is a sequence of 0s and 1s. As every computer geek knows, a 0 or 1 is called a *bit.*

As it turns out, the sequence 01001010, which stands for the letter *J*, can also stand for the number 74. The same sequence can also stand for $1.0369608636003646 \times 10^{-43}$. In fact, if the bits are interpreted as screen pixels, the same sequence can be used to represent the dots shown in Figure 4-4. The meaning of 01001010 depends on the way the software interprets this sequence of 0s and 1s.

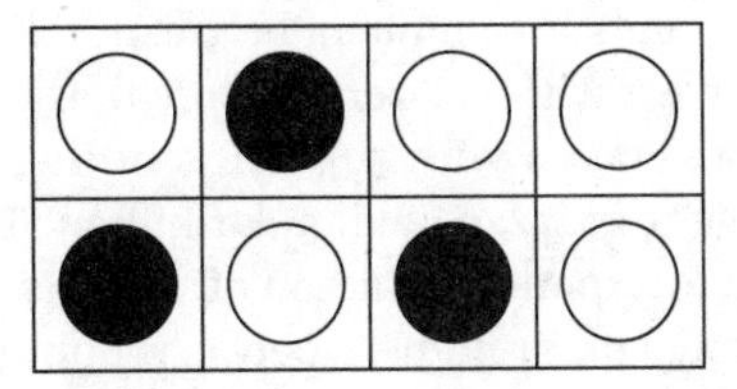

FIGURE 4-4: An extreme close-up of eight black and white screen pixels.

How do you tell the computer what 01001010 stands for? The answer is in the concept of type. The *type* of a variable is the range of values that the variable is permitted to store. I copied the lines from Listing 4-1 and put them into a complete Java program. The program is in Listing 4-2. When I run the program in Listing 4-2, I get the output shown in Figure 4-5.

LISTING 4-2: A Program Uses amountInAccount

```
public class Millionaire {

    public static void main(String[] args) {
        double amountInAccount;

        amountInAccount = 50.22;
        amountInAccount = amountInAccount + 1000000.00;

        System.out.print("You have $");
        System.out.print(amountInAccount);
        System.out.println(" in your account.");
    }
}
```

FIGURE 4-5: Running the program in Listing 4-2.

```
You have $1000050.22 in your account.
```

In Listing 4-2, look at the first line in the body of the `main` method:

```
double amountInAccount;
```

This line is called a *variable declaration.* Putting this line in your program is like saying, "I'm declaring my intention to have a variable named *amountInAccount* in my program." This line reserves the name *amountInAccount* for your use in the program.

In this variable declaration, the word *double* is a Java keyword. This word *double* tells the computer what kinds of values you intend to store in `amountInAccount`. In particular, the word *double* stands for numbers between -1.8×10^{308} and 1.8×10^{308}. (These are enormous numbers with 308 zeros before the decimal point. Only the world's richest people write checks with 308 zeros in them. The second of these numbers is one-point-eight gazazzo-zillion-kaskillion. The number 1.8×10^{308}, a constant defined by the International Bureau of Weights and Measures, is the number of eccentric computer programmers between Sunnyvale, California, and the M31 Andromeda Galaxy.)

More important than the humongous range of the `double` keyword's numbers is the fact that a `double` value can have digits beyond the decimal point. After you declare `amountInAccount` to be of type `double`, you can store all sorts of numbers in `amountInAccount`. You can store `50.22`, `0.02398479`, or `-3.0`. In Listing 4-2, if I hadn't declared `amountInAccount` to be of type `double`, I may not have been able to store `50.22`. Instead, I would have had to store plain old `50`, with no digits beyond the decimal point.

Another type — type `float` — also allows you to have digits beyond the decimal point. But `float` values aren't as accurate as `double` values.

TIP

In many situations, you have a choice. You can declare certain values to be either `float` values or `double` values. But don't sweat the choice between `float` and `double`. For most programs, just use `double`. With today's fancy processors, the space you save using the `float` type is almost never worth the loss of accuracy. (For more details, see the nearby sidebar, "To the decimal point and beyond!")

The big million-dollar jackpot in Listing 4-2 is impressive. But Listing 4-2 doesn't illustrate the best way to deal with dollar amounts. In a Java program, the best way to represent currency is to shun the `double` and `float` types and opt instead for a type named `BigDecimal`. For more information, see this book's website (`http://JavaForDummies.allmycode.com`).

TO THE DECIMAL POINT AND BEYOND!

Java has two different types that have digits beyond the decimal point: type `double` and type `float`. So, what's the difference? When you declare a variable to be of type `double`, you're telling the computer to keep track of 64 bits when it stores the variable's values. When you declare a variable to be of type `float`, the computer keeps track of only 32 bits.

You could change Listing 4-2 and declare `amountInAccount` to be of type `float`:

```
float amountInAccount;
```

Surely, 32 bits are enough to store a small number like 50.22, right? Well, they are and they aren't. You could easily store 50.00 with only 32 bits. Heck, you could store 50.00 with only 6 bits. The size of the number doesn't matter. The accuracy matters. In a 64-bit double variable, you're using most of the bits to store stuff beyond the decimal point. To store the .22 part of 50.22, you need more than the measly 32 bits that you get with type `float`.

Do you really believe what you just read — that it takes more than 32 bits to store . 22? To help convince you, I made a few changes to the code in Listing 4-2. I made `amountInAccount` be of type `float`. Then I changed the first three statements inside the main method:

```
float amountInAccount;
amountInAccount = 50.22F;
amountInAccount = amountInAccount + 1000000.00F;
```

(To understand why I used the letter `F` in `50.22F` and `1000000.00F`, see Table 4-1, later in this chapter.) The output I got was

```
You have $1000050.25 in your account.
```

Compare this with the output in Figure 4-5. When I switch from type `double` to type `float`, Charles has an extra three cents in his account. By changing to the 32-bit `float` type, I've clobbered the accuracy in the `amountInAccount` variable's hundredths place. That's bad.

Another difficulty with `float` values is purely cosmetic. Look again at the literals, `50.22` and `1000000.00`, in Listing 4-2. The Laws of Java say that literals like these take up 64 bits each. So, if you declare `amountInAccount` to be of type `float`, you'll run into

(continued)

(continued)

trouble. You'll have trouble stuffing those 64-bit literals into your little 32-bit `amountInAccount` variable. To compensate, you can switch from `double` literals to `float` literals by adding an F to each `double` literal, but a number with an extra F at the end looks funny.

```
float amountInAccount;
amountInAccount = 50.22F;
amountInAccount = amountInAccount + 1000000.00F;
```

To experiment with numbers, visit `http://babbage.cs.qc.cuny.edu/IEEE-754.old/Decimal.html`. The page takes any number you enter and shows you how the number would be represented as 32 bits and as 64 bits.

How to hold the line

The last three statements in Listing 4-2 use a neat formatting trick. You want to display several different items on a single line on the screen. You put these items in separate statements. All but the last of the statements are calls to `System.out.print`. (The last statement is a call to `System.out.println`.) Calls to `System.out.print` display text on part of a line and then leave the cursor at the end of the current line. After executing `System.out.print`, the cursor is still at the end of the same line, so the next `System.out.`*`whatever`* can continue printing on that same line. With several calls to print capped off by a single call to `println`, the result is just one nice-looking line of output. (Refer to Figure 4-5.)

REMEMBER

A call to `System.out.print` writes some things and leaves the cursor sitting at the end of the line of output. A call to `System.out.println` writes things and then finishes the job by moving the cursor to the start of a brand-new line of output.

TRY IT OUT

Run the code in Listing 4-2 to make sure that it runs correctly on your computer. Then see what happens when you make the following changes:

NUMBER FORMAT EXPERIMENTS

- Add thousands-separators to the number `1000000.00` in the code. For example, if you live in the United States, where the thousands-separator is a comma, change the number to `1,000,000.00` and see what happens. (***Hint:*** Nothing good happens.)
- Try using underscores as thousands-separators in the code. That is, change `1000000.00` to `1_000_000.00` and see what happens.

» Add a currency symbol to the number `50.22` in the code. For example, if you live in the United States, where the currency symbol is $, see what happens when you change the first assignment statement to `amountInAccount = $50.22`.

HOW TO DISPLAY VALUES

» Listing 4-2 has two `System.out.print` statements and one `System.out.println` statement. Change all three to `System.out.println` statements and then run the program.

» The code in Listing 4-2 displays one line of text in its output. Using the `amountInAccount` variable, add statements to the program so that it displays a second line of text. Have the second line of text be "Now you have even more! You have 2000000.00 in your account."

Numbers without decimal points

"In 1995, the average family had 2.3 children."

At this point, a wise guy always remarks that no real family has exactly 2.3 children. Clearly, whole numbers have a role in this world. Therefore, in Java, you can declare a variable to store nothing but whole numbers. Listing 4-3 shows a program that uses whole number variables.

LISTING 4-3: Using the int Type

```
public class ElevatorFitter {

    public static void main(String[] args) {
        int weightOfAPerson;
        int elevatorWeightLimit;
        int numberOfPeople;

        weightOfAPerson = 150;
        elevatorWeightLimit = 1400;
        numberOfPeople = elevatorWeightLimit / weightOfAPerson;

        System.out.print("You can fit ");
        System.out.print(numberOfPeople);
        System.out.println(" people on the elevator.");
    }
}
```

The story behind the program in Listing 4-3 takes some heavy-duty explaining. Here goes:

You have a hotel elevator whose weight capacity is 1,400 pounds. One weekend the hotel hosts the Brickenchicker family reunion. A certain branch of the Brickenchicker family has been blessed with identical dectuplets (ten siblings, all with the same physical characteristics). Normally, each of the Brickenchicker dectuplets weighs exactly 145 pounds. But on Saturday the family has a big catered lunch, and, because lunch included strawberry shortcake, each of the Brickenchicker dectuplets now weighs 150 pounds. Immediately after lunch, all ten of the Brickenchicker dectuplets arrive at the elevator at exactly the same time. (Why not? All ten of them think alike.) So, the question is, how many of the dectuplets can fit on the elevator?

Now remember, if you put one ounce more than 1,400 pounds of weight on the elevator, the elevator cable breaks, plunging all dectuplets on the elevator to their sudden (and costly) deaths.

The answer to the Brickenchicker riddle (the output of the program of Listing 4-3) is shown in Figure 4-6.

```
You can fit 9 people on the elevator.
```

FIGURE 4-6: Save the Brickenchickers.

At the core of the Brickenchicker elevator problem, you have whole numbers — numbers with no digits beyond the decimal point. When you divide 1,400 by 150, you get 9⅓, but you shouldn't take the ⅓ seriously. No matter how hard you try, you can't squeeze an extra 50 pounds' worth of Brickenchicker dectuplet onto the elevator. This fact is reflected nicely in Java. In Listing 4-3, all three variables (`weightOfAPerson`, `elevatorWeightLimit`, and `numberOfPeople`) are of type `int`. An `int` value is a whole number. When you divide one `int` value by another (as you do with the slash in Listing 4-3), you get another `int`. When you divide 1,400 by 150, you get 9 — not 9⅓. You see this in Figure 4-6. Taken together, the following statements display 9 onscreen:

```
numberOfPeople = elevatorWeightLimit / weightOfAPerson;

System.out.print(numberOfPeople);
```

TAKE A FLYING LEAP

TRY IT OUT

My wife and I were married on February 29, so we have one anniversary every four years. Write a program with a variable named `years`. Based on the value of the `years` variable, the program displays the number of anniversaries we've had. For example, if the value of `years` is 4, the program displays the sentence `Number of anniversaries: 1`. If the value of `years` is 7, the program still displays `Number of anniversaries: 1`. But if the value of years is 8, the program displays `Number of anniversaries: 2`.

Combining declarations and initializing variables

Look back at Listing 4-3. In that listing, you see three variable declarations — one for each of the program's three `int` variables. I could have done the same thing with just one declaration:

```
int weightOfAPerson, elevatorWeightLimit, numberOfPeople;
```

WARNING

If two variables have completely different types, you can't create both variables in the same declaration. For instance, to create an `int` variable named *weightOfFred* and a `double` variable named *amountInFredsAccount*, you need two separate variable declarations.

You can give variables their starting values in a declaration. In Listing 4-3, for instance, one declaration can replace several lines in the `main` method (all but the calls to `print` and `println`):

```
int weightOfAPerson = 150, elevatorWeightLimit = 1400,
    numberOfPeople = elevatorWeightLimit/weightOfAPerson;
```

FOUR WAYS TO STORE WHOLE NUMBERS

Java has four types of whole numbers. The types are `byte`, `short`, `int`, and `long`. Unlike the complicated story about the accuracy of types `float` and `double`, the only thing that matters when you choose among the whole number types is the size of the number you're trying to store. If you want to use numbers larger than 127, don't use `byte`. To store numbers larger than 32767, don't use `short`.

Most of the time, you'll use `int`. But if you need to store numbers larger than 2147483647, forsake `int` in favor of `long`. (A `long` number can be as big as 9223372036854775807.) For the whole story, see Table 4-1, a little earlier in this chapter.

When you do this, you don't say that you're assigning values to variables. The pieces of the declarations with equal signs in them aren't really called assignment statements. Instead, you say that you're *initializing* the variables. Believe it or not, keeping this distinction in mind is helpful.

For example, when you initialize a variable inside a program's `main` method, you don't have to specify the new variable's type. You can start Listing 4-2 like this:

```
public class Millionaire {

    public static void main(String[] args) {
        var amountInAccount = 0.0;
```

If you want, you can start Listing 4-3 this way:

```
public class ElevatorFitter {

    public static void main(String[] args) {
        var weightOfAPerson = 150;
        var elevatorWeightLimit = 1400;
        var numberOfPeople = elevatorWeightLimit/weightOfAPerson;
```

In either case, the word `var` replaces the name of a type. This trick works because Java is smart. When you write

```
var amountInAccount = 0.0;
```

Java looks at the number `0.0` and realizes that `amountInAccount` is a `double` value. It's a number with digits to the right of the decimal point. In the same way, Java sees

```
var weightOfAPerson = 150;
```

and figures out on its own that `weightOfAPerson` has to be `int` value.

This `var` business doesn't work if you don't initialize your new variable. For example, the lines

```
// BAD CODE:
var numberOfCats;
numberOfCats = 3;
```

are unacceptable as far as Java is concerned. Java can't wait until the assignment statement `numberOfCats = 3` to decide what type of variable `numberOfCats` is. Java wants to know immediately.

WARNING

There are other situations in which the use of `var` is illegal, even with an initialization. For details, see Chapter 7.

Like everything else in life, initializing a variable has advantages and disadvantages:

- **When you combine six lines of Listing 4-3 into just one declaration, the code becomes more concise.** Sometimes concise code is easier to read. Sometimes it's not. As a programmer, it's your judgment call.
- **By initializing a variable, you might automatically avoid certain programming errors.** For an example, see Chapter 7.
- **In some situations, you have no choice. The nature of your code forces you either to initialize or not to initialize.** For an example that doesn't lend itself to variable initialization, see the deleting-evidence program in Chapter 6.

Experimenting with JShell

The programs in Listings 4-2 and 4-3 both begin with the same old, tiresome refrain:

```
public class SomethingOrOther {

    public static void main(String[] args) {
```

A Java program requires this verbose introduction because

- In Java, the entire program is a class.
- The `main` method is called into action automatically when the program begins running.

I explain all of this in Chapter 3.

Anyway, retyping this boilerplate code into an editor window can be annoying, especially when your goal is to test the effect of executing a few simple statements. To fix this problem, the stewards of Java came up with a new tool in Java 9. They call it *JShell.*

Instructions for launching JShell differ from one computer to the next. For instructions that work on your computer, visit this book's website (`http://JavaForDummies.allmycode.com`).

When you use JShell, you hardly ever type an entire program. Instead, you type a Java statement, and then JShell responds to your statement, and then you type a second statement, and then JShell responds to your second statement, and then you type a third statement, and so on. A single statement is enough to get a response from JShell.

TECHNICAL STUFF

Some folks have tweaked JShell to make its behavior a bit different from what I describe in this book. If you're running IntelliJ IDEA's JShell Console or some other specialized JShell variant, be sure to check the vendor's documentation.

JShell is only one example of a language's *Read Evaluate Print Loop* (REPL). Many programming languages have REPLs and, with Java 9, the Java language finally has a REPL of its own.

In Figure 4-7, I use JShell to find out how Java responds to the assignment statements in Listings 4-2 and 4-3.

```
jshell> double amountInAccount
amountInAccount ==> 0.0

jshell> amountInAccount = 50.22
amountInAccount ==> 50.22

jshell> amountInAccount = amountInAccount + 1000000.00
amountInAccount ==> 1000050.22

jshell> int weightOfAPerson, elevatorWeightLimit
weightOfAPerson ==> 0
elevatorWeightLimit ==> 0

jshell> weightOfAPerson = 150;
weightOfAPerson ==> 150

jshell> elevatorWeightLimit = 1400
elevatorWeightLimit ==> 1400

jshell> elevatorWeightLimit / weightOfAPerson
$8 ==> 9

jshell> $8 + 1
$9 ==> 10

jshell> 42 + 7
$10 ==> 49

jshell>
```

FIGURE 4-7: An intimate conversation between JShell and me.

When you run JShell, the dialogue goes something like this:

```
jshell> You type a statement
JShell responds

jshell> You type another statement
JShell responds
```

For example, in Figure 4-7, I type `double amountInAccount` and then press Enter. JShell responds by displaying

```
amountInAccount ==> 0.0
```

Here are a few things to notice about JShell:

- **You don't have to type an entire Java program.**

  Typing a few statements such as

  ```
  double amountInAccount
  amountInAccount = 50.22
  amountInAccount = amountInAccount + 1000000.00
  ```

  does the trick. It's like running the code snippet in Listing 4-1 (except that Listing 4-1 doesn't declare `amountInAccount` to be a `double`).

- **In JShell, semicolons are (to a large extent) optional.**

  In Figure 4-7, I type a semicolon at the end of only one of my nine lines.

CROSS REFERENCE

  For some advice about using semicolons in JShell, see Chapter 5.

- **JShell responds immediately after you type each line.**

  After I declare `amountInAccount` to be `double`, JShell responds by telling me that the `amountInAccount` variable has the value `0.0`. After I type `amountInAccount = amountInAccount + 1000000.00`, JShell tells me that the new value of `amountInAccount` is `1000050.22`.

- **You can mix statements from many different Java programs.**

  In Figure 4-7, I mix statements from the programs in Listings 4-2 and 4-3. JShell doesn't care.

- **You can ask JShell for the value of an expression.**

  You don't have to assign the expression's value to a variable. For example, in Figure 4-7, I type

  ```
  elevatorWeightLimit / weightOfAPerson
  ```

JShell responds by telling me that the value of `elevatorWeightLimit / weightOfAPerson` is `9`. JShell makes up a temporary name for that value. In Figure 4-7, the name happens to be `$8`. So, on the next line in Figure 4-7, I ask for the value of `$8 +1`, and JShell gives me the answer `10`.

» **You can even get answers from JShell without using variables.**

On the last line in Figure 4-7, I ask for the value of `42 + 7`, and JShell generously answers with the value `49`.

TIP

While you're running JShell, you don't have to retype commands that you've already typed. You don't even have to copy and paste commands. If you press the up-arrow key once, JShell shows you the command that you typed most recently. If you press the up-arrow key twice, JShell shows you the next-to-last command that you typed. And so on. When JShell shows you a command, you can press the left- and right-arrow keys to move to any character in the middle of the command. You can modify characters in the command. Finally, when you press Enter, JShell executes the newly modified command.

To end your run of JShell, you type **/exit** (starting with a slash). But `/exit` is only one of many commands you can give to JShell. To ask JShell what other kinds of commands you can use, type **/help**.

With JShell, you can test your statements before you put them into a full-blown Java program. That makes JShell a truly useful tool.

SHELL GAME

TRY IT OUT

Visit this book's website (`http://JavaForDummies.allmycode.com`) for instructions on launching JShell on your computer. After launching JShell, type a few lines of code from Figure 4-7. See what happens when you type some slightly different lines.

What Happened to All the Cool Visual Effects?

The programs in Listings 4-2 and 4-3 are text-based. A *text-based* program has no windows, no dialog boxes — nothing of that kind. All you see is line after line of plain, unformatted text. The user types something, and the computer displays a response beneath each line of input.

The opposite of a text-based program is a *graphical user interface (GUI)* program. A GUI program has windows, text fields, buttons, and other visual goodies.

As visually unexciting as text-based programs are, they contain the basic concepts for all computer programming. Also, text-based programs are easier for the novice programmer to read, write, and understand than the corresponding GUI programs. So, in this book I take a three-pronged approach:

- **Text-based examples:** I introduce most of the new concepts with these examples.
- **The `DummiesFrame` class:** Alongside the text-based examples, I present GUI versions using the `DummiesFrame` class, which I created especially for this book. (I introduce the `DummiesFrame` class in Chapter 7.)
- **GUI programming techniques:** I describe some of the well-known techniques in Chapters 9, 10, 14, and 16. I even have a tiny GUI example in this chapter. (See the later section "The Molecules and Compounds: Reference Types.")

With this careful balance of drab programs and sparkly programs, you're sure to learn Java.

The Atoms: Java's Primitive Types

The words *int* and *double* that I describe in the previous sections are examples of *primitive types* (also known as *simple* types) in Java. The Java language has exactly eight primitive types. As a newcomer to Java, you can pretty much ignore all but four of these types. (As programming languages go, Java is nice and compact that way.) Table 4-1 shows the complete list of primitive types.

The types that you shouldn't ignore are `int`, `double`, `char`, and `boolean`. Previous sections in this chapter cover the `int` and `double` types. So the next two sections cover `char` and `boolean` types.

The char type

Several decades ago, people thought computers existed only for doing big number-crunching calculations. Nowadays, nobody thinks that way. So, if you haven't been in a cryogenic freezing chamber for the past 20 years, you know that computers store letters, punctuation symbols, and other characters.

TABLE 4-1 **Java's Primitive Types**

Type Name	What a Literal Looks Like	Range of Values
Whole number types		
`byte`	`(byte)42`	–128 to 127
`short`	`(short)42`	–32768 to 32767
`int`	`42`	–2147483648 to 2147483647
`long`	`42L`	–9223372036854775808 to 9223372036854775807
Decimal number types		
`float`	`42.0F`	-3.4×10^{38} to 3.4×10^{38}
`Double`	`42.0`	-1.8×10^{308} to 1.8×10^{308}
Character type		
`Char`	`'A'`	Thousands of characters, glyphs, and symbols
Logical type		
`Boolean`	`true`	true, false

The Java type that's used to store characters is called *char.* Listing 4-4 has a simple program that uses the `char` type. Figure 4-8 shows the output of the program in Listing 4-4.

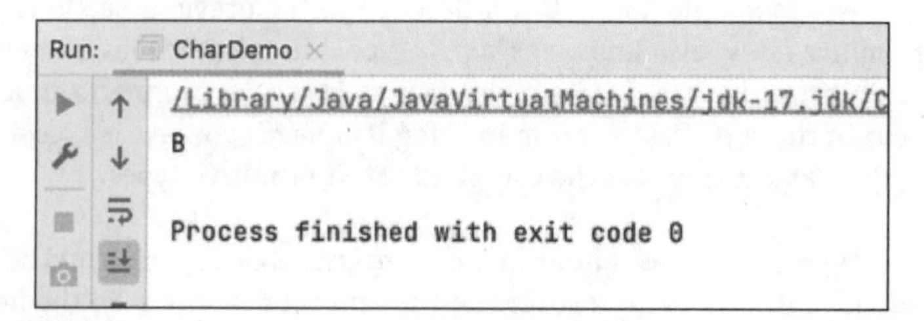
Run:
CharDemo
/Library/Java/JavaVirtualMachines/jdk-17.jdk/C
B
Process finished with exit code 0

FIGURE 4-8: An exciting run of the program of Listing 4-4 as it appears in IntelliJ's Run tool window.

LISTING 4-4: **Using the char Type**

```
public class CharDemo {

  public static void main(String[] args) {
     char myLittleChar = 'b';
     char myBigChar = Character.toUpperCase(myLittleChar);
```

```
        System.out.println(myBigChar);
    }
}
```

In Listing 4-4, the first initialization stores the letter *b* in the variable `myLittleChar`. In the initialization, notice how *b* is surrounded by single quote marks. In Java, every `char` literal starts and ends with a single quote mark.

REMEMBER

In a Java program, single quote marks surround the letter in a `char` literal.

If you need help sorting out the terms *assignment, declaration*, and *initialization*, see the section "Combining declarations and initializing variables," earlier in this chapter.

In the second initialization of Listing 4-4, the program calls an API method whose name is *Character.toUpperCase*. The `Character.toUpperCase` method does just what its name suggests — the method produces the uppercase equivalent of the letter *b*. This uppercase equivalent (the letter *B*) is assigned to the `myBigChar` variable, and the *B* that's stored in `myBigChar` appears on the screen.

CROSS REFERENCE

For an introduction to the Java application programming interface (API), see Chapter 3.

If you're tempted to write the following statement,

```
char myLittleChars = 'barry'; //Don't do this
```

please resist the temptation. You can't store more than one letter at a time in a `char` variable, and you can't put more than one letter between a pair of single quotes. If you're trying to store words or sentences (not just single letters), you need to use something called a String.

For a look at Java's `String` type, see the section "The Molecules and Compounds: Reference Types," later in this chapter.

WARNING

If you're used to writing programs in other languages, you may be aware of something called ASCII character encoding. Most languages use ASCII; Java uses Unicode. In the old ASCII representation, each character takes up only 8 bits, but in Unicode, each character takes up 8, 16, or 32 bits. Whereas ASCII stores the letters of the Roman (English) alphabet, Unicode has room for characters from most of the world's commonly spoken languages. The only problem is that some of the Java API methods are geared specially toward 16-bit Unicode. Occasionally, this bites you in the back (or it bytes you in the back, as the case may be). If you're using a method to write `Hello` on the screen and `H e l l o` shows up instead, check the method's documentation for mention of Unicode characters.

It's worth noticing that the two methods, `Character.toUpperCase` and `System.out.println`, are used quite differently in Listing 4-4. The method `Character.toUpperCase` is called as part of either an initialization or an assignment statement, but the method `System.out.println` is called on its own. To find out more about this topic, see the explanation of return values in Chapter 7.

The boolean type

A variable of type `boolean` stores one of two values: `true` or `false`. Listing 4-5 demonstrates the use of a `boolean` variable. Figure 4-9 shows the output of the program in Listing 4-5.

LISTING 4-5: Using the boolean Type

```
public class ElevatorFitter2 {

    public static void main(String[] args) {
        System.out.println("True or False?");
        System.out.println("You can fit all ten of the");
        System.out.println("Brickenchicker dectuplets");
        System.out.println("on the elevator:");
        System.out.println();

        int weightOfAPerson = 150;
        int elevatorWeightLimit = 1400;
        int numberOfPeople = elevatorWeightLimit / weightOfAPerson;

        boolean allTenOkay = numberOfPeople >= 10;

        System.out.println(allTenOkay);
    }
}
```

```
True or False?
You can fit all ten of the
Brickenchicker dectuplets
on the elevator:

false
```

FIGURE 4-9: The Brickenchicker dectuplets strike again.

In Listing 4-5, the allTenOkay variable is of type boolean. To find a value for the allTenOkay variable, the program checks to see whether numberOfPeople is greater than or equal to ten. (The symbols >= stand for *greater than or equal to.*)

At this point, it pays to be fussy about terminology. Any part of a Java program that has a value is an *expression.* If you write

```
weightOfAPerson = 150;
```

then 150 is an expression (an expression whose value is the quantity 150). If you write

```
numberOfEggs = 2 + 2;
```

then 2 + 2 is an expression (because 2 + 2 has the value 4). If you write

```
int numberOfPeople = elevatorWeightLimit / weightOfAPerson;
```

then elevatorWeightLimit / weightOfAPerson is an expression. (The value of the expression elevatorWeightLimit / weightOfAPerson depends on whatever values the variables elevatorWeightLimit and weightOfAPerson have when the code containing the expression is executed.)

REMEMBER

Any part of a Java program that has a value is an *expression.*

In Listing 4-5, the code numberOfPeople >= 10 is an expression. The expression's value depends on the value stored in the numberOfPeople variable. But, as you know from seeing the strawberry shortcake at the Brickenchicker family's catered lunch, the value of numberOfPeople isn't greater than or equal to ten. As a result, the value of numberOfPeople >= 10 is false. So, in the statement in Listing 4-5, in which allTenOkay is assigned a value, the allTenOkay variable is assigned a false value.

TIP

In Listing 4-5, I call System.out.println() with nothing inside the parentheses. When I do this, Java adds a line break to the program's output. In Listing 4-5, System.out.println() tells the program to display a blank line.

The Molecules and Compounds: Reference Types

By combining simple things, you get more complicated things. That's the way things always go. Take some of Java's primitive types, whip them together to make a primitive type stew, and what do you get? You get a more complicated type called a *reference type.*

The program in Listing 4-6 uses reference types. Figure 4-10 shows you what happens when you run the program in Listing 4-6.

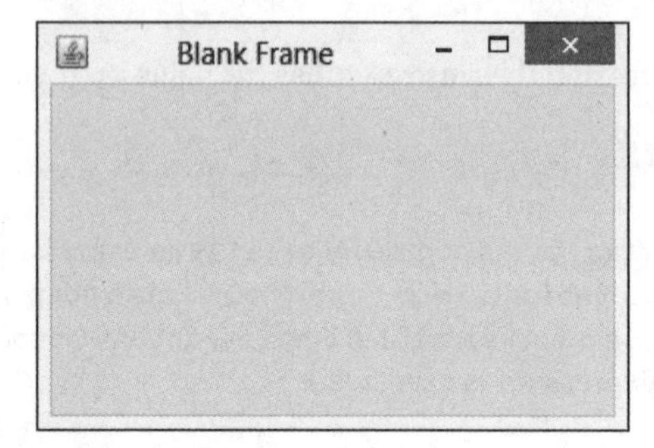

FIGURE 4-10: An empty frame.

LISTING 4-6: Using Reference Types

```
import javax.swing.JFrame;

public class ShowAFrame {

    public static void main(String[] args) {
        JFrame myFrame = new JFrame();
        String myTitle = "Blank Frame";

        myFrame.setTitle(myTitle);
        myFrame.setSize(300, 200);
        myFrame.setDefaultCloseOperation(JFrame.EXIT_ON_CLOSE);
        myFrame.setVisible(true);
    }
}
```

The program in Listing 4-6 uses two references types. Both types are defined in the Java API. One of the types (the one that you'll use all the time) is called *String.* The other type (the one that you can use to create GUIs) is called *JFrame.*

A `String` is a bunch of characters. It's like having several `char` values in a row. So, with the `myTitle` variable declared to be of type `String`, assigning `"Blank Frame"` to the `myTitle` variable makes sense in Listing 4-6. The `String` class is declared in the Java API.

REMEMBER

In a Java program, double quote marks surround the letters in a `String` literal.

A Java *JFrame* is a lot like a window. (The only difference is that you call it a JFrame instead of a window.) To keep Listing 4-6 short and sweet, I decided not to put anything in my frame — no buttons, no fields, nothing.

Even with a completely empty frame, Listing 4-6 uses tricks that I don't describe until later in this book. So don't try reading and interpreting every word of Listing 4-6. The big thing to get from Listing 4-6 is that the program has two variable declarations. In writing the program, I made up two variable names: `myTitle` and `myFrame`. According to the declarations, `myTitle` is of type `String`, and `myFrame` is of type `JFrame`.

You can look up `String` and `JFrame` in Java's API documentation. But, even before you do, I can tell you what you'll find. You'll find that `String` and `JFrame` are the names of Java classes. So that's the big news. Every class is the name of a reference type. You can reserve `amountInAccount` for `double` values by writing

```
double amountInAccount;
```

or by writing

```
double amountInAccount = 50.22;
```

You can also reserve `myFrame` for a `JFrame` value by writing

```
JFrame myFrame;
```

or by writing

```
JFrame myFrame = new JFrame();
```

or even

```
var myFrame = new JFrame();
```

To review the notion of a Java class, see the sections on object-oriented programming (OOP) in Chapter 1.

Every Java class is a reference type. If you declare a variable to have some type that's not a primitive type, the variable's type is (most of the time) the name of a Java class.

Now, when you declare a variable to have type `int`, you can visualize what that declaration means in a fairly straightforward way. It means that, somewhere inside the computer's memory, a storage location is reserved for that variable's value. In the storage location is a bunch of bits. The arrangement of the bits ensures that a certain whole number is represented.

That explanation is fine for primitive types like `int` or `double`, but what does it mean when you declare a variable to have a reference type? What does it mean to declare variable `myFrame` to be of type `JFrame`?

Well, what does it mean to declare *i thank You God* to be an E. E. Cummings poem? What would it mean to write the following declaration?

```
EECummingsPoem ithankYouGod;
```

It means that a class of things is `EECummingsPoem`, and `ithankYouGod` refers to an instance of that class. In other words, `ithankYouGod` is an object belonging to the `EECummingsPoem` class.

Because `JFrame` is a class, you can create objects from that class. (If you don't believe me, read some of my paragraphs about classes and objects in Chapter 1.) Each object (each instance of the `JFrame` class) is an actual frame — a window that appears on the screen when you run the code in Listing 4-6. By declaring the variable `myFrame` to be of type `JFrame`, you're reserving the use of the name `myFrame`. This reservation tells the computer that `myFrame` can refer to an actual `JFrame`-type object. In other words, `myFrame` can become a nickname for one of the windows that appears on the computer screen. Figure 4-11 illustrates the situation.

When you declare *`ClassName variableName;`*, you're saying that a certain variable can refer to an instance of a particular class.

In Listing 4-6, the phrase `JFrame myFrame` reserves the use of the name `myFrame`. On that same line of code, the phrase new `JFrame()` creates a new object (an instance of the `JFrame` class). Finally, that line's equal sign makes `myFrame` refer to the new object. Knowing that the two words `new JFrame()` create an object can be vitally important. For a more thorough explanation of objects, see Chapter 7.

Try these experiments:

TRY IT OUT

RUN IT TWICE

- Run the code in Listing 4-6 on your computer.
- Run the code in Listing 4-6 again. But before running the code, comment out the `myFrame.setVisible(true)` statement by inserting two forward slashes (`//`) immediately to the left of the statement. Does anything happen when you run the modified code?

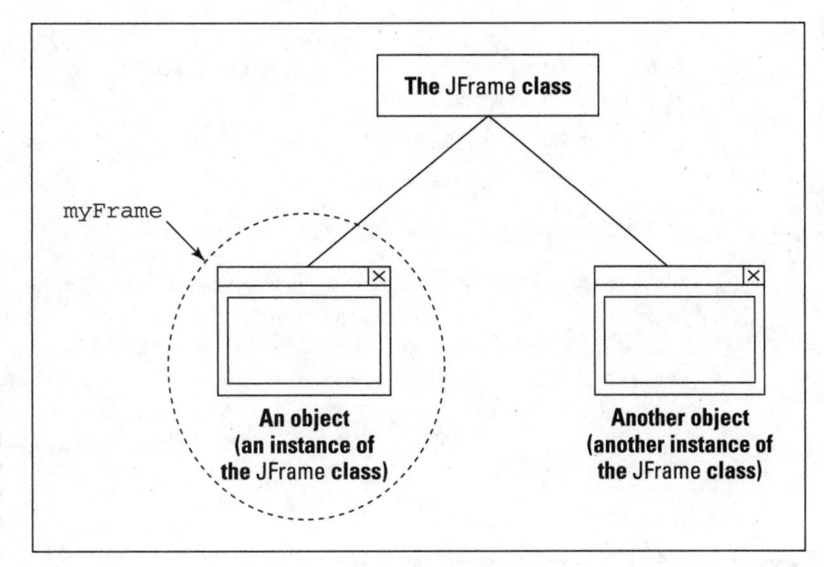

FIGURE 4-11: The variable `myFrame` refers to an instance of the `JFrame` class.

SHUFFLE PLAY

Experiment with the code in Listing 4-6 by changing the order of the statements inside the body of the `main` method. Which rearrangements of these statements are okay, and which aren't?

PRIMITIVE-TYPE STEW

While I'm on the subject of frames, what's a frame, anyway? A *frame* is a window that has a certain height and width and a certain location on your computer's screen. Therefore, deep inside the declaration of the `Frame` class, you can find variable declarations that look something like this:

```
int width;
int height;
int x;
int y;
```

Here's another example: `Time`. An instance of the `Time` class may have an hour (a number from 1 through 12), a number of minutes (from 0 through 59), and a letter (*a* for A.M.; *p* for P.M.):

```
int hour;
int minutes;
char amOrPm;
```

Notice that this high-and-mighty thing called a Java API class is neither high nor mighty. A class is just a collection of declarations. Some of those declarations are the declarations of variables. Some of those variable declarations use primitive types, and other variable declarations use reference types. These reference types, however, come from other classes, and the declarations of those classes have variables. The chain goes on and on. Ultimately, everything comes, in one way or another, from the primitive types.

An Import Declaration

It's always good to announce your intentions upfront. Consider the following classroom lecture:

> *Today, in our History of Film course, we'll discuss the career of actor* ***Lionel Herbert Blythe Barrymore****.*
>
> *Born in Philadelphia,* ***Barrymore*** *appeared in more than 200 films, including* It's a Wonderful Life, Key Largo, and Dr. Kildare's Wedding Day. *In addition,* ***Barrymore*** *was a writer, composer, and director. Barrymore did the voice of Ebenezer Scrooge every year on radio. . . .*

Interesting stuff, heh? Now compare these paragraphs with a lecture in which the instructor doesn't begin by introducing the subject:

> *Welcome once again to the History of Film.*
>
> *Born in Philadelphia,* ***Lionel Barrymore*** *appeared in more than 200 films, including* It's a Wonderful Life, Key Largo, *and* Dr. Kildare's Wedding Day. *In addition,* ***Barrymore (not Ethel, John, or Drew)*** *was a writer, composer, and director.* ***Lionel Barrymore*** *did the voice of Ebenezer Scrooge every year on radio. . . .*

Without a proper introduction, a speaker may have to remind you repeatedly that the discussion is about Lionel Barrymore and not about any other Barrymore. The same is true in a Java program. Look again at Listing 4-6:

```
import javax.swing.JFrame;

public class ShowAFrame {

    public static void main(String[] args) {
        JFrame myFrame = new JFrame();
```

In Listing 4-6, you announce in the introduction (in the import declaration) that you're using `JFrame` in your Java class. You clarify what you mean by `JFrame` with the full name `javax.swing.JFrame`. (Hey! Didn't the first lecturer clarify with the full name "Lionel Herbert Blythe Barrymore"?) After announcing your intentions in the import declaration, you can use the abbreviated name JFrame in your Java class code.

If you don't use an import declaration, you have to repeat the full `javax.swing.JFrame` name wherever you use the name JFrame in your code. For example, without an import declaration, the code of Listing 4-6 would look like this:

```
public class ShowAFrame {

    public static void main(String[] args) {
        javax.swing.JFrame myFrame = new javax.swing.JFrame();
        String myTitle = "Blank Frame";

        myFrame.setTitle(myTitle);
        myFrame.setSize(3200, 200);
        myFrame.setDefaultCloseOperation(javax.swing.JFrame.EXIT_ON_CLOSE);
        myFrame.setVisible(true);
    }
}
```

TIP

The details of this import stuff can be pretty nasty. But fortunately, many IDEs have convenient helper features for import declarations. In your IDE of choice, look for menus such as Code⇨Optimize Imports or Source⇨Organize Imports.

No single section in this book can present the entire story about import declarations. To begin untangling some of the import declaration's subtleties, see the sidebar entitled "Import declarations: The ugly truth" later in this chapter. See also Chapters 5 and 7.

Creating New Values by Applying Operators

What could be more comforting than your old friend the plus sign? It was the first topic you learned about in elementary school math. Almost everybody knows how to add 2 and 2. In fact, in English usage, adding 2 and 2 is a metaphor for something that's easy to do. Whenever you see a plus sign, a cell in your brain says, "Thank goodness — it could be something much more complicated."

Java has a plus sign. You can use it for several purposes. You can use the plus sign to add two numbers, like this:

```
int apples, oranges, fruit;
apples = 5;
oranges = 16;
fruit = apples + oranges;
```

You can also use the plus sign to paste `String` values together:

```
String startOfChapter =
    "It's three in the morning. I'm dreaming about the" +
    "history course that I failed in high school.";
System.out.println(startOfChapter);
```

This can be handy because, in Java, you can't make an ordinary `String` straddle from one line to another. In other words, the following code wouldn't work:

```
String thisIsBadCode =
    "It's three in the morning. I'm dreaming about the
     history course that I failed in high school.";
System.out.println(thisIsBadCode);
```

If you want a single string to span several lines, you need another trick. For details, see Chapter 5.

The correct way to say that you're pasting `String` values together is to say that you're *concatenating* `String` values.

You can even use the plus sign to paste numbers next to `String` values:

```
int apples, oranges, fruit;
apples = 5;
oranges = 16;
fruit = apples + oranges;
System.out.println("You have" + fruit + "pieces of fruit.");
```

Of course, the old minus sign is available, too (but not for `String` values):

```
apples = fruit - oranges;
```

Use an asterisk (*) for multiplication and a slash (/) for division:

```
double rate, pay;
int hours;

rate = 6.25;
hours = 35;
pay = rate * hours;
System.out.println(pay);
```

For an example using division, refer to Listing 4-3.

When you divide an `int` value by another `int` value, you get an `int` value. The computer doesn't round. Instead, the computer chops off any remainder. If you put `System.out.println(11 / 4)` in your program, the computer prints `2`, not `2.75`. To get past this, make either (or both) of the numbers you're dividing `double` values. If you put `System.out.println(11.0 / 4)` in your program, the computer prints `2.75`.

Another useful arithmetic operator is called the *remainder* operator. The symbol for the remainder operator is the percent sign (`%`). When you put `System.out.println(11 % 4)` in your program, the computer prints `3`. It does this because 4 goes into 11 who-cares-how-many times with a remainder of 3. The remainder operator turns out to be fairly useful. Listing 4-7 has an example.

LISTING 4-7: Making Change

```
import static java.lang.System.out;

public class MakeChange {

    public static void main(String[] args) {
        int total = 248;
        int quarters = total / 25;
        int whatsLeft = total % 25;

        int dimes = whatsLeft / 10;
        whatsLeft = whatsLeft % 10;

        int nickels = whatsLeft / 5;
        whatsLeft = whatsLeft % 5;

        int cents = whatsLeft;

        out.println("From " + total + " cents you get");
        out.println(quarters + " quarters");
        out.println(dimes + " dimes");
        out.println(nickels + " nickels");
        out.println(cents + " cents");
    }
}
```

Figure 4-12 shows a run of the code in Listing 4-7. You start with a total of 248 cents. Then

```
quarters = total / 25
```

divides 248 by 25, giving 9. That means you can make 9 quarters from 248 cents. Next,

```
whatsLeft = total % 25
```

divides 248 by 25 again and puts only the remainder, 23, into `whatsLeft`. Now you're ready for the next step, which is to take as many dimes as you can out of cents.

```
From 248 cents you get
9 quarters
2 dimes
0 nickels
3 cents
```

FIGURE 4-12: Change for $2.48.

TECHNICAL STUFF

The code in Listing 4-7 makes change in US currency with the following coin denominations: 1 cent, 5 cents (one nickel), 10 cents (one dime), and 25 cents (one quarter). With these denominations, the `MakeChange` class gives you more than simply a set of coins adding up to 248 cents. The `MakeChange` class gives you the *smallest number of coins* that add up to 248 cents. With some minor tweaking, you can make the code work in any country's coinage. You can always get a set of coins adding up to a total. But, for some kinds of coinage, you won't always get the *smallest number of coins* that add up to a total. For example, in the mid-1970s, England had coins with values 25p, 20p, 10p, and 5p. To put 40p together, a program like the one in Listing 4-7 would suggest 25p + 10p + 5p. But you could use fewer coins by shelling out 20p + 20p.

TRY IT OUT

SMOOTH OPERATORS

Find the values of the following expressions by typing each expression in JShell:

```
» 5 / 4
» 5 / 4.0
» 5.0 / 4
» 5.0 / 4.0
» "5" + "4"
» 5 + 4
» " " + 5 + 4
```

Initialize once, assign often

Listing 4-7 has three lines that put values into the variable `whatsLeft`:

```
int whatsLeft = total % 25;

whatsLeft = whatsLeft % 10;

whatsLeft = whatsLeft % 5;
```

Only one of these lines is a declaration. The other two lines are assignment statements. That's good because you can't declare the same variable more than once (not without creating something called a *block*). If you goof and write

```
int whatsLeft = total % 25;

int whatsLeft = whatsLeft % 10;
```

IMPORT DECLARATIONS: THE UGLY TRUTH

Notice the import declaration at the top of Listing 4-7:

```
import static java.lang.System.out;
```

Compare this with the import declaration at the top of Listing 4-6:

```
import javax.swing.JFrame;
```

By adding the `import static java.lang.System.out;` line to Listing 4-7, I can make the rest of the code a bit easier to read, and I can avoid having long Java statements that start on one line and continue on another. But you never have to do that. If you remove the `import static java.lang.System.out;` line and pepper the code liberally with `System.out.println`, the code works just fine.

Here's a question: Why does one declaration include the word *static* and the other declaration doesn't? Well, to be honest, I wish I hadn't asked!

For the real story about *static,* you have to read part of Chapter 10. And frankly, I don't recommend skipping ahead to that chapter's *static* section if you take medicine for a heart condition, if you're pregnant or nursing, or if you have no previous experience with object-oriented programming. For now, rest assured that Chapter 10 is easy to read after you've made the journey through Part 3 of this book. And when you have to decide whether to use the word *static* in an import declaration, remember these hints:

- The vast majority of import declarations in Java program do not use the word *static.*
- In this book, I never use *import static* to import anything except `System.out`. (Well, almost never. . . .)
- Most import declarations don't use the word *static* because most declarations import classes. Unfortunately, `System.out` is not the name of a class.

in Listing 4-7, you see an error message (such as `Duplicate variable whatsLeft` or `Variable 'whatsLeft' is already defined`) when you try to compile your code.

CROSS REFERENCE

To find out what a block is, see Chapter 5. Then, for some honest talk about redeclaring variables, see Chapter 10.

The increment and decrement operators

Java has some neat little operators that make life easier (for the computer's processor, for your brain, and for your fingers). Altogether, four such operators exist — two increment operators and two decrement operators. The increment operators add 1, and the decrement operators subtract 1. The increment operators use double plus signs (++), and the decrement operators use double minus signs (--). To see how they work, you need some examples. The first example is shown in Figure 4-13.

```
import static java.lang.System.out;
public class preIncrementDemo {
        public static void main(String args[]) {
                int numberOfBunnies = 27;

                ++numberOfBunnies;
                out.println(numberOfBunnies);
                out.println(++numberOfBunnies);
                out.println(numberOfBunnies);
        }
}
```

numberOfBunnies becomes 28.
28 gets printed.
numberOfBunnies becomes 29, and 29 gets printed
29 gets printed again.

FIGURE 4-13: Using preincrement.

Figure 4-14 shows a run of the program in Figure 4-13. In this horribly uneventful run, the count of bunnies prints three times.

```
28
29
29
```

FIGURE 4-14: A run of the code in Figure 4-13.

The double plus signs go by two names, depending on where you put them. When you put the ++ before a variable, the ++ is called the *preincrement* operator. (The *pre* stands for *before.*)

The word *before* has two meanings:

» You put ++ before the variable.

» The computer adds 1 to the variable's value before the variable is used in any other part of the statement.

To understand this, look at the bold line in Figure 4-13. The computer adds 1 to numberOfBunnies (raising the value of numberOfBunnies to 29) and then prints 29 onscreen.

REMEMBER

With out.println(++numberOfBunnies), the computer adds 1 to numberOfBunnies before printing the new value of numberOfBunnies onscreen.

An alternative to preincrement is *postincrement*. (The *post* stands for *after*.) The word *after* has two different meanings:

- You put ++ after the variable.
- The computer adds 1 to the variable's value after the variable is used in any other part of the statement.

To see more clearly how postincrement works, look at the bold line in Figure 4-15. The computer prints the old value of numberOfBunnies (which is 28) on the screen, and then the computer adds 1 to numberOfBunnies, which raises the value of numberOfBunnies to 29.

```
import static java.lang.System.out;
public class postIncrementDemo {
      public static void main(String args[]) {
            int numberOfBunnies = 27;

            numberOfBunnies++;
            out.println(numberOfBunnies);
            out.println(numberOfBunnies++);
            out.println(numberOfBunnies);
      }
}
```

numberOfBunnies becomes 28.
28 gets printed.
28 gets printed, and then numberOfBunnies becomes 29.
29 gets printed.

FIGURE 4-15: Using postincrement.

REMEMBER

With out.println(numberOfBunnies++), the computer adds 1 to numberOfBunnies after printing the old value that numberOfBunnies already had.

Figure 4-16 shows a run of the code in Figure 4-15. Compare Figure 4-16 with the run in Figure 4-14:

- With preincrement in Figure 4-14, the second number is 29.
- With postincrement in Figure 4-16, the second number is 28.

  In Figure 4-16, 29 doesn't show onscreen until the end of the run, when the computer executes one last out.println(numberOfBunnies).

```
28
28
29
```

FIGURE 4-16: A run of the code in Figure 4-15.

TIP

Are you trying to decide between using preincrement or postincrement? Try no longer. Most programmers use postincrement. In a typical Java program, you often see things like `numberOfBunnies++`. You seldom see things like `++numberOfBunnies`.

In addition to preincrement and postincrement, Java has two operators that use `--`. These operators are called *predecrement* and *postdecrement:*

» With **predecrement** (`--numberOfBunnies`), the computer subtracts 1 from the variable's value before the variable is used in the rest of the statement.

» With **postdecrement** (`numberOfBunnies--`), the computer subtracts 1 from the variable's value after the variable is used in the rest of the statement.

STATEMENTS AND EXPRESSIONS

You can describe the pre- and postincrement and pre- and postdecrement operators in two ways: the way everyone understands them and the right way. The way that I explain the concept in most of this section (in terms of time, with *before* and *after*) is the way that everyone understands it. Unfortunately, the way everyone understands the concept isn't really the right way. When you see ++ or --, you can think in terms of time sequence. But occasionally a programmer uses ++ or -- in a convoluted way, and the notions of *before* and *after* break down. So, if you're ever in a tight spot, think about these operators in terms of statements and expressions.

First, remember that a statement tells the computer to do something, and an expression has a value. (I discuss statements in Chapter 3, and I describe expressions elsewhere in this chapter.) Which category does `numberOfBunnies++` belong to? The surprising answer is both — the Java code `numberOfBunnies++` is both a statement and an expression.

Assume that, before the computer executes the code `out.println(numberOfBunnies++)`, the value of `numberOfBunnies` is 28:

- As a statement, `numberOfBunnies++` tells the computer to add 1 to `numberOfBunnies`.
- As an expression, the value of `numberOfBunnies++` is 28, not 29.

(continued)

(continued)

So, even though the computer adds 1 to `numberOfBunnies`, the code `out.println(numberOfBunnies++)` really means `out.println(28)`.

Now, almost everything you just read about `numberOfBunnies++` is true about `++numberOfBunnies`. The only difference is that as an expression, `++numberOfBunnies` behaves in a more intuitive way:

- As a statement, `++numberOfBunnies` tells the computer to add 1 to `numberOfBunnies`.
- As an expression, the value of `++numberOfBunnies` is 29.

So, with `out.println(++numberOfBunnies)`, the computer adds 1 to the variable `numberOfBunnies`, and the code `out.println(++numberOfBunnies)` really means `out.println(29)`.

TECHNICAL STUFF

Rather than write `++numberOfBunnies`, you can achieve the same effect by writing `numberOfBunnies = numberOfBunnies + 1`. So, some people conclude that Java's `++` and `--` operators are for saving keystrokes — to keep those poor fingers from overworking themselves. This is entirely incorrect. The best reason for using `++` is to avoid the inefficient and error-prone practice of writing the same variable name, such as `numberOfBunnies`, twice in the same statement. If you write `numberOfBunnies` only once (as you do when you use `++` or `--`), the computer has to figure out what `numberOfBunnies` means only once. On top of that, when you write `numberOfBunnies` only once, you have only one chance (instead of two chances) to type the variable name incorrectly. With simple expressions like `numberOfBunnies++`, these advantages hardly make a difference. But with more complicated expressions, such as `inventoryItems[(quantityReceived--*itemsPerBox+17)]++`, the efficiency and accuracy that you gain by using `++` and `--` are significant.

PROGNOSTICATION GAME

TRY IT OUT

Before you run the following code, try to predict what the code's output will be. Then run the code to find out whether your prediction is correct:

```
public class Main {

  public static void main(String[] args) {
    int i = 10;
    System.out.println(i++);
    System.out.println(--i);
```

```
        --i;
        i--;
        System.out.println(i);
        System.out.println(++i);
        System.out.println(i--);
        System.out.println(i);
        i++;
        i = i++ + ++i;
        System.out.println(i);
        i = i++ + i++;
        System.out.println(i);
    }
}
```

SEE PLUS PLUS

Type the boldface text, one line after another, into JShell and see how JShell responds:

```
int i = 8
i++
i
i
i++
i
++i
i + i++
i++ + i
```

Assignment operators

If you read the preceding section, which is about operators that add 1, you may be wondering whether you can manipulate these operators to add 2 or add 5 or add 1000000. Can you write `numberOfBunnies++++` and still call yourself a Java programmer? Well, you can't. If you try it, an error message appears when you try to compile your code.

What can you do? As luck would have it, Java has plenty of assignment operators you can use. With an *assignment operator*, you can add, subtract, multiply, or divide by anything you want. You can do other cool operations, too. Listing 4-8 has a smorgasbord of assignment operators (the ones with equal signs). Figure 4-17 shows the output from running Listing 4-8.

```
28
33
86
172
165
100
```

FIGURE 4-17: A run of the code in Listing 4-8.

LISTING 4-8: Assignment Operators

```
public class UseAssignmentOperators {

    public static void main(String[] args) {
        int numberOfBunnies = 27;
        int numberExtra = 53;

        numberOfBunnies += 1;
        System.out.println(numberOfBunnies);

        numberOfBunnies += 5;
        System.out.println(numberOfBunnies);

        numberOfBunnies += numberExtra;
        System.out.println(numberOfBunnies);

        numberOfBunnies *= 2;
        System.out.println(numberOfBunnies);

        System.out.println(numberOfBunnies -= 7);

        System.out.println(numberOfBunnies = 100);
    }
}
```

Listing 4-8 shows how versatile Java's assignment operators are. With the assignment operators, you can add, subtract, multiply, or divide a variable by any number. Notice how `+= 5` adds 5 to `numberOfBunnies`, and how `*= 2` multiplies `numberOfBunnies` by 2. You can even use another expression's value (in Listing 4-8, `numberExtra`) as the number to be applied.

The last two lines in Listing 4-8 demonstrate a special feature of Java's assignment operators. You can use an assignment operator as part of a larger Java statement. In the next-to-last line of Listing 4-8, the operator subtracts 7 from `numberOfBunnies`, decreasing the value of `numberOfBunnies` from `172` to `165`. Then the whole assignment business is stuffed into a call to `System.out.println`, so `165` prints onscreen.

Lo and behold, the last line of Listing 4-8 shows how you can do the same thing with Java's plain old equal sign. The thing that I call an assignment statement near the start of this chapter is really one of the assignment operators that I describe in this section. Therefore, whenever you assign a value to something, you can make that assignment be part of a larger statement.

TECHNICAL STUFF

Each use of an assignment operator does double duty as a statement and an expression. In all cases, the expression's value equals whatever value you assign. For example, before executing the code `System.out.println(numberOfBunnies -= 7)`, the value of `numberOfBunnies` is `172`. As a statement, `numberOfBunnies -= 7` tells the computer to subtract 7 from `numberOfBunnies` (so the value of `numberOfBunnies` goes from `172` to `165`). As an expression, the value of `numberOfBunnies -= 7` is `165`. So the code `System.out.println(numberOfBunnies -= 7)` really means `System.out.println(165)`. The number `165` displays on the computer screen.

For a richer explanation of this kind of thing, see the sidebar "Statements and expressions," earlier in this chapter.

THE OPERATION IS A SUCCESS

TRY IT OUT

Before you run the following code, try to predict what the code's output will be. Then run the code to find out whether your prediction is correct:

```
public class Main {

  public static void main(String[] args) {
    int i = 10;

    i += 2;
    i -= 5;
    i *= 6;

    System.out.println(i);
    System.out.println(i += 3);
    System.out.println(i /= 2);
  }
}
```

IN THIS CHAPTER

» Forming logical conditions

» Putting statements inside one another

» Choosing among many alternatives

Chapter 5
Controlling Program Flow with Decision-Making Statements

The TV show *Dennis the Menace* aired on CBS from 1959 to 1963. I remember one episode in which Mr. Wilson was having trouble making an important decision. I think it was something about changing jobs or moving to a new town. Anyway, I can still see that shot of Mr. Wilson sitting in his yard, sipping lemonade, and staring into nowhere for the whole afternoon. Of course, the annoying character Dennis was continually interrupting Mr. Wilson's peace and quiet. That's what made this situation funny.

What impressed me about this episode (the reason I remember it clearly, even now) was Mr. Wilson's dogged intent in making the decision. This guy wasn't going about his everyday business, roaming around the neighborhood while thoughts about the decision wandered in and out of his mind. He was sitting quietly in his yard, making marks carefully and logically on his mental balance sheet. How many people actually make decisions this way?

At that time, I was still pretty young. I'd never faced the responsibility of making a big decision that affected my family and me. But I wondered what such a decision-making process would be like. Would it help to sit there like a stump for hours on end? Would I make my decisions by the careful weighing and tallying of options? Or would I shoot in the dark, take risks, and act on impulse? Only time would tell.

Making Decisions (Java if Statements)

When you're writing computer programs, you're constantly hitting forks in roads. Did the user correctly type the password? If yes, let the user work; if no, kick the bum out. So the Java programming language needs a way of making a program branch in one of two directions. Fortunately, the language has a way: It's called an `if` statement.

Guess the number

Listing 5-1 illustrates the use of an `if` statement. Two runs of the program in Listing 5-1 are shown in Figure 5-1.

LISTING 5-1: A Guessing Game

```
import java.util.Random;
import java.util.Scanner;

import static java.lang.System.out;

public class GuessingGame {

    public static void main(String[] args) {
        Scanner keyboard = new Scanner(System.in);

        out.print("Enter an int from 1 to 10: ");

        int inputNumber = keyboard.nextInt();
        int randomNumber = new Random().nextInt(10) + 1;

        if (inputNumber == randomNumber) {
            out.println("**********");
            out.println("*You win.*");
            out.println("**********");
```

```
        } else {
            out.println("You lose.");
            out.print("The random number was ");
            out.println(randomNumber + ".");
        }

        out.println("Thank you for playing.");

        keyboard.close();
    }
}
```

The program in Listing 5-1 plays a guessing game with the user. The program gets a number (a guess) from the user and then generates a random number between 1 and 10. If the number that the user entered is the same as the random number, the user wins. Otherwise, the user loses and the program tells the user what the random number was.

```
Enter an int from 1 to 10: 2
**********
*You win.*
**********
Thank you for playing.


Enter an int from 1 to 10: 4
You lose.
The random number was 10.
Thank you for playing.
```

FIGURE 5-1: Two runs of the guessing game.

She controlled keystrokes from the keyboard

Taken together, the lines

```
import java.util.Scanner;

        Scanner keyboard = new Scanner(System.in);

        int inputNumber = keyboard.nextInt();
```

in Listing 5-1 get whatever number the user types on the computer's keyboard. The last of the three lines puts this number into a variable named *inputNumber*. If these lines look complicated, don't worry: You can copy these lines almost word-for-word whenever you want to read from the keyboard. Include the first two lines (the `import` and `Scanner` lines) just once in your program. Later in your program, wherever the user types an `int` value, include a line with a call to `nextInt` (as in the last of the preceding three lines of code).

Of all the names in these three lines of code, the only two names I coined myself are *inputNumber* and *keyboard*. All the other names are part of Java. So, if I want to be creative, I can write the lines this way:

```
import java.util.Scanner;

        Scanner readingThingie = new Scanner(System.in);

        int valueTypedIn = readingThingie.nextInt();
```

I can also beef up my program's import declarations, as I do later on, in Listings 5-2 and 5-3. Other than that, I have very little leeway.

As you read on in this book, you'll start recognizing the patterns behind these three lines of code, so I don't clutter up this section with all the details. For now, you can just copy these three lines and keep the following guidelines in mind:

» **When you import** `java.util.Scanner`, you don't use the word ***static.***

But importing `Scanner` is different from importing `System.out`. When you import `java.lang.System.out`, you use the word `static`. (Refer to Listing 5-1.) The difference creeps into the code because `Scanner` is the name of a class and `System.out` isn't the name of a class.

CROSS REFERENCE

For a quick look at the use of the word *static* in import declarations, see the sidebar in Chapter 4 about import declarations: the ugly truth. For a more complete story about the word, see Chapter 10.

» **Typically (on a desktop or laptop computer), the name *System.in* stands for the keyboard.**

To get characters from someplace other than the keyboard, you can type something other than `System.in` inside the parentheses.

CROSS REFERENCE

What else can you put inside the `new Scanner(...)` parentheses? For some ideas, see Chapter 8.

In Listing 5-1, I make the arbitrary decision to give one of my variables the name `keyboard`. The name `keyboard` reminds you, the reader, that this variable refers to a bunch of plastic buttons in front of your computer. Naming something `keyboard` tells Java nothing about plastic buttons or about user input. On the other hand, the name `System.in` always tells Java about those plastic buttons. The code `Scanner keyboard = new Scanner`

(System.in) in Listing 5-1 connects the name keyboard with the plastic buttons that we all know and love.

- **When you expect the user to type an `int` value (a whole number of some kind), use `nextInt()`.**

  If you expect the user to type a double value (a number containing a decimal point), use nextDouble(). If you expect the user to type **true** or **false**, use nextBoolean(). If you expect the user to type a word like *Barry*, *Java*, or *Hello*, use next().

WARNING

Decimal points vary from one country to another. In the United States, *10.5* (with a period) represents ten-and-a-half, but in France, *10,5* (with a comma) represents ten-and-a-half. In Switzerland, 10.50 is an amount of money, and 10,50 is an amount that's not money-related. In the Persian language, a decimal point looks like a slash (but it sits a bit lower than the digit characters). Your computer's operating system stores information about the country you live in, and Java reads that information to decide what ten-and-a-half looks like. If you run a program containing a nextDouble() method call and Java responds with an InputMismatchException, check your input. You might have input 10.5 when your country's conventions require 10,5 (or another way of representing ten-and-a-half). For more information, see the sidebar "Where on earth do you live?" in Chapter 8.

CROSS REFERENCE

For an example in which the user types a word, see Listing 5-3, later in this chapter. For an example in which the user types a single character, see Listing 6-4, in Chapter 6. For an example in which a program reads an entire line of text (all in one big gulp), see Chapter 8.

- **You can get several values from the keyboard, one after another.**

  To do this, use the keyboard.nextInt() code several times.

  To see a program that reads more than one value from the keyboard, go to Listing 5-4, later in this chapter.

- **Whenever you use Java's `Scanner`, you should call the `close` method after your last `nextInt` call (or your last `nextDouble` call or your last `next`*`Whatever`* call).**

  In Listing 5-1, the main method's last statement is

```
keyboard.close();
```

  This statement does some housekeeping to disconnect the Java program from the computer keyboard. (The amount of required housekeeping is more than you might think!) If I omit this statement from Listing 5-1, nothing terrible happens. Java's virtual machine usually cleans up after itself very nicely. But using close() to explicitly detach from the keyboard is good practice, and some IDEs display warnings if you omit the keyboard.close() statement. In this book's example, I always remember to close my Scanner variables.

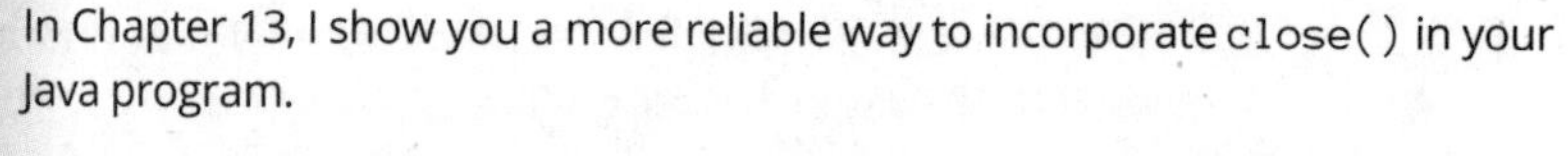
In Chapter 13, I show you a more reliable way to incorporate `close()` in your Java program.

When your program calls `System.out.println`, your program uses the computer's screen. So why don't you call a `close` method after all your `System.out.println` calls? The answer is subtle. In Listing 5-1, your own code connects to the keyboard by calling `new Scanner(System.in)`. So, later in the program, your code cleans up after itself by calling the `close` method. But with `System.out.println`, your own code doesn't create a connection to the screen. (The `out` variable refers to a `PrintStream`, but you don't call `new PrintStream()` to prepare for calling `System.out.println`.) Instead, the Java virtual machine connects to the screen on your behalf. The Java virtual machine's code (which you never have to see) contains a call to `new PrintStream()` in preparation for your calling `System.out.println`. So, because it's a well-behaved piece of code, the Java virtual machine eventually calls `out.close()` with no effort on your part.

Creating randomness

In Listing 5-1, the code `new Random().nextInt(10)` stands for a number that appears to be randomly generated — a whole number in the range from 0 to 9. With 1 added on, the expression `new Random().nextInt(10) + 1` is a number from 1 to 10.

Notice my careful wording in the previous paragraph about "a number that appears to be randomly generated." Achieving real randomness is surprisingly difficult. Mathematician Persi Diaconis says that if you flip a coin several times, always starting with the head side up, you're likely to toss heads more often than tails. If you toss several more times, always starting with the tail side up, you'll likely toss tails more often than heads. In other words, coin tossing isn't really fair.*

Computers aren't much better than coins and human thumbs. A computer mimics the generation of random sequences, but in the end the computer just does what it's told and does all of this in a purely deterministic fashion. So, in Listing 5-1, when the computer executes

```
import java.util.Random;

        int randomNumber = new Random().nextInt(10) + 1;
```

* Diaconis, Persi. "The Search for Randomness." American Association for the Advancement of Science annual meeting. Seattle. 14 Feb. 2004.

the computer appears to give a randomly generated number — a whole number between 1 and 10. But it's all a fake. The computer only follows instructions. It's not really random, but without bending a computer over backward, it's the best that anyone can do.

Once again, I ask you to take this code on blind faith. Don't worry about what new Random().nextInt really does until you have more experience with Java. Just copy this code into your own programs and have fun with it. And, if the numbers from 1 to 10 aren't in your flight plans, don't fret. To roll an imaginary die, write the statement

```
int rollEmBaby = new Random().nextInt(6) + 1;
```

With the execution of this statement, the variable rollEmBaby gets a value from 1 to 6.

The if statement

At the core of Listing 5-1 is a Java if statement. This if statement represents a fork in the road. (See Figure 5-2.) The computer follows one of two prongs: the prong that prints You win or the prong that prints You lose. The computer decides which prong to take by testing the truth or falsehood of a *condition*. In Listing 5-1, the condition being tested is

```
inputNumber == randomNumber
```

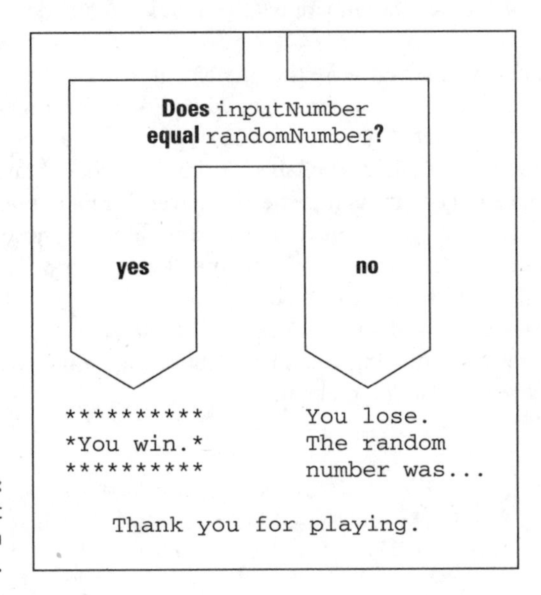

FIGURE 5-2: An if statement is like a fork in the road.

Does the value of `inputNumber` equal the value of `randomNumber`? When the condition is true, the computer does the stuff between the condition and the word *else*. When the condition turns out to be false, the computer does the stuff after the word *else*. Either way, the computer goes on to execute the last `println` call, which displays `Thank you for playing`.

REMEMBER

The condition in an `if` statement must be enclosed in parentheses. However, a line like `if (inputNumber == randomNumber)` is not a complete statement (just as "If I had a hammer" isn't a complete sentence). So, this line `if (inputNumber == randomNumber)` shouldn't end with a semicolon.

TECHNICAL STUFF

Sometimes, when I'm writing about a condition that's being tested, I slip into using the word *expression* instead of *condition*. That's okay because every condition is an expression. An expression is something that has a value and, sure enough, every condition has a value. The condition's value is either `true` or `false`. (For revealing information about expressions and values like `true` and `false`, see Chapter 4.)

Equal, equal

In Listing 5-1, in the `if` statement's condition, notice the use of the double equal sign. Comparing two numbers to see whether they're the same isn't the same as setting something equal to something else. That's why the symbol to compare for equality isn't the same as the symbol that's used in an assignment or an initialization. In an `if` statement's condition, you can't replace the double equal sign with a single equal sign. If you do, your program just won't work. (You almost always get an error message when you try to compile your code.)

On the other hand, if you never make the mistake of using a single equal sign in a condition, you're not normal. Not long ago, while I was teaching an introductory Java course, I promised that I'd swallow my laser pointer if no one made the single equal sign mistake during any of the lab sessions. This wasn't an idle promise. I knew I'd never have to keep it. As it turned out, even if I had ignored the first ten times anybody made the single equal sign mistake during those lab sessions, I would still be laser-pointer-free. Everybody mistakenly uses the single equal sign several times in a programming career.

TIP

The trick is not to avoid making the single-equal-sign mistake; the trick is to catch the mistake whenever you make it.

Brace yourself

The `if` statement in Listing 5-1 has two halves: a top half and a bottom half. I have names for these two parts of an `if` statement. I call them the *if part* (the top half) and the *else part* (the bottom half).

The `if` part in Listing 5-1 seems to have more than one statement in it. I make this happen by enclosing the three statements of the `if` part in a pair of curly braces. When I do this, I form a block. A *block* is a bunch of statements scrunched together by a pair of curly braces.

With this block, three calls to `println` are tucked away safely inside the `if` part. With the curly braces, the rows of asterisks and the words `You win` display only when the user's guess is correct.

This business with blocks and curly braces applies to the else part as well. In Listing 5-1, whenever `inputNumber` doesn't equal `randomNumber`, the computer executes three `print/println` calls. To convince the computer that all three of these calls are inside the `else` clause, I put these calls into a block. That is, I enclose these three calls in a pair of curly braces.

TECHNICAL STUFF

Strictly speaking, Listing 5-1 has only one statement between the `if` and the `else` statements and only one statement after the `else` statement. The trick is that when you place a bunch of statements inside curly braces, you get a block; and a block behaves, in all respects, like a single statement. In fact, the official Java documentation lists blocks as one of the many kinds of statements. So, in Listing 5-1, the block that prints `You win` and asterisks is a single statement that has, within it, three smaller statements.

Your intent to indent

Notice how, in Listing 5-1, the `print` and `println` calls inside the `if` statement are indented. (This includes both the `You win` and `You lose` statements. The `print` and `println` calls that come after the word `else` are still part of the `if` statement.) Strictly speaking, you don't have to indent the statements that are inside an `if` statement. For all the compiler cares, you can write your whole program on a single line or place all your statements in an artful, misshapen zigzag. The problem is that neither you nor anyone else can make sense of your code if you don't indent your statements in some logical fashion. In Listing 5-1, the indenting of the `print` and `println` statements helps your eye (and brain) see quickly that these statements are subordinate to the overall `if/else` flow.

In a small program, unindented or poorly indented code is barely tolerable. But in a complicated program, indentation that doesn't follow a neat, logical pattern is a big, ugly nightmare.

REMEMBER

Many Java IDEs have tools to indent your code automatically. In your favorite IDE, look for menus such as Source⇨Format or Code⇨Reformat Code.

WARNING

When you write `if` statements, you may be tempted to chuck out the window all the rules about curly braces and simply rely on indentation. This strategy works in other programming languages, such as Python and Haskell, but not in Java. If you indent three statements after the word `else` and forget to enclose those statements in curly braces, the computer thinks that the `else` part includes only the first of the three statements. What's worse, the indentation misleads you into believing that the `else` part includes all three statements. This makes it more difficult for you to figure out why your code isn't behaving the way you think it should. Watch those braces!

Elseless in Helsinki

Okay, so the title of this section is contrived. Big deal! The idea is that you can create an `if` statement without the `else` part. Take, for instance, the code in Listing 5-1, shown earlier. Maybe you'd rather not rub it in whenever the user loses the game. The modified code in Listing 5-2 shows you how to do this (and Figure 5-3 shows you the result).

```
Enter an int from 1 to 10: 4
*You win.*
That was a very good guess :-)
The random number was 4.
Thank you for playing.


Enter an int from 1 to 10: 4
That was a very good guess :-)
The random number was 6.
Thank you for playing.
```

FIGURE 5-3: Two runs of the game in Listing 5-2.

LISTING 5-2: A Kinder, Gentler Guessing Game

```
import java.util.Random;
import java.util.Scanner;

import static java.lang.System.in;
import static java.lang.System.out;
```

```
public class DontTellThemTheyLost {

    public static void main(String[] args) {
        Scanner keyboard = new Scanner(in);

        out.print("Enter an int from 1 to 10: ");

        int inputNumber = keyboard.nextInt();
        int randomNumber = new Random().nextInt(10) + 1;

        if (inputNumber == randomNumber) {
            out.println("*You win.*");
        }

        out.println("That was a very good guess :-)");
        out.print("The random number was ");
        out.println(randomNumber + ".");
        out.println("Thank you for playing.");

        keyboard.close();
    }
}
```

The `if` statement in Listing 5-2 has no `else` part. When `inputNumber` is the same as `randomNumber`, the computer prints `You win`. When `inputNumber` is different from `randomNumber`, the computer doesn't print `You win`.

Listing 5-2 illustrates another new idea. With an `import` declaration for `System.in`, I can reduce `new Scanner(System.in)` to the shorter `new Scanner(in)`. Adding this `import` declaration is hardly worth the effort. In fact, I do more typing with the `import` declaration than without it. Nevertheless, the code in Listing 5-2 demonstrates that it's possible to import `System.in`.

STRAIGHT TALK

TRY IT OUT

In Chapter 4, Listing 4-5 tells you whether you can or cannot fit ten people on an elevator. A run of the listing's code looks something like this:

```
True or False?
You can fit all ten of the
Brickenchicker dectuplets
on the elevator:

false
```

Use what you know about Java's `if` statements to make the program's output more natural. Depending on the value of the program's `elevatorWeightLimit` variable, the output should be either

```
You can fit all ten of the
Brickenchicker dectuplets
on the elevator.
```

or

```
You can't fit all ten of the
Brickenchicker dectuplets
on the elevator.
```

Using Blocks in JShell

Chapter 4 introduces Java 9's interactive JShell environment. You type a statement, and JShell responds immediately by executing the statement. That's fine for simple statements, but what happens when you have a statement inside of a block?

In JShell, you can start typing a statement with one or more blocks. JShell doesn't respond until you finish typing the entire statement — blocks and all. To see how it works, look over this conversation that I had recently with JShell:

```
jshell> import static java.lang.System.out

jshell> import java.util.Random

jshell> int randomNumber = new Random().nextInt(10) + 1
randomNumber ==> 4

jshell> int inputNumber = 4
inputNumber ==> 4

jshell> if (inputNumber == randomNumber) {
   ...>     out.println("*You win.*");
   ...> }
*You win.*

jshell>
```

In this dialogue, I've set the text that I type in bold. JShell's responses aren't set in bold.

When I type `if (inputNumber == randomNumber) {` and press Enter, JShell doesn't do much — it only displays a `...>` prompt, which indicates that whatever lines I've typed don't form a complete statement. I have to respond by typing the rest of the `if` statement.

When I finish the `if` statement with a close curly brace, JShell finally acknowledges that I've typed an entire statement. JShell executes the statement and (in this example) displays `*You win.*`.

Notice the semicolon at the end of the `out.println` line:

- When you type a statement that's not inside of a block, JShell lets you omit the semicolon at the end of the statement.
- When you type a statement that's inside of a block, JShell (like the plain old Java in Listing 5-2) doesn't let you omit the semicolon.

TIP

When you type a block in JShell, you always have the option of typing the entire block on one line, with no line breaks, like so:

```
if (inputNumber == randomNumber) { out.println("*You win.*"); }
```

Forming Conditions with Comparisons and Logical Operators

The Java programming language has plenty of little squiggles and doodads for your various condition-forming needs. This section tells you all about them.

Comparing numbers; comparing characters

Table 5-1 shows you the operators that you can use to compare one value with another.

You can use all of Java's comparison operators to compare numbers and characters. When you compare numbers, things go pretty much the way you think they should go. But when you compare characters, things are a little strange. Comparing uppercase letters with one another is no problem. Because the letter *B* comes alphabetically before *H*, the condition `'B' < 'H'` is true. Comparing lowercase letters with one another is also okay. What's strange is that when you compare an uppercase letter with a lowercase letter, the uppercase letter is always smaller. So, even though `'Z' < 'A'` is `false`, `'Z' < 'a'` is `true`.

TABLE 5-1 Comparison Operators

Operator Symbol	Meaning	Example
`==`	is equal to	`numberOfCows == 5`
`!=`	is not equal to	`buttonClicked != panicButton`
`<`	is less than	`numberOfCows < 5`
`>`	is greater than	`myInitial > 'B'`
`<=`	is less than or equal to	`numberOfCows <= 5`
`>=`	is greater than or equal to	`myInitial >= 'B'`

TECHNICAL STUFF

Under the hood, the letters *A* through *Z* are stored with numeric codes 65 through 90. The letters *a* through *z* are stored with codes 97 through 122. That's why each uppercase letter is smaller than each lowercase letter.

WARNING

Be careful when you compare two numbers for equality (with `==`) or inequality (with `!=`). After you do some calculations and obtain two `double` values or two `float` values, the values that you have are seldom dead-on equal to one another. (The problem comes from those pesky digits beyond the decimal point.) For instance, the Fahrenheit equivalent of 21 degrees Celsius is 69.8, and when you calculate `9.0 / 5 * 21 + 32` by hand, you get 69.8. But the condition `9.0 / 5 * 21 + 32 == 69.8` turns out to be false. That's because, when the computer calculates `9.0 / 5 * 21 + 32`, it gets 69.80000000000001, not 69.8.

Comparing objects

When you start working with objects, you find that you can use `==` and `!=` to compare objects with one another. For instance, a button you see on the computer screen is an object. You can ask whether the thing that was just mouse-clicked is a particular button on your screen. You do this with Java's equality operator:

```
if (e.getSource() == bCopy) {
clipboard.setText(which.getText());
```

CROSS REFERENCE

To find out more about responding to button clicks, read Chapter 16.

The big gotcha with Java's comparison scheme comes when you compare two strings. (For a word or two about Java's `String` type, see the section about reference types in Chapter 4.) When you compare two strings with one another, you don't want to use the double equal sign. Using the double equal sign would ask, "Is this string stored in exactly the same place in memory as that other string?" Usually, that's not what you want to ask. Instead, you usually want to ask, "Does this string have the same characters in it as that other string?" To ask the second question (the more appropriate question), Java's `String` type has a method named `equals`. (Like everything else in the known universe, this `equals` method is defined in the Java API, short for *a*pplication *p*rogramming *i*nterface.) The `equals` method compares two strings to see whether they have the same characters in them. For an example using Java's `equals` method, see Listing 5-3. (Figure 5-4 shows a run of the program in Listing 5-3.)

```
What's the password? swordfish
You typed >>swordfish<<

The word you typed is not
stored in the same place as
the real password, but that's
no big deal.

The word you typed has the
same characters as the real
password. You can use our
precious system.
```

FIGURE 5-4: The results of using == and using Java's `equals` method.

LISTING 5-3: Checking a Password

```
import java.util.Scanner;

import static java.lang.System.*;

public class CheckPassword {

    public static void main(String[] args) {

        out.print("What's the password?");

        var keyboard = new Scanner(in);
        String password = keyboard.next();

        out.println("You typed >>" + password + "<<");
        out.println();
```

(continued)

LISTING 5-3: *(continued)*

```
        if (password == "swordfish") {
            out.println("""
                    The word you typed is stored
                    in the same place as the real
                    password. You must be a hacker.""");
        } else {
            out.println("""
                    The word you typed is not
                    stored in the same place as
                    the real password, but that's
                    no big deal.""");
        }
        out.println();

        if (password.equals("swordfish")) {
            out.println("""
                    The word you typed has the
                    same characters as the real
                    password. You can use our
                    precious system.""");
        } else {
            out.println("""
                    The word you typed doesn't
                    have the same characters as
                    the real password. You can't
                    use our precious system.""");
        }

        keyboard.close();
    }
}
```

In Listing 5-3, the call `keyboard.next()` grabs whatever word the user types on the computer keyboard. The code shoves this word into the variable named *password*. Then the program's `if` statements use two different techniques to compare `password` with `"swordfish"`.

The more appropriate of the two techniques uses Java's `equals` method. The `equals` method looks funny because when you call it, you put a dot after one string and put the other string in parentheses. But that's the way you have to do it.

In calling Java's `equals` method, it doesn't matter which string gets the dot and which gets the parentheses. For instance, in Listing 5-3, you could have written

```
if ("swordfish".equals(password))
```

The method would work just as well.

TECHNICAL STUFF

A call to Java's equals method looks imbalanced, but it's not. There's a reason behind the apparent imbalance between the dot and the parentheses. The idea is that you have two objects: the password object and the "swordfish" object. Each of these two objects is of type String. (However, password is a variable of type String, and "swordfish" is a String literal.) When you write password.equals("swordfish"), you're calling an equals method that belongs to the password object. When you call that method, you're feeding "swordfish" to the method as the method's parameter (pun intended). You can read more about methods belonging to objects in Chapter 7.

TIP

In addition to its equals method, Java has an equalsIgnoreCase method. Even though "SWORDFISH".equals("swordfish") is false, its close cousin "SWORDFISH".equalsIgnoreCase("swordfish") is true.

Look!

The big new in Listing 5-3 is Java's equals method, but the listing has several other interesting features. This section describes three of them.

On the var side

In Listing 5-2, the following statement associates my made-up name keyboard with Java's well-established name Scanner:

```
Scanner keyboard = new Scanner(in);
```

The word Scanner appears twice in this statement — once as Scanner keyboard and again as new Scanner(in). You may ask whether this repetition of the name Scanner is necessary. Chapter 4 introduces Java's use of the word var, and Listing 5-3 puts var to good use. In Listing 5-3, the statement

```
var keyboard = new Scanner(in);
```

tells Java to figure out on its own that keyboard refers to a Scanner value. Believe it or not, older versions of Java couldn't jump to that conclusion. Before Java 10 came along, repeating words like Scanner, as in Listings 5-1 and 5-2, was the only option.

This newfangled var word is handy! But remember, you can't always replace Scanner keyboard with var keyboard. For details, see Chapter 7.

CROSS REFERENCE

The expression `new Scanner(in)` is an example of a constructor call. For the low-down on constructor calls, see Chapter 9.

When one line isn't enough

Listing 5-3 uses a feature that didn't become an official part of Java until September 2020 (with Java 15): A *text block* is a bunch of text surrounded on both sides by three double quotes (`"""`).

```
out.println("""
        The word you typed is stored
        in the same place as the real
        password. You must be a hacker.""");
```

A text block starts with three double quotation marks, and the remainder of that block's first line must be blank. If you mistakenly put text after those first three quotation marks, Java becomes sick to its stomach:

```
// Don't do this:
out.println("""The word you typed is stored
        in the same place as the real
        password. You must be a hacker.""");
```

Text blocks are useful because the text inside a block can straddle more than one line. Without text blocks, you may be tempted to put one quotation mark at each end, but that doesn't work. The following code, with traditional Java string notation, is forbidden:

```
// This code is incorrect:
out.println("
        The word you typed is stored
        in the same place as the real
        password. You must be a hacker.");
```

This book's seventh edition hit the shelves in 2017, before Java had text blocks. In that edition, Listing 5-3 had a truckload of `out.println` calls:

```
out.println("The word you typed is stored");
out.println("in the same place as the real");
out.println("password. You must be a");
out.println("hacker.");
```

That was some ugly code!

Importing everything in one fell swoop

The first line of Listing 5-3 illustrates a lazy way of importing both `System.out` and `System.in`. To import everything that `System` has to offer, you use the asterisk wildcard character (*). In fact, importing `java.lang.System.*` is like having about 30 separate import declarations, including `System.in`, `System.out`, `System.err`, `System.nanoTime`, and many other `System` things.

I don't use the wildcard very much in this book's examples. But for larger programs — programs that use dozens of names from the Java API — the lazy asterisk trick is handy.

TECHNICAL STUFF

You can't toss an asterisk anywhere you want inside an `import` declaration. For example, you can't import everything starting with `java` by writing `import java.*`. You can substitute an asterisk only for the name of a class or for the name of something static that's tucked away inside a class. For more information about asterisks in `import` declarations, see Chapter 7. For information about `static` things, see Chapter 10.

Java's logical operators

Mr. Spock would be pleased: Java has all the operators you need for mixing and matching logical tests. The operators are shown in Table 5-2.

TABLE 5-2 Logical Operators

Operator Symbol	What It Means	Example
&&	and	`5 < x && x < 10`
||	or	`x < 5 || 10 < x`
!	not	`!password.equals("swordfish")`

You can use these operators to form all kinds of elaborate conditions. Listing 5-4 has an example.

LISTING 5-4: Checking Username and Password

```
import javax.swing.JOptionPane;

public class Authenticator {

    public static void main(String[] args) {

        String username = JOptionPane.showInputDialog("Username:");
        String password = JOptionPane.showInputDialog("Password:");

        if (
              username != null && password != null &&
              (
                (username.equals("bburd") && password.equals("swordfish")) ||
                (username.equals("hritter") && password.equals("preakston"))
              )
           )
        {
            JOptionPane.showMessageDialog(null, "You're in.");
        } else {
            JOptionPane.showMessageDialog(null, "You're suspicious.");
        }
    }
}
```

Several runs of the program in Listing 5-4 are shown in Figure 5-5. When the username is *bburd* and the password is *swordfish* or when the username is *hritter* and the password is *preakston*, the user sees a nice message. Otherwise, the user is a bum who sees the nasty message they deserve.

Confession: Figure 5-5 is a fake! To help you read the usernames and passwords, I added an extra statement to Listing 5-4. The extra statement (`UIManager.put("TextField.font", new Font("Dialog", Font.BOLD, 14))`) enlarges each text field's font size. Yes, I modified the code before creating the figure. Shame on me!

Listing 5-4 illustrates a new way to get user input; namely, to show the user an input dialog box. The statement

```
String password = JOptionPane.showInputDialog("Password:");
```

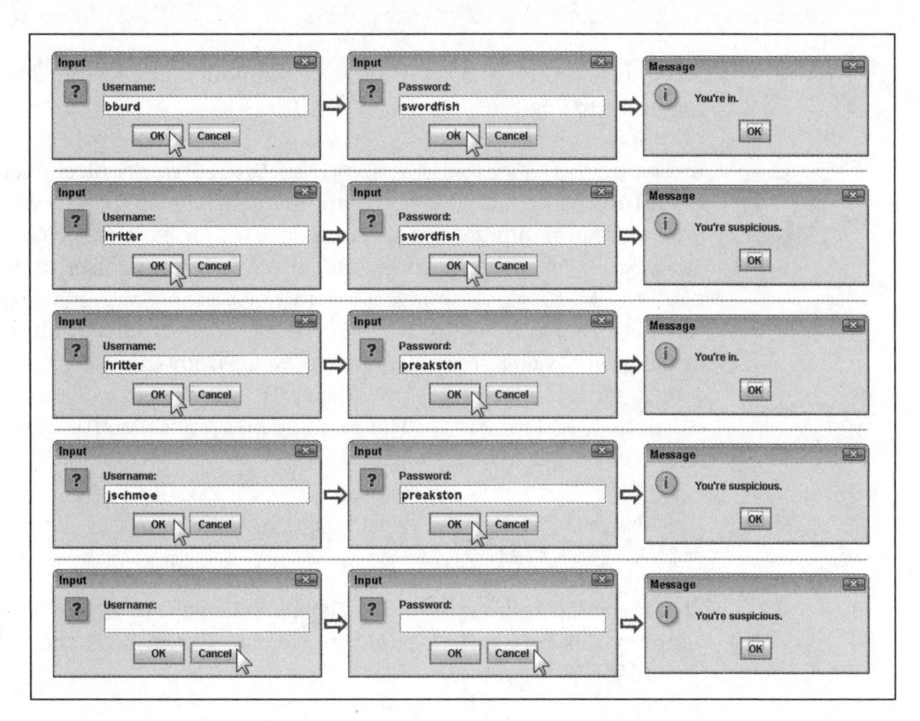

FIGURE 5-5: Several runs of the code from Listing 5-4.

in Listing 5-4 performs more or less the same task as the statement

```
String password = keyboard.next();
```

from Listing 5-3. The big difference is, while `keyboard.next()` displays dull-looking text in a console, `JOptionPane.showInputDialog("Username:")` displays a fancy dialog box containing a text field and buttons. (Compare Figures 5-4 and 5-5.) When the user clicks OK, the computer takes whatever text is in the text field and hands that text over to a variable. In fact, Listing 5-4 uses `JOptionPane.showInputDialog` twice — once to get a value for the username variable and a second time to get a value for the password variable.

Near the end of Listing 5-4, I use a slight variation on the `JOptionPane` business:

```
JOptionPane.showMessageDialog(null, "You're in.");
```

With `showMessageDialog`, I show a simple dialog box — a box with no text field. (Again, see Figure 5-5.)

Like thousands of other names, the name `JOptionPane` is defined in Java's API. (To be more specific, `JOptionPane` is defined inside something called `javax.swing`, which in turn is defined inside Java's API.) So, to use the name

JOptionPane throughout Listing 5-4, I import javax.swing.JOptionPane at the top of the listing.

TIP

In Listing 5-4, JOptionPane.showInputDialog works nicely because the user's input (username and password) are mere strings of characters. If you want the user to input a number (an int or a double, for example), you have to do some extra work. For example, to get an int value from the user, type something like int numberOfCows = Integer.parseInt(JOptionPane.showInputDialog("How many cows?")). The extra Integer.parseInt stuff forces your text field's input to be an int value. To get a double value from the user, type something like double fractionOfHolsteins = Double.parseDouble(JOptionPane.showInputDialog("Holsteins:")). The extra Double.parseDouble business forces your text field's input to be a double value.

Vive les nuls!

The French translations of *For Dummies* books are books *Pour les Nuls.* So a "dummy" in English is a "nul" in French.* But in Java, the word null means "nothing." When you see

```
if (
      username != null
```

in Listing 5-4, you can imagine that you see

```
if (
      username isn't nothing
```

or

```
if (
      username has any value at all
```

To find out how usernames can have no value, see the last row in Figure 5-5. When you click Cancel in the first dialog box, the computer hands null to your program. So, in Listing 5-4, the variable username becomes null. The comparison username != null checks to make sure that you haven't clicked Cancel in the program's first dialog box. The comparison password != null performs the same kind of check for the program's second dialog box. When you see the if statement in Listing 5-4, you can imagine that you see the following:

* In Russian, a "dummy" is a "чайник," which, when interpreted literally, means a "teapot." So, in Russian, this book is *Java For Teapots.* I've never been called a teapot, and I'm not sure how I'd react if I were.

```
if (
      you didn't press Cancel in the username dialog and
      you didn't press Cancel in the password dialog and
      (
        (you typed bburd in the username dialog and
         you typed swordfish in the password dialog) or
        (you typed hritter in the username dialog and
         you typed preakston in the password dialog)
      )
   )
```

WARNING

In Listing 5-4, the comparisons `username != null` and `password != null` are not optional. If you forget to include these and then click Cancel when the program runs, you get a nasty `NullPointerException` message, and the program comes crashing down before your eyes. The word `null` represents nothing, and in Java, you can't compare nothing to a string like `"bburd"` or `"swordfish"`. In Listing 5-4, the purpose of the comparison `username != null` is to prevent Java from moving on to check `username.equals("bburd")` whenever you happen to click Cancel. Without this preliminary `username != null` test, you're courting trouble.

TECHNICAL STUFF

The last couple of `null`s in Listing 5-4 are different from the others. In the code `JOptionPane.showMessageDialog (null, "You're in.")`, the word `null` stands for "no other dialog box." In particular, the call `showMessageDialog` tells Java to pop up a new dialog box, and the word `null` indicates that the new dialog box doesn't grow out of any existing dialog box. One way or another, Java insists that you say something about the origin of the newly popped dialog box. (For some reason, Java doesn't insist that you specify the origin of the `showInputDialog` box. Go figure!) Anyway, in Listing 5-4, having a `showMessageDialog` box pop up from nowhere is quite useful.

(Conditions in parentheses)

Keep an eye on those parentheses! When you're combining conditions with logical operators, it's better to waste typing effort and add unneeded parentheses than to goof up your result by using too few parentheses. Take, for example, the expression

```
2 < 5 || 100 < 6 && 27 < 1
```

By misreading this expression, you might conclude that the expression is false. That is, you could wrongly read the expression as meaning `(`*`something-or-other`*`) && 27 < 1`. Because `27 < 1` is false, you would conclude that the whole expression is false. The fact is that, in Java, any `&&` operator is evaluated before any `||` operator. So the expression really asks whether `2 < 5 || (`*`something-or-other`*`)`. Because `2 < 5` is true, the whole expression is true.

To change the expression's value from `true` to `false`, you can put the expression's first two comparisons in parentheses, like this:

```
(2 < 5 || 100 < 6) && 27 < 1
```

TIP

Java's `||` operator is *inclusive*. This means that you get a `true` value whenever the thing on the left side is true, the thing on the right side is true, or both things are true. For instance, the expression `2 < 10 || 20 < 30` is true.

WARNING

In Java, you can't combine comparisons the way you do in ordinary English. In English, you may say, "We'll have between three and ten people at the dinner table." But in Java, you get an error message if you write `3 <= people <= 10`. To do this comparison, you need something like `3 <= people && people <= 10`.

In Listing 5-4, the `if` statement's condition has more than a dozen parentheses. What happens if you omit two of them?

```
if (
      username != null && password != null &&
      // open parenthesis omitted
          (username.equals("bburd") && password.equals("swordfish")) ||
          (username.equals("hritter") && password.equals("preakston"))
      // close parenthesis omitted
   )
```

Java tries to interpret your wishes by grouping everything before the "or" (the `||` operator):

```
if (
      username != null && password != null &&
      (username.equals("bburd") && password.equals("swordfish"))

      ||

      (username.equals("hritter") && password.equals("preakston"))
   )
```

When the user clicks Cancel and username is `null`, Java says, "Okay! The stuff before the `||` operator is false, but maybe the stuff after the `||` operator is true. I'll check the stuff after the `||` operator to find out whether it's true." (Java often talks to itself. The psychiatrists are monitoring this situation.)

Anyway, when Java finally checks `username.equals("hritter")`, your program aborts with an ugly `NullPointerException` message. You've made Java angry by trying to apply .equals to a null username. (Psychiatrists have recommended anger management sessions for Java, but Java's insurance plan refuses to pay for the sessions.)

TRY IT OUT

Make some changes to the code in Listing 5-4.

THE RULE OF THREE

Add a third username/password combination to the list of acceptable logins.

OUT, DAMN'D NOT!

In Listing 5-4, change

```
username != null && password != null
```

to

```
!(username == null || password == null)
```

Does the program still work? Why or why not?

EQUAL BYTES

In Listing 5-4, change

```
username != null && password != null
```

to

```
!(username == null && password == null)
```

This is almost the same as the previous experiment. The only difference is the use of && instead of || between the two == null tests. Does the program still work? Why or why not?

The Nesting Habits of if Statements

Have you seen those cute Russian matryoshka nesting dolls? Open one, and another one is inside. Open the second, and a third one is inside it. You can do the same thing with Java's if statements. (Talk about fun!) Listing 5-5 shows you how.

LISTING 5-5: Nested if Statements

```
import java.util.Scanner;

import static java.lang.System.out;

public class Authenticator2 {

    public static void main(String[] args) {
        var keyboard = new Scanner(System.in);

        out.print("Username: ");
        String username = keyboard.next();

        if (username.equals("bburd")) {
            out.print("Password: ");
            String password = keyboard.next();

            if (password.equals("swordfish")) {
                out.println("You're in.");
            } else {
                out.println("Incorrect password");
            }

        } else {
            out.println("Unknown user");
        }

        keyboard.close();
    }
}
```

Figure 5-6 shows several runs of the code in Listing 5-5. The main idea is that to log on, you have to pass two tests. (In other words, two conditions must be true.) The first condition tests for a valid username; the second condition tests for the correct password. If you pass the first test (the username test), you march right into another `if` statement that performs a second test (the password test). If you fail the first test, you never make it to the second test. Figure 5-7 shows the overall plan.

WARNING

The code in Listing 5-5 does a good job with nested `if` statements, but it does a terrible job with real-world user authentication. First, never show a password in plain view (without asterisks to masquerade the password). Second, don't handle passwords without encrypting them. Third, don't tell the malicious user which of the two words (the username or the password) was entered incorrectly. Fourth . . . well, I could go on and on. The code in Listing 5-5 just isn't meant to illustrate good username/password practices.

```
Username: bburd
Password: swordfish
You're in.

Username: bburd
Password: catfish
Incorrect password

Username: jschmoe
Unknown user
```

FIGURE 5-6: Three runs of the code in Listing 5-5.

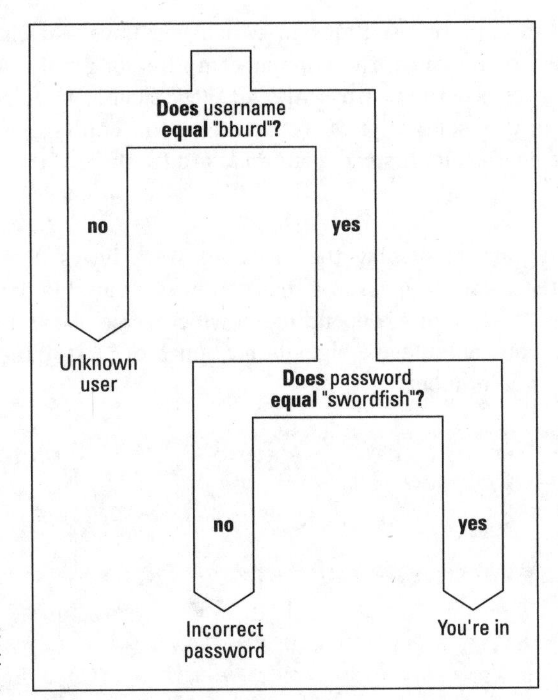

FIGURE 5-7: Don't eat with this fork.

I CHANGED MY MIND

Modify the program in Listing 5-4 so that, if the user clicks Cancel for either the username or the password, the program replies with a `Not enough information` message.

Choosing among Many Alternatives

I'm the first to admit that I hate making decisions. If things go wrong, I would rather have the problem be someone else's fault. Writing the previous sections (on making decisions with Java's `if` statement) knocked the stuffing right out of me. That's why my mind boggles as I begin this section on choosing among many alternatives. What a relief it is to have that confession out of the way!

Java's glorious switch statement

Now it's time to explore situations in which you have a decision with many branches. Take, for instance, the popular campfire song "Al's All Wet." (For a review of the lyrics, see the nearby "Al's All Wet" sidebar.) You're eager to write code that prints this song's lyrics. Fortunately, you don't have to type all the words over and over again. Instead, you can take advantage of the repetition in the lyrics.

A complete program to display the "Al's All Wet" lyrics doesn't come until Chapter 6. In the meantime, assume that you have a variable named `verse`. The value of `verse` is `1`, `2`, `3`, or `4`, depending on which verse of "Al's All Wet" you're trying to print. You could have a big, clumsy bunch of `if` statements that checks each possible `verse` number:

```
if (verse == 1) {
    out.println("That's because he has no brain.");
}
if (verse == 2) {
    out.println("That's because he is a pain.");
}
if (verse == 3) {
    out.println("'Cause this is the last refrain.");
}
```

But that approach seems wasteful. Why not create a statement that checks the value of `verse` just once and then takes an action based on the value it finds? Fortunately, just such a statement exists. It's called a *switch* statement. Listing 5-6 has an example of a `switch` statement.

"AL'S ALL WET"

Sung to the tune of "Gentille Alouette"

Al's all wet. Oh, why is Al all wet? Oh,
Al's all wet 'cause he's standing in the rain.
Why is Al out in the rain?
That's because he has no brain.
Has no brain, has no brain,
In the rain, in the rain.

Oh, oh, oh, oh

Al's all wet. Oh, why is Al all wet? Oh,
Al's all wet 'cause he's standing in the rain.
Why is Al out in the rain?
That's because he is a pain.
He's a pain, he's a pain,
Has no brain, has no brain,
In the rain, in the rain.

Oh, oh, oh, oh

Al's all wet. Oh, why is Al all wet? Oh,
Al's all wet 'cause he's standing in the rain.
Why is Al out in the rain?
'Cause this is the last refrain.
Last refrain, last refrain,
He's a pain, he's a pain,
Has no brain, has no brain,
In the rain, in the rain.

Oh, oh, oh, oh

Al's all wet. Oh, why is Al all wet? Oh,
Al's all wet 'cause he's standing in the rain.

—Harriet Ritter and Barry Burd

LISTING 5-6: A switch Statement

```
import java.util.Scanner;

import static java.lang.System.out;

public class JustSwitchIt {

    public static void main(String[] args) {
        var keyboard = new Scanner(System.in);
        out.print("Which verse? ");
        int verse = keyboard.nextInt();

        switch (verse) {
            case 1 -> out.println("That's because he has no brain.");
            case 2 -> out.println("That's because he is a pain.");
            case 3 -> out.println("'Cause this is the last refrain.");
            default -> out.println("No such verse. Please try again.");
        }

        out.println("Oh, oh, oh, oh");

        keyboard.close();
    }
}
```

Figure 5-8 shows two runs of the program in Listing 5-6. (Figure 5-9 illustrates the program's overall idea.) First, the user types a number, like the number 2. Then execution of the program reaches the top of the `switch` statement. The computer checks the value of the `verse` variable. When the computer determines that the `verse` variable's value is `2`, the computer checks each case of the `switch` statement. The value `2` doesn't match the topmost case, so the computer proceeds to the middle of the three cases. The value posted for the middle case (the number 2) matches the value of the `verse` variable, so the computer executes the statement in case 2:

```
out.println("That's because he is a pain.");
```

If the pesky user asks for verse 6, the computer bypasses cases 1, 2, and 3. The computer goes straight to the default. In the default, the computer displays `No such verse. Please try again` and then jumps out of the `switch` statement. After the computer is out of the `switch` statement, the computer displays `Oh, oh, oh, oh`.

```
Which verse? 2
That's because he is a pain.
Oh, oh, oh, oh

Which verse? 6
No such verse. Please try again.
Oh, oh, oh, oh
```

FIGURE 5-8: Running the code from Listing 5-6 two times.

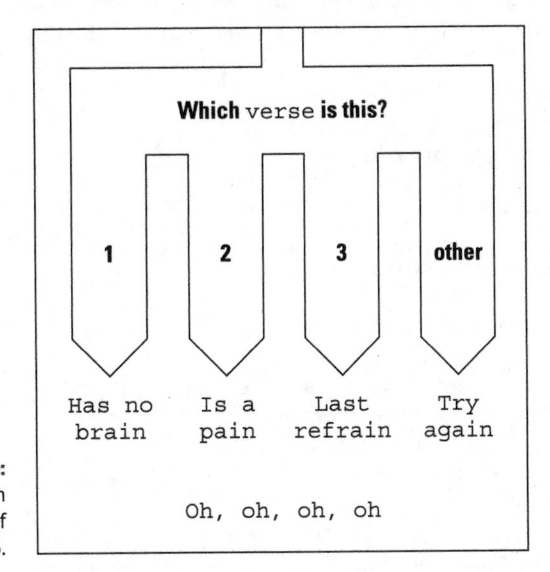

FIGURE 5-9: The big fork in the code of Listing 5-6.

TIP

In newer versions of Java, you can put more than one value in each case of a switch statement. For example, you can write the following friendly code:

```
var keyboard = new Scanner(System.in);
out.print("Will you pay me? ");
String reply = keyboard.next();

switch (reply) {
    case "Yes", "YES", "Y", "OK" -> out.println("Thank you!");
    case "No", "NO", "n" -> out.println("Thanks for nothing!");
}
```

A switch in time saves 14

In 2020, with the release of Java 14, the Java world "switched gears" (pun intended). The stewards of Java introduced a brand-new feature — namely, the *switch expression*. Here's the story:

If you look again at Listing 5-6, you wonder why someone took so much delight in typing the words `out.println`. Those words appear four times in the `switch` statement, and the only difference is the choice of a sarcastic verse. This repeated use of `out.println` seems wasteful. Why not have only one call to `out.println` for all the different answers the program can display?

Java's `switch` expression addresses this issue. In fact, runs of the code in Listing 5-7 look exactly the same as the runs of Listing 5-6. (Refer to Figure 5-8.)

LISTING 5-7: Out with the Old out.println!

```
import java.util.Scanner;

import static java.lang.System.out;

public class MakingAGoodExpression {

    public static void main(String[] args) {
        var keyboard = new Scanner(System.in);
        out.print("Which verse? ");
        int verse = keyboard.nextInt();
        String line;

        line = switch (verse) {
            case 1 -> "That's because he has no brain.";
            case 2 -> "That's because he is a pain.";
            case 3 -> "'Cause this is the last refrain.";
            default -> "No such verse. Please try again.";
        };

        out.println(line);
        out.println("Oh, oh, oh, oh");

        keyboard.close();
    }
}
```

WARNING

If you're using an older version of Java, don't bother trying to run the code in Listing 5-7. It won't work.

In any computer language, the word *expression* refers to a bunch of code that stands for a value. In Listing 5-7, the entire `switch` expression stands for a string of characters. For example, when `verse` is 2, the code

```
        switch (verse) {
    case 1 -> "That's because he has no brain.";
    case 2 -> "That's because he is a pain.";
    case 3 -> "'Cause this is the last refrain.";
    default -> "No such verse. Please try again.";
};
```

stands for the string `"That's because he is a pain."` So, in that scenario, Listing 5-7 does the same thing as this code:

```
line = "What part of 'no' don't you understand?";
out.println(line);
```

In this small snippet of code, notice how the first line ends with a semicolon. That's why there's a semicolon at the end of the big `switch` expression in Listing 5-7.

You can squeeze even more goodness out of Java's `switch` expression. For example, you don't have to assign the expression's value to a variable. You can bypass the variable and print the `switch` expression's value:

```
out.println(switch (verse) {
    case 1 -> "That's because he has no brain.";
    case 2 -> "That's because he is a pain.";
    case 3 -> "'Cause this is the last refrain.";
    default -> "No such verse. Please try again.";
});
```

But remember that a `switch` expression can't stand independently on its own. The following code makes no sense:

```
// Bad code!
int verse = keyboard.nextInt();
String line;
"That's because he is a pain.";    // This isn't a complete statement.
out.println(line);
```

The code with a freestanding `switch` expression doesn't work, either.

```
// Bad code!
int verse = keyboard.nextInt();
String line;

switch (verse) {                                // This isn't a complete statement.
```

```
        case 1 -> "That's because he has no brain.";
        case 2 -> "That's because he is a pain.";
        case 3 -> "'Cause this is the last refrain.";
        default -> "No such verse. Please try again.";
    };

    out.println(line);
```

Your grandparents' switch statement

In versions of Java before Java 14, the code in Listings 5-6 and 5-7 would have failed. That's because Java's `switch` statement came originally from the `switch` statement in the C/C++ language family, and the C/C++ `switch` statement was quite clunky. The code in Listing 5-8 does the same thing as the code in Listings 5-6 and 5-7, but the Listing 5-8 code works in all versions of Java — old and new.

LISTING 5-8: A la recherche du temps perdu

```
import java.util.Scanner;

import static java.lang.System.out;

public class TheVeryOldSwitcheroo {

    public static void main(String[] args) {
        var keyboard = new Scanner(System.in);
        out.print("Which verse? ");
        int verse = keyboard.nextInt();

        switch (verse) {
            case 1:
                out.println("That's because he has no brain.");
                break;
            case 2:
                out.println("That's because he is a pain.");
                break;
            case 3:
                out.println("'Cause this is the last refrain.");
                break;
            default:
                out.println("No such verse. Please try again.");
                break;
        }
```

```
        out.println("Oh, oh, oh, oh");

        keyboard.close();
    }
}
```

Runs of the code in Listing 5-8 look exactly the same as runs of Listings 5-6 and 5-7. (Refer to Figure 5-8.) The big difference is that Listing 5-8 uses the old form of the `switch` statement. In the old form:

- You use colons (:) instead of arrows (->).
- You need `break` statements to bypass any remaining cases.

When Java encounters a `break` statement, it jumps out of whatever `switch` statement it's in. For example, imagine that `verse` has the value `2`. In Listing 5-6, Java prints `That's because he is a pain`. Then, because of the `break` statement, Java skips right past the case that would display `'Cause this is the last refrain`. In fact, Java jumps out of the entire `switch` statement and goes directly to the statement just after the end of the `switch` statement. The computer displays `Oh, oh, oh, oh` because that's what the statement after the `switch` statement tells Java to do.

You don't really need to put a `break` at the end of an old `switch` statement. In Listing 5-8, the last `break` (the `break` that's part of the default) is just for the sake of overall tidiness.

The need for `break` inside the old `switch` statement is a Java programmer's nightmare. Everyone forgets to add these `break` statements, and when you forget to add `break` statements, you get fall-though. With *fall-through*, execution of the code falls right through from one case to the next. Execution keeps falling through until you eventually reach a `break` statement or the end of the entire `switch` statement.

Free fall

Usually, when you're using a `switch` statement, you don't want fall-through, so you pepper `break` statements throughout the `switch` statements. But, occasionally, fall-through is just the thing you need. Take, for instance, the "Al's All Wet" song. (The classy lyrics are shown in the sidebar bearing the song's name.) Each verse of "Al's All Wet" adds new lines in addition to the lines from previous verses. This situation (accumulating lines from one verse to another) cries out for a `switch` statement with fall-through. Listing 5-9 demonstrates the idea.

LISTING 5-9: A switch Statement with Fall-Through

```
import java.util.Scanner;

import static java.lang.System.out;

public class FallingForYou {

    public static void main(String[] args) {
        var keyboard = new Scanner(System.in);
        out.print("Which verse? (one, two or three) ");
        String verse = keyboard.next();

        switch (verse) {
            case "three":
                out.print("Last refrain, ");
                out.println("last refrain,");
            case "two":
                out.print("He's a pain, ");
                out.println("he's a pain,");
            case "one":
                out.print("Has no brain, ");
                out.println("has no brain,");
        }

        out.print("In the rain, ");
        out.println("in the rain.");

        out.println("Oh, oh, oh, oh");
        out.println();

        keyboard.close();
    }
}
```

Figure 5-10 shows several runs of the program in Listing 5-9. Because the `switch` has no `break` statements in it, fall-through happens all over the place. For instance, when the user enters the word `two`, Java starts by executing the two statements in `case "two"`:

```
out.print("He's a pain, ");
out.println("he's a pain,");
```

Then Java marches right on to execute the two statements in `case "one"`:

```
out.print("Has no brain, ");
out.println("has no brain,");
```

```
Which verse? (one, two or three) one
Has no brain, has no brain,
In the rain, in the rain.
Oh, oh, oh, oh

Which verse? (one, two or three) two
He's a pain, he's a pain,
Has no brain, has no brain,
In the rain, in the rain.
Oh, oh, oh, oh

Which verse? (one, two or three) three
Last refrain, last refrain,
He's a pain, he's a pain,
Has no brain, has no brain,
In the rain, in the rain.
Oh, oh, oh, oh

Which verse? (one, two or three) six
In the rain, in the rain.
Oh, oh, oh, oh
```

FIGURE 5-10: Running the code of Listing 5-9 four times.

That's good because the song's second verse has all these lines in it.

Notice what happens when the user asks for verse six. The `switch` statement in Listing 5-9 has no `case "six"` and no `default`, so none of the actions inside the `switch` statement is executed. Even so, with statements that print `In the rain, in the rain` and `Oh, oh, oh, oh` right after the `switch` statement, the computer displays something when the user asks for verse `"six"`.

TECHNICAL STUFF

If you travel back in time to the year 2010, you can't run the code in Listing 5-9. It wasn't until July 2011, with the release of Java 7, that strings became acceptable as Java `case` values. In Java 6, `case 3:` was okay, but `case "three":` was strictly forbidden.

TRY IT OUT

Get some practice with `if` statements and `switch` statements!

MONTH-TO-MONTH RESUSCITATION

Write a program that inputs the name of a month and outputs the number of days in that month. In this first version of the program, assume that February always has 28 days. Use Java's `switch` statement — the kind I used in Listing 5-6.

LEAPING TO CONCLUSIONS

Make changes to your code from the earlier section "Month-to-Month Resuscitation." Have the user input a month name and also `yes` or `no` in response to the question `Is it a leap year?`

HOW ABOUT A DATE?

Make changes to your code from the earlier section "Leaping to Conclusions." Rather than mimic Listing 5-6, use a `switch` expression of the kind you find in Listing 5-7. Then use Java's old-style `switch` statement like the one you see in Listing 5-8.

THE YIELD APP

Take this code for a spin and see what it does:

```
out.println(switch (month) {
    case "January", "March", "May", "July",
            "August", "October", "December" -> 31;
    case "April", "June", "September", "November" -> 30;
    case "February" -> {
        out.print("Leap year (true/false)? ");
        isLeapYear = keyboard.nextBoolean();
        if (isLeapYear) {
            yield 29;
        } else {
            yield 28;
        }
    }
    default -> 0;
} + " days");
```

» Counting as you loop

» Repeating relentlessly (until the user gives you a clear answer)

Chapter 6
Controlling Program Flow with Loops

In 1966, the company that brings you Head & Shoulders shampoo made history. On the back of the bottle, the directions for using the shampoo read, "LATHER–RINSE–REPEAT." Never before had a complete set of directions (for doing anything, let alone shampooing your hair) been summarized so succinctly. People in the direction-writing business hailed this as a monumental achievement. Directions like these stood in stark contrast to others of the time. (For instance, the first sentence on a can of bug spray read, "Turn this can so that it points away from your face." Duh!)

Aside from their brevity, the thing that made the Head & Shoulders directions so cool was that, with three simple words, they managed to capture a notion that's at the heart of all instruction-giving: the notion of repetition. That last word, *REPEAT*, took an otherwise bland instructional drone and turned it into a sophisticated recipe for action.

The fundamental idea is that when you're following directions, you don't just follow one instruction after another. Instead, you take turns in the road. You make decisions ("If HAIR IS DRY, then USE CONDITIONER") and you go into loops ("LATHER–RINSE, and then LATHER–RINSE again."). In computer programming, you use decision-making and looping all the time. This chapter explores looping in Java.

Repeating Instructions Over and Over Again (Java while Statements)

Here's a guessing game for you. The computer generates a random number from 1 to 10. The computer asks you to guess the number. If you guess incorrectly, the game continues. As soon as you guess correctly, the game is over. Listing 6-1 shows the program to play the game, and Figure 6-1 shows a round of play.

```
      ************
Welcome to the Guessing Game
      ************

Enter an int from 1 to 10: 2

Try again...
Enter an int from 1 to 10: 5

Try again...
Enter an int from 1 to 10: 8

Try again...
Enter an int from 1 to 10: 3
You win after 4 guesses.
```

FIGURE 6-1: Play until you drop.

LISTING 6-1: **A Repeating Guessing Game**

```
import java.util.Random;
import java.util.Scanner;

import static java.lang.System.out;

public class GuessAgain {

    public static void main(String[] args) {
        var keyboard = new Scanner(System.in);

        int numGuesses = 0;
        int randomNumber = new Random().nextInt(10) + 1;

        out.println(" ************ ");
        out.println("Welcome to the Guessing Game");
        out.println(" ************ ");
        out.println();

        out.print("Enter an int from 1 to 10: ");
        int inputNumber = keyboard.nextInt();
        numGuesses++;
```

```
        while (inputNumber != randomNumber) {
            out.println();
            out.println("Try again...");
            out.print("Enter an int from 1 to 10: ");
            inputNumber = keyboard.nextInt();
            numGuesses++;
        }
        out.print("You win after ");
        out.println(numGuesses + " guesses.");

        keyboard.close();
    }
}
```

In Figure 6-1, the user makes four guesses. Each time around, the computer checks to see whether the guess is correct. An incorrect guess generates a request to try again. For a correct guess, the user gets a rousing `You win`, along with a tally of the number of guesses they made. The computer repeats several statements, checking each time through to see whether the user's guess is the same as a certain randomly generated number. Each time the user makes a guess, the computer adds 1 to its tally of guesses. When the user makes the correct guess, the computer displays that tally. Figure 6-2 illustrates the flow of action.

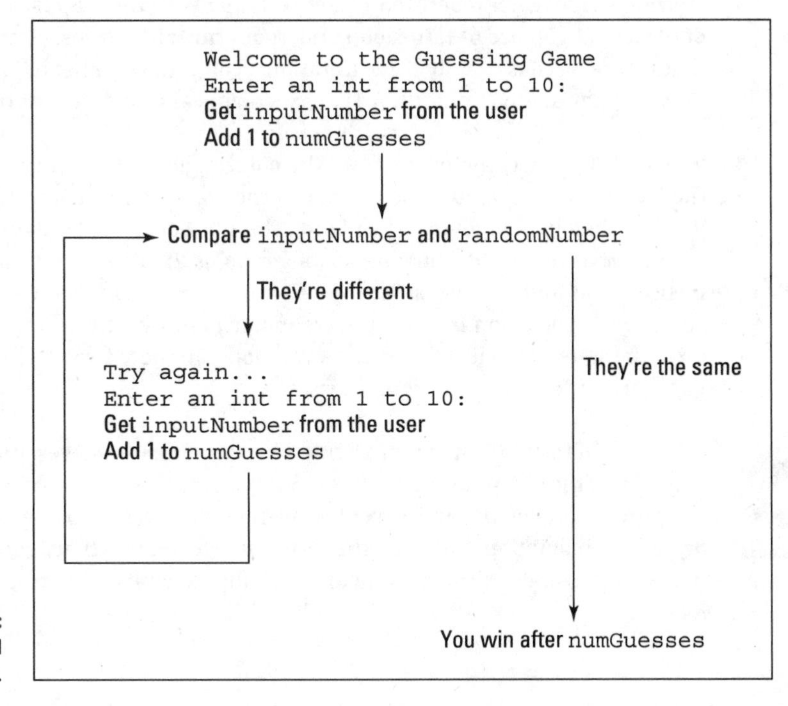

FIGURE 6-2: Around and around you go.

When you look over Listing 6-1, you see the code that does all this work. At the core of the code is a thing called a *while statement* (also known as a *while loop*). Rephrased in English, the `while` statement says:

```
while the inputNumber is not equal to the randomNumber
keep doing all the stuff in curly braces: {

}
```

The stuff in curly braces (the stuff that repeats) is the code that prints `Try again` and `Enter an int ...`, gets a value from the keyboard, and adds 1 to the count of the user's guesses.

TIP

When you're dealing with counters, like `numGuesses` in Listing 6-1, you may easily become confused and be off by 1 in either direction. You can avoid this headache by making sure that the ++ statements stay close to the statements whose events you're counting. For example, in Listing 6-1, the variable `numGuesses` starts with a value of `0`. That's because, when the program starts running, the user hasn't made any guesses. Later in the program, right after each call to `keyboard.nextInt`, is a `numGuesses++` statement. That's how you do it — you increment the counter as soon as the user enters another guess.

The statements in curly braces are repeated as long as `inputNumber != randomNumber` is true. Each repetition of the statements in the loop is called an *iteration* of the loop. In Figure 6-1, the loop undergoes three iterations. (If you don't believe that Figure 6-1 has exactly three iterations, count the number of `Try again` printings in the program's output. A `Try again` appears for each incorrect guess.)

When, at long last, the user enters the correct guess, the computer goes back to the top of the `while` statement, checks the condition in parentheses, and finds itself in double double-negative land. The not equal (`!=`) relationship between `inputNumber` and `randomNumber` no longer holds. In other words, the `while` statement's condition, `inputNumber != randomNumber`, is false. Because the `while` statement's condition is false, the computer jumps past the `while` loop and goes on to the statements just below the `while` loop. In these two statements, the computer prints `You win` after four guesses.

REMEMBER

With code of the kind shown in Listing 6-1, the computer never jumps out in mid-loop. When the computer finds that `inputNumber` isn't equal to `randomNumber`, the computer marches on and executes all five statements inside the loop's curly braces. The computer performs the test again (to see whether `inputNumber` is still not equal to `randomNumber`) only after it fully executes all five statements in the loop.

TRY IT OUT

I have two things for you to try:

BIGGER AND BETTER

Modify the program in Listing 6-1 so that the randomly generated number is a number from 1 through 100. To make life bearable for the game player, have the program give a hint whenever the player guesses incorrectly. Hints such as `Try a higher number` or `Try a lower number` are helpful.

NO NEGATIVITY

Write a program in which the user types `int` values, one after another. The program stops looping when the user types a number that isn't positive (for example, the number 0 or the number −17). After all the looping, the program displays the largest number that the user typed. For example, if the user types the numbers

```
7
25
3
9
0
```

the program displays the number `25`.

Count On Me

"Write *I will not talk in class* on the blackboard 100 times."

What your teacher really meant was this:

```
Set the count to 0.
As long as the count is less than 100,
   Write 'I will not talk in class' on the blackboard,
   Add 1 to the count.
```

Fortunately, you didn't know about loops and counters at the time. If you pointed out all this stuff to your teacher, you'd have gotten into a lot more trouble than you were already in.

One way or another, life is filled with examples of counting loops. And computer programming mirrors life — or is it the other way around? When you tell a computer what to do, you're often telling the computer to print three lines, process ten accounts, dial a million phone numbers, or whatever. Because counting loops is so common in programming, the people who create programming languages have developed statements just for loops of this kind. In Java, the statement that repeats something a certain number of times is called a *for statement.* Listings 6-2 and 6-3 illustrate the use of the `for` statement. Listing 6-2 has a rock-bottom simple example and Listing 6-3 has a more exotic example. Take your pick.

LISTING 6-2: The World's Most Boring for Loop

```
import static java.lang.System.out;

public class Yawn {

    public static void main(String[] args) {

        for (int count = 1; count <= 10; count++) {
            out.print("The value of count is ");
            out.print(count);
            out.println(".");
        }

        out.println("Done!");
    }
}
```

Figure 6-3 shows you what you get when you run the program from Listing 6-2. (You get exactly what you deserve.) The `for` statement in Listing 6-2 starts by setting the `count` variable to `1`. Then the statement tests to make sure that `count` is less than or equal to 10 (which it certainly is). Then the `for` statement dives ahead and executes the printing statements between the curly braces. (At this early stage of the game, the computer prints `The value of count is 1.`) Finally, the `for` statement does that last thing inside its parentheses — it adds 1 to the value of `count`.

```
The value of count is 1.
The value of count is 2.
The value of count is 3.
The value of count is 4.
The value of count is 5.
The value of count is 6.
The value of count is 7.
The value of count is 8.
The value of count is 9.
The value of count is 10.
Done!
```

FIGURE 6-3: Counting to ten.

With count now equal to 2, the for statement checks again to make sure that count is less than or equal to 10. (Yes, 2 is smaller than 10.) Because the test turns out okay, the for statement marches back into the curly-braced statements and prints The value of count is 2 on the screen. Finally, the for statement does that last thing inside its parentheses — it adds 1 to the value of count, increasing the value of count to 3.

And so on. This whole business repeats until, after ten iterations, the value of count finally reaches 11. When this happens, the check for count being less than or equal to ten fails and the loop's execution ends. The computer jumps to whatever statement comes immediately after the for statement. In Listing 6-2, the computer prints Done! as its output. Figure 6-4 illustrates the whole process.

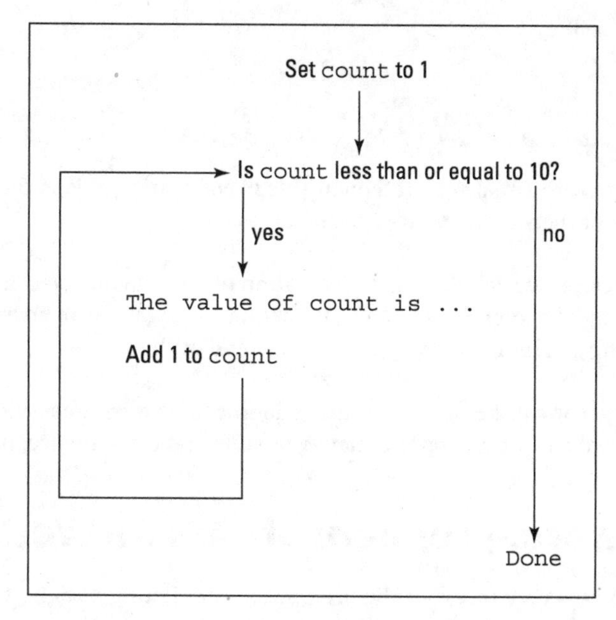

FIGURE 6-4: The action of the for loop in Listing 6-2.

The anatomy of a for statement

After the word *for*, you always put three things in parentheses. The first of these three is called an *initialization*, the second is an *expression*, and the third is an *update:*

```
for ( initialization ; expression ; update )
```

Each of the three items in parentheses plays its own, distinct role:

- The **initialization** is executed once, when the run of your program first reaches the for statement.
- The **expression** is evaluated several times (before each iteration).
- The **update** is also evaluated several times (at the end of each iteration).

If it helps, think of the loop as though its text is shifted all around:

```
int count = 1
for count <= 10 {
    out.print("The value of count is ");
    out.print(count);
    out.println(".");
    count++;
}
```

You can't write a real for statement this way. Even so, this is the order in which the parts of the statement are executed.

WARNING

If you declare a variable in the initialization of a for loop, you can't use that variable outside the loop. For instance, in Listing 6-2, you see an error message if you try putting out.println(count) after the end of the loop.

TIP

Anything that can be done with a for loop can also be done with a while loop. Choosing to use a for loop is a matter of style and convenience, not necessity.

The world premiere of "Al's All Wet"

Listing 6-2 is very nice, but the program in that listing does nothing interesting. For a more eye-catching example, see Listing 6-3. In Listing 6-3, I make good on a promise I make in Chapter 5: The program in Listing 6-3 prints all the lyrics of the hit single "Al's All Wet." (You can find the lyrics in Chapter 5.)

LISTING 6-3: The Unabridged "Al's All Wet" Song

```
import static java.lang.System.out;

public class AlsAllWet {
```

```
    public static void main(String[] args) {
        String intro = """
                Al's all wet. Oh, why is Al all wet? Oh,
                Al's all wet 'cause he's standing in the rain.""";

        for (int verse = 1; verse <= 3; verse++) {
            out.println(intro);
            out.println("Why is Al out in the rain?");

            out.println(switch (verse) {
                case 1 -> "That's because he has no brain.";
                case 2 -> "That's because he is a pain.";
                case 3 -> "'Cause this is the last refrain.";
                default -> "";
            });

            switch (verse) {
                case 3:
                    out.println("Last refrain, last refrain,");
                case 2:
                    out.println("He's a pain, he's a pain,");
                case 1:
                    out.println("Has no brain, has no brain,");
            }

            out.println("""
                    In the rain, in the rain.
                    Oh, oh, oh, oh""");
            out.println();
        }

        out.println(intro);
    }
}
```

Listing 6-3 is nice because it combines many of the ideas from Chapters 5 and 6. Listing 6-3 has two switches nested inside a `for` loop. One of them is a `switch` expression; the other is a `switch` statement with fall-through. As the value of the `for` loop's counter variable (`verse`) goes from `1` to `2` and then to `3`, Java executes all the parts in both of these switches. When the program is near the end of its run and execution has dropped out of the `for` loop, the program's last statement prints the song's final verse.

TECHNICAL STUFF

When I boldly declare that a `for` statement is for counting, I'm stretching the truth just a bit. Java's `for` statement is *versatile:* You can use a `for` statement in situations that have nothing to do with counting. For instance, a statement with no update part, such as `for (i = 0; i < 10; )`, just keeps going. The looping ends when some action inside the loop assigns a big number to the variable `i`. You can even create a `for` statement with nothing inside the parentheses. The loop `for ( ; ; )` runs forever, which is good if the loop controls a serious piece of machinery. Usually, when you write a `for` statement, you're counting how many times to repeat something. But, in truth, you can do just about any kind of repetition with a `for` statement.

TRY IT OUT

Would you like some practice? Try these experiments and challenges:

FOR AMOUR

A `for` statement's initialization may have several parts. A `for` statement's update may also have several parts. To find out how, enter the following lines in Java's JShell, or add the lines to a small Java program:

```
import static java.lang.System.out
for (int i = 0, j = 10; i < j; i++, j--) {out.println(i + " " + j);}
```

COLLECTING VALUES

What's the output of the following code?

```
int total = 0;
for (int i = 0; i < 10; i++) {
    total += i;
}
System.out.println(total);
```

In this code, the variable `total` is called an *accumulator* because it accumulates (adds up) a bunch of values inside the loop.

FACTORIAL

In mathematics, the exclamation point (!) means *factorial* — the number you get when you multiply all the positive `int` values up to and including a certain number. For example, 3! is $1 \times 2 \times 3$, which is 6. And 5! is $1 \times 2 \times 3 \times 4 \times 5$, which is 120.

Write a program in which the user types a positive `int` value (call it *n*), and Java displays the value of *n*! as its output.

SEEING STARS

Without running the following code, try to predict the code's output:

```
for (int row = 0; row < 5; row++) {
    for (int column = 0; column < 5; column++) {
        System.out.print("*");
    }
    System.out.println();
}
```

After making your prediction, run the code to find out whether your prediction is correct.

SEEING MORE AND MORE STARS

The code in this experiment is a slight variation on the code in the previous experiment. First, try to predict what the code will output. Then run the code to find out whether your prediction is correct:

```
for (int row = 0; row < 5; row++) {
    for (int column = 0; column <= row; column++) {
        System.out.print("*");
    }
    System.out.println();
}
```

THREE TRIANGLES

Write a program that uses loops to display three copies of the following pattern, one after another:

```
*****
****
***
**
*
```

You Can Always Get What You Want

Fools rush in where angels fear to tread.

—ALEXANDER POPE

Today, I want to be young and foolish (or, at the very least, foolish). Look back at Figure 6-2 and notice how Java's `while` loop works. When execution enters a while loop, the computer checks to make sure that the loop's condition is true. If the condition isn't true, the statements inside the loop are never executed — not even once. In fact, you can easily cook up a while loop whose statements are never executed (although I can't think of a reason you would ever want to do it):

```
int twoPlusTwo = 2 + 2;

while (twoPlusTwo == 5) {
    out.println("""
            Are you kidding?
            2 + 2 doesn't equal 5.
            Everyone knows that 2 + 2 equals 3.""");
}
```

In spite of this silly twoPlusTwo example, the while statement turns out to be the most versatile of Java's looping constructs. In particular, the while loop is good for situations in which you must look before you leap. For example, "While money is in my account, write a mortgage check every month." When you first encounter this statement, if your account has a zero balance, you don't want to write a mortgage check — not even a single check.

But at times (not many), you want to leap before you look. Take, for instance, the situation in which you're asking the user for a response. Maybe the user's response makes sense, but maybe it doesn't. If it doesn't, you want to ask again. Maybe the user's finger slipped, or perhaps the user didn't understand the question.

Figure 6-5 shows some runs of a program to delete a file. Before deleting the file, the program asks the user whether making the deletion is okay. If the user answers *y* or *n*, the program proceeds according to the user's wishes. But if the user enters any other character (any digit, uppercase letter, punctuation symbol, or whatever), the program asks the user for another response.

```
Delete evidence? (y/n) n
Sorry, buddy. Just asking.


Delete evidence? (y/n) u
Delete evidence? (y/n) Y
Delete evidence? (y/n) L
Delete evidence? (y/n) 8
Delete evidence? (y/n) .
Delete evidence? (y/n) y
Okay, here goes...
The evidence has been deleted.
```

FIGURE 6-5: Two runs of the code in Listing 6-4.

To write this program, you need a loop — a loop that repeatedly asks the user whether the file should be deleted. The loop keeps asking until the user gives a meaningful response. Now, the thing to notice is that the loop doesn't need to check anything before asking the user the first time. Indeed, before the user gives the first response, the loop has nothing to check. The loop doesn't start with "As long as such-and-such is true, then get a response from the user." Instead, the loop just leaps ahead, gets a response from the user, and then checks the response to see whether it makes sense.

That's why the program in Listing 6-4 has a *do* loop (also known as a *do . . . while* loop). With a do loop, the program jumps right in, takes action, and then checks a condition to see whether the result of the action makes sense. If the result makes sense, execution of the loop is done. If not, the program goes back to the top of the loop for another go-round.

LISTING 6-4: To Delete or Not to Delete

```
import java.io.File;
import java.util.Scanner;

import static java.lang.System.out;

public class DeleteEvidence {

    public static void main(String[] args) {
        var evidence = new File("cookedBooks.txt");
        var keyboard = new Scanner(System.in);
        char reply;

        do {
            out.print("Delete evidence? (y/n) ");
            reply = keyboard.findWithinHorizon(".", 0).charAt(0);
        } while (reply != 'y' && reply != 'n');

        if (reply == 'y') {
            out.println("Okay, here goes...");
            evidence.delete();
            out.println("The evidence has been deleted.");
        } else {
            out.println("Sorry, buddy. Just asking.");
        }
        keyboard.close();
    }
}
```

Figure 6-5 shows two runs of the code in Listing 6-4. The program accepts lowercase letters *y* and *n*, but not the uppercase letters *Y* and *N*. To make the program accept uppercase letters, change the conditions in the code as follows:

```
do {
    out.print("Delete evidence? (y/n) ");
    reply = keyboard.findWithinHorizon(".", 0).charAt(0);
} while (reply != 'y' && reply != 'Y' && reply != 'n' && reply!='N');

if (reply == 'y' || reply == 'Y') {
```

Figure 6-6 shows the flow of control in the loop of Listing 6-4. With a do loop, the situation in the twoPlusTwo program (shown at the beginning of this section) can never happen. Because the do loop carries out its first action without testing a condition, every do loop is guaranteed to perform at least one iteration.

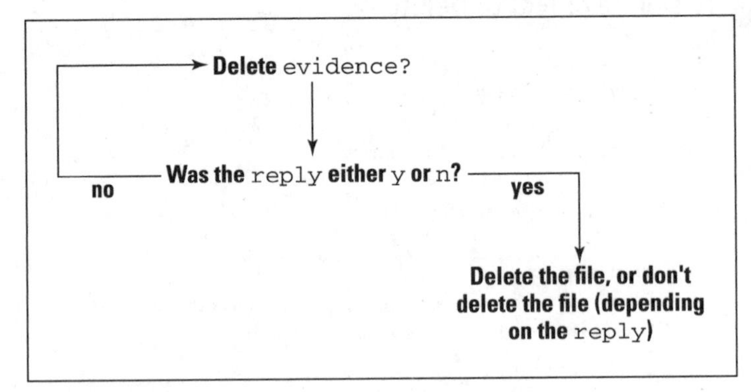

FIGURE 6-6: Here we go loop, do loop.

The root of the matter

The location of Listing 6-4's cookedBooks.txt file on your computer's hard drive depends on several factors. If you create a cookedBooks.txt file in the wrong directory, the code in Listing 6-4 cannot delete your file. (More precisely, if cookedBooks.txt is in the wrong directory on your hard drive, the code in Listing 6-4 can't find the cookedBooks.txt file in preparation for deleting the file.)

In most settings, you start testing Listing 6-4 by creating a project within your IDE. The new project lives in a folder on your hard drive. That folder is called the project's *root folder*. With the code in Listing 6-4, the cookedBooks.txt file belongs directly inside your project's root folder.

For example, I have a project named 06-04. That project lives on my hard drive in a folder named 06-04 — the project's root folder. Directly inside that 06-04 root folder, I have a file named cookedBooks.txt. Also, inside the 06-04 folder, I have a src subfolder. The src subfolder contains my DeleteEvidence.java file.

TIP

If you have trouble deciding where the cookedBooks.txt file should be, add the following code to Listing 6-4 immediately after the new File statement:

```
try {
    out.println("Looking for " + evidence.getCanonicalPath());
} catch (java.io.IOException e) {
    e.printStackTrace();
}
```

When you run the code, Java tells you where, on your hard drive, the cooked-Books.txt file should be.

CROSS REFERENCE

For more information about files and their folders, see Chapter 8.

Reading a single character

Over in Listing 5-3, from Chapter 5, the user types a word on the keyboard. The keyboard.next method grabs the word and places the word into a String variable named *password*. Everything works nicely because a String variable can store many characters at a time, and the next method can read many characters at a time.

But in Listing 6-4, you're not interested in reading several characters. You expect the user to press one letter — either y or n. So you don't create a String variable to store the user's response. Instead, you create a char variable — a variable that stores just one symbol at a time.

The Java API has no nextChar method. To read something suitable for storage in a char variable, you have to improvise. In Listing 6-4, the improvisation looks like this:

```
keyboard.findWithinHorizon(".", 0).charAt(0)
```

You can use this code exactly as it appears in Listing 6-4 whenever you want to read a single character.

REMEMBER

A String variable can contain many characters or just one. But a String variable that contains only one character isn't the same as a char variable. No matter what you put in a String variable, String variables and char variables have to be treated differently.

File handling in Java

In Listing 6-4, the actual file-handling statements deserve some attention. These statements involve the use of classes, objects, and methods. Many of the meaty details about these things are in other chapters, like Chapters 7 and 9. Even so, I can't do any harm by touching on some highlights right here.

So you can find a class in the Java language API named *java.io.File*. The statement

```
var evidence = new File("cookedBooks.txt");
```

creates a new object in the computer's memory. This object, formed from the `java.io.File` class, describes everything that the program needs to know about the disk file `cookedBooks.txt`. From this point on in Listing 6-4, the variable `evidence` refers to the disk file `cookedBooks.txt`.

The `evidence` object, as an instance of the `java.io.File` class, has a `delete` method. (What can I say? It's in the API documentation.) When you call `evidence.delete`, the computer gets rid of the file for you.

Of course, you can't get rid of something that doesn't already exist. When the computer executes

```
var evidence = new File("cookedBooks.txt");
```

Java doesn't check to make sure that you have a file named `cookedBooks.txt`. To force Java to do the checking, you have a few options. The simplest is to call the `exists` method. When you call `evidence.exists()`, the method looks in the folder where Java expects to find `cookedBooks.txt`. The call `evidence.exists()` returns true if Java finds `cookedBooks.txt` inside that folder. Otherwise, the call `evidence.exists()` returns false. Here's how it works:

```
if (evidence.exists()) {
    var keyboard = new Scanner(System.in);
    char reply;

    do {
        out.print("Delete evidence? (y/n) ");
        reply = keyboard.findWithinHorizon(".", 0).charAt(0);
    } while (reply != 'y' && reply != 'n');

    if (reply == 'y') {
        out.println("Okay, here goes...");
        evidence.delete();
```

```
        out.println("The evidence has been deleted.");
    } else {
        out.println("Sorry, buddy. Just asking.");
    }

    keyboard.close();
}
```

Block on the while side

A bunch of statements surrounded by curly braces forms a *block.* If you declare a variable inside a block, you generally can't use that variable outside the block. For instance, in Listing 6-4, you see an error message if you make the following change:

```
do {
    out.print("Delete evidence? (y/n) ");
    char reply = keyboard.findWithinHorizon(".", 0).charAt(0);
} while (reply != 'y' && reply != 'n');

if (reply == 'y')
```

With the declaration `char reply` inside the loop's curly braces, no use of the name `reply` makes sense anywhere outside the braces. When you try to compile this code, you get three error messages — two for the `reply` words in `while (reply != 'y' && reply != 'n')` and a third for the `if` statement's `reply`.

So, in Listing 6-4, your hands are tied. The program's first real use of the `reply` variable is inside the loop. But to make that variable available after the loop, you have to declare `reply` before the loop. In this situation, you're best off declaring the `reply` variable without initializing the variable. Very interesting!

CROSS REFERENCE

To read more about variable initializations, see Chapter 4. To find out more about blocks, see Chapter 5.

All versions of Java have the three kinds of loops described in this chapter (`while` loops, `for` loops, and `do ... while` loops). But newer Java versions (namely, Java 5 and beyond) have yet another kind of loop, called an *enhanced for loop.* For a look at Java's `enhanced for` loop, see Chapter 11.

MISSED OPPORTUNITY

TRY IT OUT

Copy the code from Listing 6-1, but with the following change:

```
out.print("Enter an int from 1 to 10: ");
int inputNumber = keyboard.nextInt();
numGuesses++;

do {
    out.println();
    out.println("Try again...");
    out.print("Enter an int from 1 to 10: ");
    inputNumber = keyboard.nextInt();
    numGuesses++;
} while (inputNumber != randomNumber);

out.print("You win after ");
out.println(numGuesses + " guesses.");
```

The code in Listing 6-1 has a `while` loop, but this modified code has a `do` loop. Does this modified code work correctly? Why or why not?

LET'S BUST OUT OF HERE

In Chapter 5, you use `break` statements to jump out of a `switch`. But a `break` statement can also play a role inside a loop. To find out how it works, run a program containing the following code:

```
var keyboard = new Scanner(System.in);
while (true) {
    System.out.print("Enter an int value: ");
    int i = keyboard.nextInt();
    if (i == 0) {
        break;
    }
    System.out.println(i);
}
System.out.println("Done!");
keyboard.close();
```

The loop's condition is always `true`. It's like starting a loop with the line

```
while (1 + 1 == 2)
```

If it weren't for the break statement, the loop would run forever. Fortunately, when you execute the break statement, Java jumps to the code immediately after the loop.

CARRY ON AND KEEP CODING

In addition to its break statement, Java has a continue statement. When you execute a continue statement, Java skips to the end of its loop and begins the next iteration of that loop. To see it in action, run a program containing the following code:

```
var keyboard = new Scanner(System.in);

while (true) {
    System.out.print("Enter an int value: ");
    int i = keyboard.nextInt();
    if (i > 10) {
        continue;
    }
    if (i == 0) {
        break;
    }
    System.out.println(i);
}

System.out.println("Done!");
keyboard.close();
```

3

Working with the Big Picture: Object-Oriented Programming

IN THIS PART . . .

Find out what classes and objects are (without bending your brain out of shape).

Learn how to reuse existing code (saving time and money).

Be the architect of your virtual world by constructing brand-new objects.

» Passing values to and from methods

» Hiding details in your object-oriented code

Chapter 7
The Inside scOOP

As a computer book author, I've been told, over and over again: Don't expect people to read sections and chapters in their logical order. People jump around, picking what they need and skipping what they don't feel like reading. With that in mind, I realize that you may have skipped Chapter 1. If that's the case, please don't feel guilty. You can compensate in just 60 seconds by reading the following information, culled from Chapter 1:

> *Because Java is an object-oriented programming language, your primary goal is to describe classes and objects. A class is the idea behind a certain kind of thing. An object is a concrete instance of a class. The programmer defines a class, and from the class definition, Java makes individual objects.*

Of course, you can certainly choose to skip over the 60-second summary paragraph. If that's the case, you may want to recoup some of your losses. You can do that by reading the following two-word summary of Chapter 1:

> *Classes; objects.*

Defining a Class (What It Means to Be an Account)

What distinguishes one bank account from another? If you ask a banker this question, you hear a long sales pitch. The banker describes interest rates, fees, penalties — the whole routine. Fortunately for you, I'm not interested in all that. Instead, I want to know how my account is different from your account. After all, my account is named *Barry Burd, trading as Burd Brain Consulting,* and your account is named *Jane Q. Reader, trading as Budding Java Expert.* My account has $24.02 in it. How about yours?

When you come right down to it, the differences between one account and another can be summarized as values of variables. Maybe there's a variable named `balance`. For me, the value of `balance` is `24.02`. For you, the value of `balance` is `55.63`. The question is, when writing a computer program to deal with accounts, how do I separate my `balance` variable from your `balance` variable?

The answer is to create two separate objects. Let one `balance` variable live inside one object and let the other `balance` variable live inside the other object. While you're at it, put a `name` variable and an `address` variable in each of the objects. And there you have it: two objects, and each object represents an account. More precisely, each object is an instance of the `Account` class. (See Figure 7-1.)

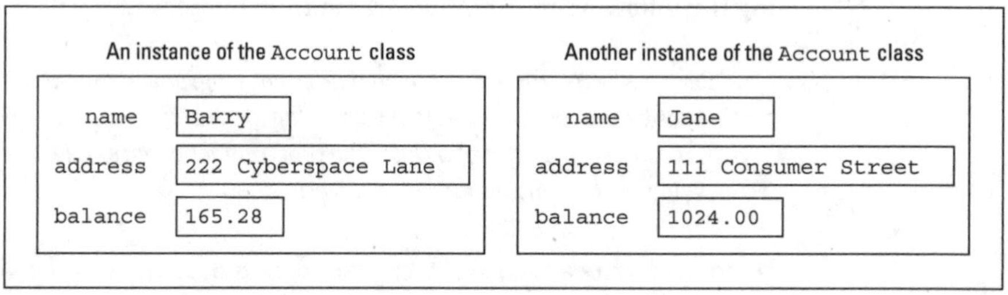

FIGURE 7-1: Two objects.

So far, so good. However, you still haven't solved the original problem. In your computer program, how do you refer to my `balance` variable as opposed to your `balance` variable? Well, you have two objects sitting around, so maybe you have variables to refer to these two objects. Create one variable named *myAccount* and another variable named *yourAccount.* The `myAccount` variable refers to my object (my instance of the `Account` class) with all the stuff that's inside it. To refer to my balance, write

```
myAccount.balance
```

To refer to my name, write

```
myAccount.name
```

For want of any better terminology, this way of referring to an object's fields is called *dot notation*.

Then `yourAccount.balance` refers to the value in your object's `balance` variable, and `yourAccount.name` refers to the value of your object's `name` variable. To tell Java how much I have in my account, you can write

```
myAccount.balance = 24.02;
```

To display your name on the screen, you can write

```
out.println(yourAccount.name);
```

These ideas come together in Listings 7-1 and 7-2. Here's Listing 7-1:

LISTING 7-1: What It Means to Be an Account

```
package com.example.accounts;

public class Account {
    String name;
    String address;
    double balance;
}
```

The `Account` class in Listing 7-1 defines what it means to be an `Account`. In particular, Listing 7-1 tells you that each of the `Account` class's instances has three variables: `name`, `address`, and `balance`. This is consistent with the information in Figure 7-1. Java programmers have a special name for variables of this kind (variables that belong to instances of classes). Each of these variables — `name`, `address`, and `balance` — is called a *field*.

REMEMBER

A variable declared inside a class but not inside any particular method is a *field*. In Listing 7-1, the variables `name`, `address`, and `balance` are fields. Another name for a field is an *instance variable*.

WARNING

Chapter 4 tells you that, when you initialize a variable in a method's body, you can replace the variable's type with the word var. In other words, you can write

```
public static void main(String[] args) {
    var greeting = "Good morning!";  //Okay!
    // ... and so on.
```

Yes, this rule about var applies to the variables that you declare inside your method bodies. But it *doesn't* apply to a class's fields. I illustrate this concept with a few lines of incorrect code:

```
public class Account {
    var name = "Alan"; //Forgive me for typing this!
    // ... and so on.
```

Honestly, it makes me nervous to type these three lines, even while I warn you that this is bad code.

If you've been grappling with the material in Chapters 4 through 6, the code for class Account (refer to Listing 7-1) may come as a big shock to you. Can you really define a complete Java class with only four lines of code (give or take a curly brace)? You certainly can. A class is a grouping of existing things. In the Account class of Listing 7-1, those existing things are two String values and a double value.

The field declarations in Listing 7-1 have *default access*, which means that I didn't add a word before the type name String. The alternatives to default access are *public*, *protected*, and *private* access:

```
public String name;
protected String address;
private double balance;
```

Professional programmers shun the use of default access because default access doesn't shield a field from accidental misuse. But in my experience, you learn best when you learn about the simplest stuff first, and in Java, default access is the simplest stuff. In this book, I delay the discussion of private access until this chapter's "Hide-and-Seek" section. And I delay the discussion of protected access until Chapter 14. As you read this chapter's examples, please keep in mind that default access isn't the best thing to use in a Java program. And, if a professional programmer asks you where you learned to use default access, please lie and blame someone else's book.

WARNING

Your IDE won't offer to run the code in Listing 7-1. That's because Listing 7-1 has no `main` method. The run of every Java program must begin with the `main` method, so the code in Listing 7-1 has no starting point. To use the listing's `Account` class, your project must contain at least two Java files: one file with the code in Listing 7-1, and another file with a `main` method. In this chapter, Listing 7-2 has a handy-dandy `main` method, so the `Account` class in Listing 7-1 won't go to waste.

But wait! You need to know another tidbit before you can use the code in Listing 7-1. The listing begins with the line `package com.example.accounts`. That line says, "This file is one of possibly many files that are inside a grouping named `com.example.accounts`." When a Java file belongs to such a grouping, the file must live in a specially named folder on your computer's hard drive. This chapter's "Package deal" sidebar has some details.

PACKAGE DEAL

Java has a feature that lets you lump classes into groups. Each group of classes is called a *package*.

The Java API defines about 4,500 classes. Each class belongs to one of about 220 packages. When you write `import java.util.Scanner`, you're referring to a class named `Scanner`, which belongs to the package named `java.util`.

But that's not all! The grouping doesn't stop with packages. Packages can be lumped into bigger things called *modules*. Each package in the Java API belongs to one of about 70 different modules. When you write `import java.util.Scanner`, you're referring to something in the class named `Scanner`, which belongs to the package named `java.util`, which in turn belongs to the module named `java.base`.

The `Account` class in Listing 7-1 belongs to a package that I named `com.example.accounts`. The first line in Listing 7-1 makes it so. That first line is called a *package declaration*.

In the Java world, programmers customarily give packages long, dot-filled names. For instance, because I've registered the domain name *allmycode.com,* I may name a package `com.allmycode.dummiesframe`. In truth, I can name a package almost anything I want — like `com.example.accounts`, `fish.and.chips`, or simply `hamburger`. In any case, informative package names like `com.allmycode.dummiesframe` are better than silly names like `midnight.in.akron` or `do.not.run.this.code`.

(continued)

When I created Listing 7-1, I didn't have to start with a package declaration. None of the listings in Chapters 1–6 starts with package declarations, so each of those classes belongs to a catchall *default package*. also known as Java's *unnamed package*. Professional Java programmers avoid using this unnamed package. So, for Listing 7-1, I decided to straighten up and start flying right.

How many classes can you define? The `Account` class in Listing 7-1 is one of them. You can also create a `Customer` class, a `Company` class, classes named `Address`, `Agency`, `Contract`, `Buyer`, `Seller`, `Consumer`, `Sale`, `BalanceSheet`, `IncomeStatement`, `Inventory`, `Lease`, `Product`, `Loan`, `Discount`, and many more. Of course, you may want to separate these classes from your other classes named `Song`, `Novel`, `Film`, `Painting`, and `Video`. So, you might create two different packages: one named `com.example.accounts` and another named `com.example.arts`.

You can start any Java file with a package declaration, but you have to remember a few things:

- **A Java file's package name dictates the names of some folders and subfolders.**

  The code in Listing 7-1 begins with `package com.example.accounts`. So your hard drive must have a folder named `com`, which has a subfolder named `example`, which in turn has its own subfolder named `accounts`. Your `Account.java` file must be inside this `accounts` folder. For example, on my computer, I have a project whose root folder is named `07-01`. My copy of Listing 7-1 is inside a `07-01/src/com/example/accounts` folder.

- **Many IDEs have to be told explicitly to create the needed folders.**

  Before you copy the code from Listing 7-1 into your IDE, you probably have to tell the IDE to create a package named `com.example.accounts`. When you do, your IDE will create the `com/example/accounts` folders for you. Then you can copy the `Account.java` file into the `com/example/accounts` folder.

  How do you tell your IDE to create a package? When you start a new project, your IDE might prompt you for a new package name. The IDE creates the package as part of the new project. When you work on an existing project, creating a package might mean fishing around for New ⇨ Package among your IDE's menus. If you can't find a way to do that, look for some hints at this book's website (`http://javafordummies.allmycode.com`).

After telling you about packages, I can clear up some of the confusion about `import` declarations: Any `import` declaration that doesn't use the word `static` must start with the name of a package and must end with either of the following:

- The name of a class within that package
- An asterisk (indicating all classes within that package)

For example, the declaration

```
import java.util.Scanner;
```

is valid because java.util is the name of a package in the Java API, and Scanner is the name of a class in the java.util package. The dotted name java.util.Scanner is called the *fully qualified name* of the Scanner class. A class's fully qualified name includes the name of the package in which the class is defined. (You can find out all this stuff about java.util and Scanner by reading Java's API documentation. For tips on reading the documentation, see Chapter 3 and this book's website.)

Here's another example. The declaration

```
import javax.swing.*;
```

is valid because javax.swing is the name of a package in the Java API, and the asterisk refers to all classes in the javax.swing package. With this import declaration at the top of your Java code, you can use abbreviated names for classes in the javax.swing package — names like JFrame, JButton, JMenuBar, JCheckBox, and many others.

Here's one more example. A line like

```
import javax.*; //Bad!!
```

is *not* a valid import declaration. The Java API has no package with the one-word name javax. You may think that this line allows you to abbreviate all names beginning with javax (names like javax.swing.JFrame, and javax.sound.midi), but that's not the way the import declaration works. Because javax isn't the name of a package, the line import javax.* just angers the Java compiler.

After all this fuss about packages, you may wonder what difference it makes that a class is in one package or another. For some insight, see Chapter 14.

Declaring variables and creating objects

A young fellow approaches me while I'm walking down the street. He tells me to print "You'll love Java!" so I print those words. If you must know, I print them with chalk on the sidewalk. But where I print the words doesn't matter. What matters is that some guy issues instructions, and I follow the instructions.

Later that day, an elderly woman sits next to me on a park bench. She says, "An account has a name, an address, and a balance." And I say, "That's fine, but what do you want me to do about it?" In response she just stares at me, so I don't do anything about her *account* pronouncement. I just sit there, she sits there, and we both do absolutely nothing.

Listing 7-1, shown earlier, is like the elderly woman. This listing defines what it means to be an `Account`, but the listing doesn't tell me to do anything with my account, or with anyone else's account. In order to do something, I need a second piece of code. I need another class — a class that contains a `main` method. Fortunately, while the woman and I sit quietly on the park bench, a young child comes by with Listing 7-2.

LISTING 7-2: Dealing with Account Objects

```
package com.example.accounts;

import static java.lang.System.out;

public class UseAccount {

    public static void main(String[] args) {
        Account myAccount;
        Account yourAccount;

        myAccount = new Account();
        yourAccount = new Account();

        myAccount.name = "Barry Burd";
        myAccount.address = "222 Cyberspace Lane";
        myAccount.balance = 24.02;

        yourAccount.name = "Jane Q. Public";
        yourAccount.address = "111 Consumer Street";
        yourAccount.balance = 55.63;

        out.print(myAccount.name);
        out.print(" (");
        out.print(myAccount.address);
        out.print(") has $");
        out.print(myAccount.balance);
        out.println();

        out.print(yourAccount.name);
        out.print(" (");
        out.print(yourAccount.address);
```

```
        out.print(") has $");
        out.print(yourAccount.balance);
    }
}
```

Taken together, the two classes — Account and UseAccount — form one complete program. The code in Listing 7-2 defines the UseAccount class, and the UseAccount class has a main method. This main method has variables of its own — yourAccount and myAccount.

In a way, the first two lines inside the main method of Listing 7-2 are misleading. Some people read Account yourAccount as if it's supposed to mean that "yourAccount is an Account" or that "the variable yourAccount refers to an instance of the Account class." That's not really what this first line means. Instead, the line Account yourAccount means, "If and when I make the variable yourAccount refer to something, that something will be an instance of the Account class." So, what's the difference?

The difference is that simply declaring Account yourAccount doesn't make the yourAccount variable refer to an object. All the declaration does is reserve the variable name *yourAccount* so that the name can eventually refer to an instance of the Account class. The creation of an actual object doesn't come until later in the code, when Java executes new Account().

CROSS REFERENCE

Technically, when Java executes new Account(), you're creating an object by calling the Account class's constructor. When you see Java's new keyword, think "constructor call." I have a lot more to say about constructors and constructor calls in Chapter 9.

When Java executes the assignment yourAccount = new Account(), Java creates a new object (a new instance of the Account class) and makes the variable yourAccount refer to that new object. (The equal sign makes the variable refer to the new object.) Figure 7-2 illustrates the situation.

To test the claim that I made in the last few paragraphs, I added an extra line to the code of Listing 7-2. I tried to print yourAccount.name after declaring yourAccount but before calling new Account():

```
Account myAccount;
Account yourAccount;

out.println(yourAccount.name);

myAccount = new Account();
yourAccount = new Account();
```

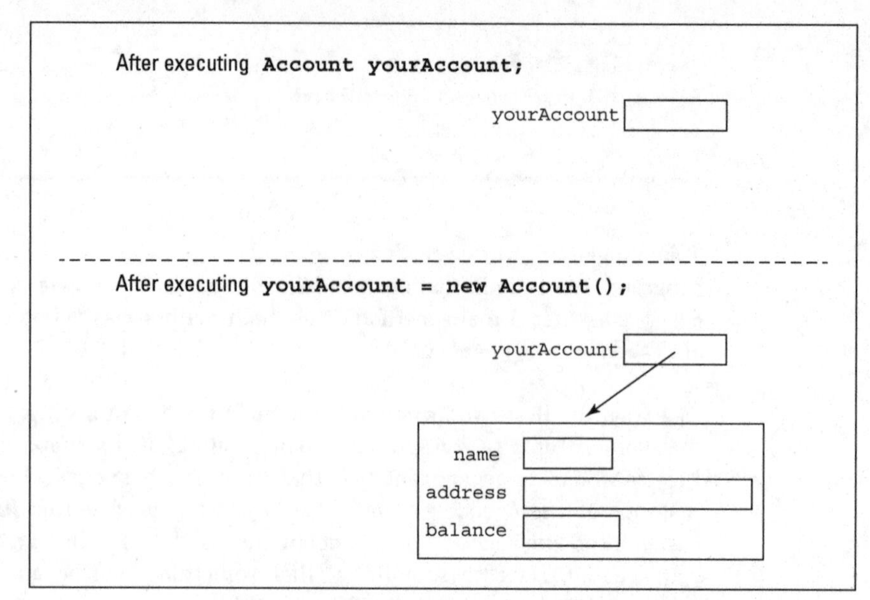

FIGURE 7-2: Before and after a constructor is called.

When I tried to compile the new code, I got this error message: `variable yourAccount might not have been initialized`. That settles it. Before you do `new Account()`, you can't print the name variable of an object because an object doesn't exist.

When a variable has a reference type, simply declaring the variable isn't enough. You don't get an object until you call a constructor and use the keyword `new`.

For information about reference types, see Chapter 4.

Initializing a variable

In Chapter 4, I announce that you can initialize a primitive type variable as part of the variable's declaration:

```
int weightOfAPerson = 150;
```

You can do the same thing with reference type variables, such as `myAccount` and `yourAccount` in Listing 7-2. You can combine the first four lines in the listing's main method into just two lines, like this:

```
Account myAccount = new Account();
Account yourAccount = new Account();
```

If you combine lines this way, you automatically avoid the `variable might not have been initialized` error that I describe in the preceding section. Sometimes you find a situation in which you can't initialize a variable. But when you can initialize, it's usually a plus.

Using an object's fields

After you've bitten off and chewed the `main` method's first four lines, the rest of the code in Listing 7-2 is sensible and straightforward. You have three lines that put values in the `myAccount` object's fields, three lines that put values in the `yourAccount` object's fields, and four lines that do some printing. Figure 7-3 shows the program's output.

FIGURE 7-3: Running the code in Listings 7-1 and 7-2.

```
Barry Burd (222 Cyberspace Lane) has $24.02
Jane Q. Public (111 Consumer Street) has $55.63
```

One program; several classes

Each program in Chapters 3–6 consists of a single class. That's great for a book's introductory chapters. But in real life, a typical program consists of hundreds or even thousands of classes. The program that spans Listings 7-1 and 7-2 consists of two classes. Sure, having two classes isn't like having thousands of classes, but it's a step in that direction.

In practice, most programmers put each class in a file of its own. When you create a program, such as the one in Listings 7-1 and 7-2, you create two files on your computer's hard drive. Therefore, when you download this section's example from the web, you get two separate files — `Account.java` and `UseAccount.java`.

For information about running a program consisting of more than one `.java` file in Eclipse, NetBeans, and IntelliJ IDEA, visit this book's website (`http://javafordummies.allmycode.com`).

Declaring a public class

The first line of Listing 7-1 is

```
public class Account {
```

The Account class is public. A public class is available for use by all other classes. For example, if you write an ATMController program in some remote corner of cyberspace, then your ATMController program can contain code, such as `myAccount.balance = 24.02`, making use of the Account class declared in Listing 7-1. (Of course, your code has to know where in cyberspace I've stored the code in Listing 7-1, but that's another story.)

Listing 7-2 contains the code `myAccount.balance = 24.02`. You might say to yourself, "The Account class has to be public because another class (the code in Listing 7-2) uses the Account class." Unfortunately, the real lowdown about public classes is a bit more complicated. In fact, when the planets align themselves correctly, one class can make use of another class's code, even though the other class isn't public. (I describe how this works in Chapter 14.)

The dirty secret in this chapter's code is that declaring certain classes to be public simply makes me feel good. Yes, programmers do certain things to feel good. In Listing 7-1, my esthetic sense of goodness comes from the fact that an Account class is useful to many other programmers. When I create a class that declares something useful and nameable — an Account, an Engine, a Customer, a BrainWave, a Headache, or a SevenLayerCake class — I declare the class to be public.

The UseAccount class in Listing 7-2 is also public. When a class contains a main method, Java programmers tend to make the class public without thinking too much about who uses the class. So even if no other class makes use of my main method, I declare the UseAccount class to be public. Most of the classes in this book contain main methods, so most of the classes in this book are public.

WARNING

When you declare a class to be public, you must declare the class in a file whose name is exactly the same as the name of the class (but with the .java extension added). For example, if you declare `public class MyImportantCode`, you must put the class's code in a file named `MyImportantCode.java`, with uppercase letters M, I, and C and all other letters lowercase. This filenaming rule has an important consequence: If your code declares two public classes, your code must consist of at least two .java files. In other words, you can't declare two public classes in one .java file.

For more news about the word *public* and other such words, see Chapter 14.

TRY IT OUT

In this section, I create an Account class. You can create classes too.

GETTING ORGANIZED

An Organization has a name (such as *XYZ Company*), an annual revenue (such as \$100,000.00), and a boolean value indicating whether the organization is or is not a profit-making organization. Companies that manufacture and sell products are generally profit-making organizations; groups that provide aid to victims of natural disasters are generally not profit-making organizations.

Declare your own `Organization` class. Declare another class that creates organizations and displays information about those organizations.

YUMMY FOODS

A product for sale in a food store has several characteristics: a type of food (peach slices), a weight (500 grams), a cost ($1.83), a number of servings (4), and a number of calories per serving (70).

Declare a `FoodProduct` class. Declare another class that creates `FoodProduct` instances and displays information about those instances.

Defining a Method within a Class (Displaying an Account)

Imagine a table containing the information about two accounts. (If you have trouble imagining such a thing, just look at Table 7-1.)

TABLE 7-1 Without Object-Oriented Programming

Name	Address	Balance
Barry Burd	222 Cyberspace Lane	24.02
Jane Q. Public	111 Consumer Street	55.63

In Table 7-1, each account has three things a name, an address, and a balance. That's how things were done before object-oriented programming came along. But object-oriented programming involved a big shift in thinking. With object-oriented programming, each account can have a name, an address, a balance, and a way of being displayed.

In object-oriented programming, each object has its own built-in functionality. An account knows how to display itself. A string can tell you whether it has the same characters inside it as another string has. A `PrintStream` instance, such as `System.out`, knows how to do `println`. In object-oriented programming, each object has its own methods. These methods are little subprograms that you can call to have an object do things to (or for) itself.

And why is this a good idea? It's good because you're making pieces of data take responsibility for themselves. With object-oriented programming, all the functionality that's associated with an account is collected inside the code for the `Account` class. Everything you have to know about a string is located in the file `String.java`. Anything having to do with year numbers (whether they have two or four digits, for instance) is handled right inside the `Year` class. Therefore, if anybody has problems with your `Account` class or your `Year` class, they know just where to look for all the code. That's great!

Imagine an enhanced account table. In this new table, each object has built-in functionality. Each account knows how to display itself on the screen. Each row of the table has its own copy of a `display` method. Of course, you don't need much imagination to picture this table. I just happen to have a table you can look at. It's Table 7-2.

An account that displays itself

In Table 7-2, each account object has four things — a name, an address, a balance, and a way of displaying itself on the screen. After you make the jump to object-oriented thinking, you'll never turn back. Listings 7-3 and 7-4 show programs that implement the ideas in Table 7-2.

TABLE 7-2 **The Object-Oriented Way**

Name	Address	Balance	Display
Barry Burd	222 Cyberspace Lane	24.02	`out.print ...`
Jane Q. Public	111 Consumer Street	55.63	`out.print ...`

LISTING 7-3: **An Account Displays Itself**

```
package com.example.accounts;

import static java.lang.System.out;

public class Account {
    String name;
    String address;
    double balance;
```

```
    public void display() {
        out.print(name);
        out.print(" (");
        out.print(address);
        out.print(") has $");
        out.print(balance);
    }
}
```

LISTING 7-4: Using the Improved Account Class

```
package com.example.accounts;

public class UseAccount {

    public static void main(String[] args) {
        var myAccount = new Account();
        var yourAccount = new Account();

        myAccount.name = "Barry Burd";
        myAccount.address = "222 Cyberspace Lane";
        myAccount.balance = 24.02;

        yourAccount.name = "Jane Q. Public";
        yourAccount.address = "111 Consumer Street";
        yourAccount.balance = 55.63;

        myAccount.display();
        System.out.println();
        yourAccount.display();
    }
}
```

A run of the code in Listings 7-3 and 7-4 looks just like a run of Listings 7-1 and 7-2. You can see the action earlier, in Figure 7-3.

In Listing 7-3, the `Account` class has four things in it: a name, an address, a balance, and a display method. These things match up with the four columns in Table 7-2. So each instance of the `Account` class has a name, an address, a balance, and a way of displaying itself. The way you call these things is nice and uniform. To refer to the name stored in `myAccount`, you write

```
myAccount.name
```

To get `myAccount` to display itself on the screen, you write

```
myAccount.display()
```

Both expressions use dot notation. The only structural difference is in the parentheses.

When you call a method, you put parentheses after the method's name.

The display method's header

Look again at Listings 7-3 and 7-4. A call to the `display` method is inside the `UseAccount` class's `main` method, but the declaration of the `display` method is up in the `Account` class. The declaration has a header and a body. (See Chapter 3.) The header has three words and some parentheses:

- **The word *public* serves roughly the same purpose as the word *public* in Listing 7-1.** Roughly speaking, any code can contain a call to a public method, even if the calling code and the public method belong to two different classes. In this section's example, the decision to make the `display` method public is a matter of taste. Normally, when I create a method that's useful in a wide variety of applications, I declare the method to be public.
- **The word *void* tells Java that when the `display` method is called, the `display` method doesn't return anything to the place that called it.** To see a method that does return something to the place that called it, see the next section.
- **The word *display* is the method's name.** Every method must have a name. Otherwise, you don't have a way to call the method.
- **The parentheses contain all the things you're going to pass to the method when you call it.** When you call a method, you can pass information to that method on the fly. The `display` method in Listing 7-3 looks strange because the parentheses in the method's header have nothing inside them. This nothingness indicates that no information is passed to the `display` method when you call it. For a meatier example, see the next section.

Listing 7-3 contains the `display` method's declaration and Listing 7-4 contains a call to the `display` method. Although Listings 7-3 and 7-4 contain different classes, both uses of `public` in Listing 7-3 are optional. To find out why, check out Chapter 14.

SHOW OFF

In the previous section, you create `Organization` and `FoodProduct` classes. Add `display` methods to both of these classes and create separate classes that make use of these `display` methods.

Sending Values to and from Methods (Calculating Interest)

Think about sending someone to the supermarket to buy bread. When you do this, you say, "Go to the supermarket and buy some bread." (Try it at home. You'll have a fresh loaf of bread in no time at all!) Of course, at some other time, you send that same person to the supermarket to buy bananas. You say, "Go to the supermarket and buy some bananas." And what's the point of all of this? Well, you have a method, and you have some on-the-fly information that you pass to the method when you call it. The method is named *goToTheSupermarketAndBuySome.* The on-the-fly information is either *bread* or *bananas,* depending on your culinary needs. In Java, the method calls would look like this:

```
goToTheSupermarketAndBuySome(bread);
goToTheSupermarketAndBuySome(bananas);
```

The things in parentheses are called *parameters,* or *parameter lists.* With parameters, your methods become much more versatile. Instead of buying the same item each time, you can send somebody to the supermarket to buy bread one time, bananas another time, and birdseed the third time. When you call your `goToTheSupermarketAndBuySome` method, you decide right then and there what you'll ask your pal to buy.

And what happens when your friend returns from the supermarket? "Here's the bread you asked me to buy," says your friend. By carrying out your wishes, your friend returns something to you. You make a method call, and the method returns information (or a loaf of bread).

The thing returned to you is called the method's *return value.* The general type of thing that is returned to you is called the method's *return type.* These concepts are made more concrete in Listings 7-5 and 7-6.

LISTING 7-5: An Account That Calculates Its Own Interest

```
package com.example.accounts;

import static java.lang.System.out;

public class Account {
    String name;
    String address;
    double balance;

    public void display() {
        out.print(name);
        out.print(" (");
        out.print(address);
        out.print(") has $");
        out.print(balance);
    }

    public double getInterest(double percentageRate) {
        return balance * percentageRate / 100.00;
    }
}
```

LISTING 7-6: Calculating Interest

```
package com.example.accounts;

import static java.lang.System.out;

public class UseAccount {

    public static void main(String[] args) {
        var myAccount = new Account();
        var yourAccount = new Account();

        myAccount.name = "Barry Burd";
        myAccount.address = "222 Cyberspace Lane";
        myAccount.balance = 24.02;

        yourAccount.name = "Jane Q. Public";
        yourAccount.address = "111 Consumer Street";
        yourAccount.balance = 55.63;
```

```
        myAccount.display();
        out.print(" plus $");
        out.print(myAccount.getInterest(5.00));
        out.println(" interest ");

        yourAccount.display();

        double yourInterestRate = 7.00;
        out.print(" plus $");
        double yourInterestAmount = yourAccount.getInterest(yourInterestRate);
        out.print(yourInterestAmount);
        out.println(" interest ");
    }
}
```

Figure 7-4 shows the output of the code in Listings 7-5 and 7-6. In Listing 7-5, the `Account` class has a `getInterest` method. This `getInterest` method is called twice from the `main` method in Listing 7-6; the actual account balances and interest rates are different each time:

- **In the first call, the balance is 24.02 and the interest rate is 5.00.** The first call, `myAccount.getInterest(5.00)`, refers to the `myAccount` object and to the values stored in the `myAccount` object's fields. (See Figure 7-5.) When this call is made, the expression `balance * percentageRate / 100.00` stands for 24.02 * 5.00 / 100.00.
- **In the second call, the balance is 55.63, and the interest rate is 7.00.** In the `main` method, just before this second call is made, the variable `yourInterestRate` is assigned the value `7.00`. The call itself, `yourAccount.getInterest(yourInterestRate)`, refers to the `yourAccount` object and to the values stored in the `yourAccount` object's fields. (Again, see Figure 7-5.) So, when the call is made, the expression `balance * percentageRate / 100.00` stands for 55.63 * 7.00 / 100.00.

FIGURE 7-4: Running the code in Listings 7-5 and 7-6.

```
Barry Burd (222 Cyberspace Lane) has $24.02 plus $1.2009999999999998 interest
Jane Q. Public (111 Consumer Street) has $55.63 plus $3.8941000000000003 interest
```

By the way, the `main` method in Listing 7-6 contains two calls to `getInterest`. One call has the literal `5.00` in its parameter list; the other call has the variable `yourInterestRate` in its parameter list. Why does one call use a literal and the other call use a variable? No reason. I just want to show you that you can do it either way.

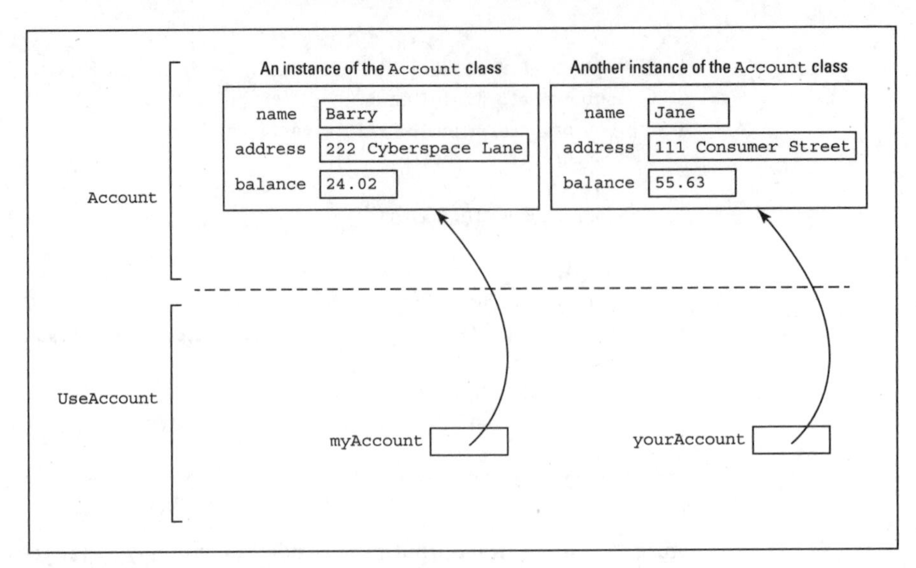

FIGURE 7-5: My account and your account.

Passing a value to a method

Take a look at the `getInterest` method's header (as you read the explanation in the next few bullets, you can follow some of the ideas visually with the diagram in Figure 7-6):

- **The word *double* tells Java that when the `getInterest` method is called, the `getInterest` method returns a `double` value back to the place that called it.** The statement in the `getInterest` method's body confirms this. The statement says `return balance * percentageRate / 100.00`, and the expression `balance * percentageRate / 100.00` has type `double`. (That's because all the things in the expression — `balance`, `percentageRate`, and `100.00` — have type `double`.)

  When the `getInterest` method is called, the `return` statement calculates `balance * percentageRate / 100.00` and hands the calculation's result back to the code that called the method.

- **The word *getInterest* is the method's name.** That's the name you use to call the method when you're writing the code for the `UseAccount` class.

- **The parentheses contain all the things that you pass to the method when you call it.** When you call a method, you can pass information to that method on the fly. This information is the method's parameter list. The `getInterest` method's header says that the `getInterest` method takes one piece of information and that piece of information must be of type `double`:

  ```
  public double getInterest(double percentageRate)
  ```

Sure enough, if you look at the first call to `getInterest` (down in the useAccount class's main method), that call has the number `5.00` in it. And `5.00` is a `double` literal. When I call `getInterest`, I'm giving the method a value of type `double`.

If you don't remember what a literal is, see Chapter 4.

The same story holds true for the second call to `getInterest`. Down near the bottom of Listing 7-6, I call `getInterest` and feed the variable `yourInterestRate` to the method in its parameter list. Luckily for me, I declared `yourInterestRate` to be of type `double` just a few lines before that.

When you run the code in Listings 7-5 and 7-6, the flow of action isn't from top to bottom. The action goes from `main` to `getInterest`, and then back to `main`, and then back to `getInterest`, and, finally, back to `main` again. Figure 7-7 shows the whole business.

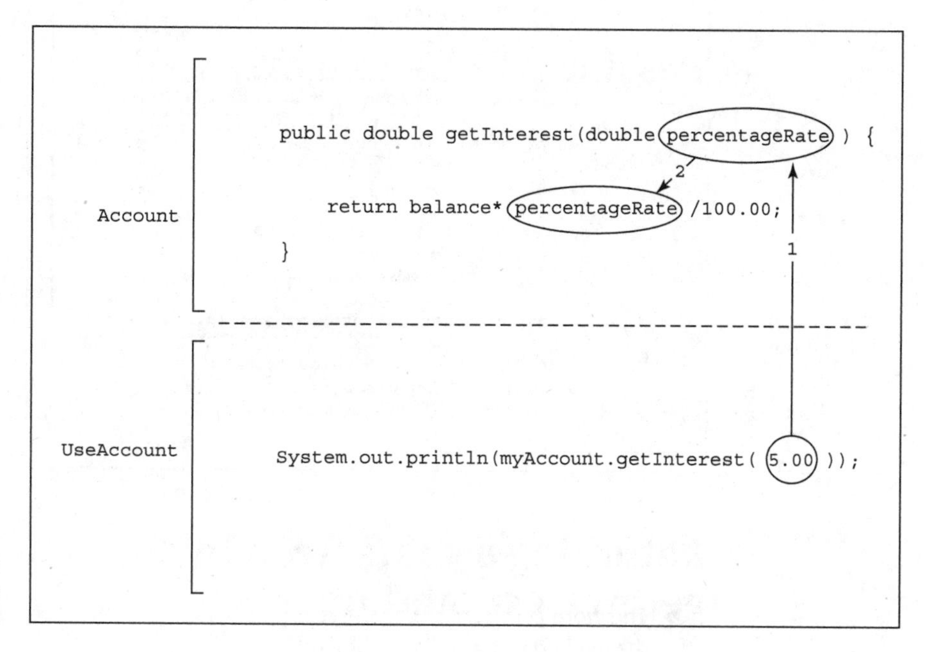

FIGURE 7-6: Passing a value to a method.

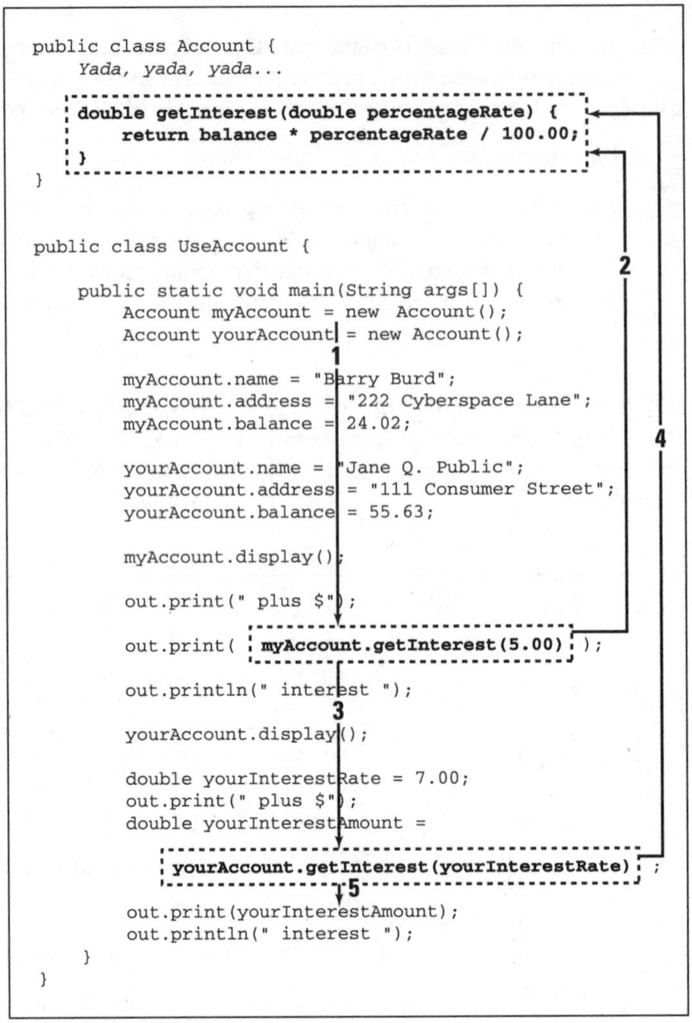

```
public class Account {
    Yada, yada, yada...

    double getInterest(double percentageRate) {
        return balance * percentageRate / 100.00;
    }
}


public class UseAccount {

    public static void main(String args[]) {
        Account myAccount = new Account();
        Account yourAccount = new Account();

        myAccount.name = "Barry Burd";
        myAccount.address = "222 Cyberspace Lane";
        myAccount.balance = 24.02;

        yourAccount.name = "Jane Q. Public";
        yourAccount.address = "111 Consumer Street";
        yourAccount.balance = 55.63;

        myAccount.display();

        out.print(" plus $");

        out.print( myAccount.getInterest(5.00) );

        out.println(" interest ");

        yourAccount.display();

        double yourInterestRate = 7.00;
        out.print(" plus $");
        double yourInterestAmount =
            yourAccount.getInterest(yourInterestRate) ;

        out.print(yourInterestAmount);
        out.println(" interest ");
    }
}
```

FIGURE 7-7: The flow of control in Listings 7-5 and 7-6.

Returning a value from the getInterest method

When the `getInterest` method is called, the method executes the one statement that's in the method's body: a `return` statement. The `return` statement computes the value of `balance * percentageRate / 100.00`. If `balance` happens to be `24.02`, and `percentageRate` is `5.00`, the value of the expression is `1.201` — around $1.20. (Because the computer works exclusively with 0s and 1s, Java gets this number wrong by an ever-so-tiny amount. Java gets 1.2009999999999998. That's just something that humans have to live with.)

Anyway, after this value is calculated, Java executes the `return`, which sends the value back to the place in `main` where `getInterest` was called. At that point in the process, the entire method call — `myAccount.getInterest(5.00)` — takes on the value 1.2009999999999998. The call itself is inside a `println`:

```
out.println(myAccount.getInterest(5.00));
```

So the `println` ends up with the following meaning:

```
out.println(1.2009999999999998);
```

The whole process, in which a value is passed back to the method call, is illustrated in Figure 7-8.

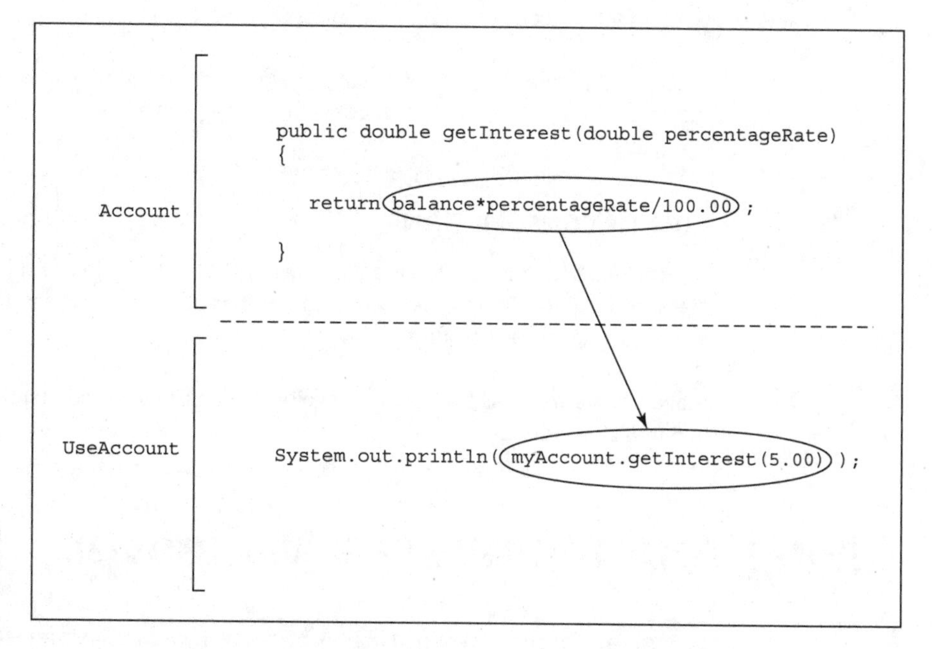

FIGURE 7-8: A method call is an expression with a value.

REMEMBER

If a method returns anything, a call to the method is an expression with a value. That value can be printed, assigned to a variable, added to something else, or whatever. Anything you can do with any other kind of value, you can do with a method call.

You might use the `Account` class in Listing 7-5 to solve a real problem. You'd call the `Account` class's `display` and `getInterest` methods in the course of an actual banking application. But the `UseAccount` class in Listing 7-6 is artificial. The `Use-Account` code creates some fake account data and then calls some `Account` class methods to convince you that the `Account` class's code works correctly. (You don't seriously think that a bank has depositors named "Jane Q. Public" and "Barry Burd," do you?) The `UseAccount` class in Listing 7-6 is a *test case* — a short-lived class whose sole purpose is to test another class's code. Like the code in Listing 7-6, each test case in this book is an ordinary class — a free-form class containing its own `main` method. Free-form classes are okay, but they're not optimal. Java developers have something better — a more disciplined way of writing test cases. The "better way" is called *JUnit*, and it's described at `https://junit.org/junit5/docs/current/user-guide/`.

TAX DAY

TRY IT OUT

In previous sections, you create your own `Organization` class. Add a method to the class that computes the amount of tax the organization pays. A profit-making organization pays 10 percent of its revenue in tax, but a nonprofit organization pays only 2 percent of its revenue in tax.

Make a separate class that creates two or three organizations and displays information about each organization, including the amount of tax the organization pays.

COST OF CONSUMPTION

In previous sections, you create your own `FoodProduct` class. Add methods to the class to compute the cost per 100 grams, the cost per serving, and the total number of calories in the product.

Make a separate class that creates two or three products and displays information about each product.

Giving Your Numbers a Makeover

Looking at Figure 7-4 again, you may be concerned that the interest on my account is only $1.2009999999999998. Seemingly, the bank is cheating me out of two hundred-trillionths of a cent. I should go straight to the bank and demand my fair interest. Maybe you and I should go together. We'll kick up some fur at that old bank and bust this scam right open. If my guess is correct, this is part of a big salami scam. In a *salami scam*, someone shaves tiny amounts off millions of accounts. People don't notice their tiny little losses, but the person doing the shaving collects enough for a quick escape to Barbados (or for a whole truckload of salami).

Wait a minute! What about you? In Listing 7-6, you have yourAccount. And in Figure 7-4, your name is Jane Q. Public. Nothing is motivating you to come with me to the bank. Checking Figure 7-4 again, I see that you're way ahead of the game. According to my calculations, the program overpays you by three hundred-trillionths of a cent. Between the two of us, we're ahead by a hundred-trillionth of a cent. What gives?

Well, because computers use 0s (zeros) and 1s and don't have an infinite amount of space to do calculations, such inaccuracies as the ones shown in Figure 7-4 are normal. The quickest solution is to display the inaccurate numbers in a more sensible fashion. You can round the numbers and display only two digits beyond the decimal point, and some handy tools from Java's API (application programming interface) can help. Listing 7-7 shows the code, and Figure 7-9 displays the pleasant result.

LISTING 7-7: **Making Your Numbers Look Right**

```
package com.example.accounts;

import static java.lang.System.out;

public class UseAccount {

    public static void main(String[] args) {
        var myAccount = new Account();
        var yourAccount = new Account();

        myAccount.balance = 24.02;
        yourAccount.balance = 55.63;

        double myInterest = myAccount.getInterest(5.00);
        double yourInterest = yourAccount.getInterest(7.00);

        out.printf("$%4.2f\n", myInterest);
        out.printf("$%5.2f\n", myInterest);
        out.printf("$%.2f\n", myInterest);
        out.printf("$%3.2f\n", myInterest);
        out.printf("$%.2f $%.2f", myInterest, yourInterest);
    }
}
```

```
$1.20
$ 1.20
$1.20
$1.20
$1.20 $3.89
```

FIGURE 7-9: Numbers that look like dollar amounts.

The inaccurate numbers in Figure 7-4 come from the computer's use of 0s and 1s. A mythical computer whose circuits were wired to use digits 0, 1, 2, 3, 4, 5, 6, 7, 8, and 9 wouldn't suffer from the same inaccuracies. So, to make things better, Java provides its own, special way around the computer's inaccurate calculations. Java's API has a class named `BigDecimal` — a class that bypasses the computer's strange 0s and 1s and uses ordinary decimal digits to perform arithmetic calculations. For more information, visit this book's website (`http://javafordummies.allmycode.com`).

Listing 7-7 uses a handy method named `printf`. When you call `printf`, you always put at least two parameters inside the call's parentheses:

- **The first parameter is a *format string*.**

  The format string uses funny-looking codes to describe exactly how the other parameters are displayed.

- **All the other parameters (after the first) are values to be displayed.**

Look at the last `printf` call of Listing 7-7. The first parameter's format string has two placeholders for numbers. The first placeholder (`%.2f`) describes the display of `myInterest`. The second placeholder (another `%.2f`) describes the display of `yourInterest`. To find out exactly how these format strings work, see Figures 7-10, 7-11, 7-12, 7-13, and 7-14.

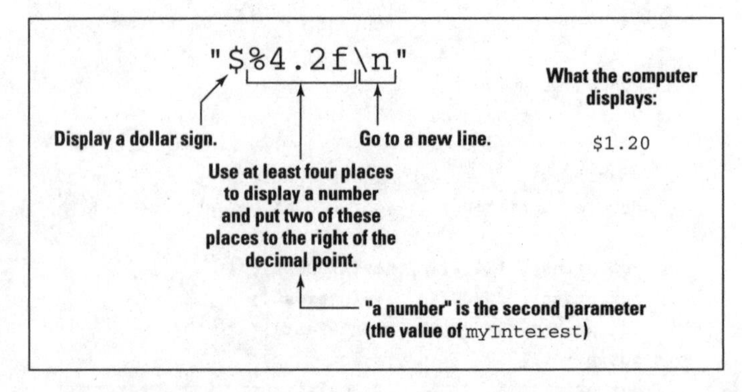

FIGURE 7-10: Using a format string.

For more examples using the `printf` method and its format strings, see Chapters 8 and 9. For a complete list of options associated with the `printf` method's format string, see the `java.util.Formatter` page of Java's API documentation at `https://docs.oracle.com/en/java/javase/17/docs/api/java.base/java/util/Formatter.html`.

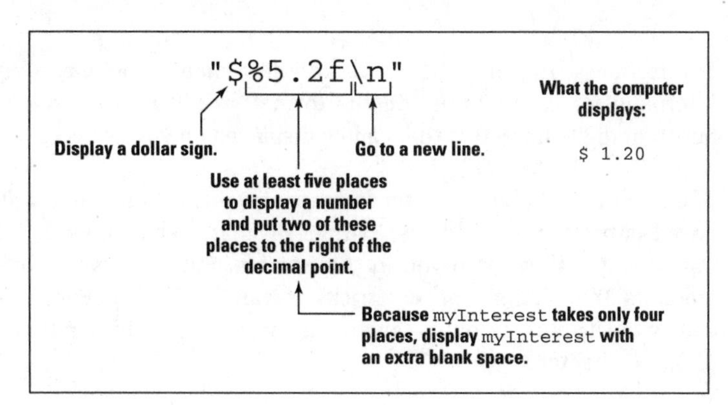

FIGURE 7-11: Adding extra places to display a value.

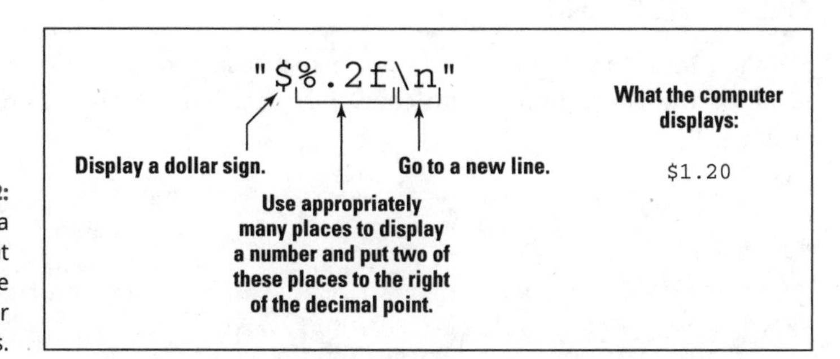

FIGURE 7-12: Displaying a value without specifying the exact number of places.

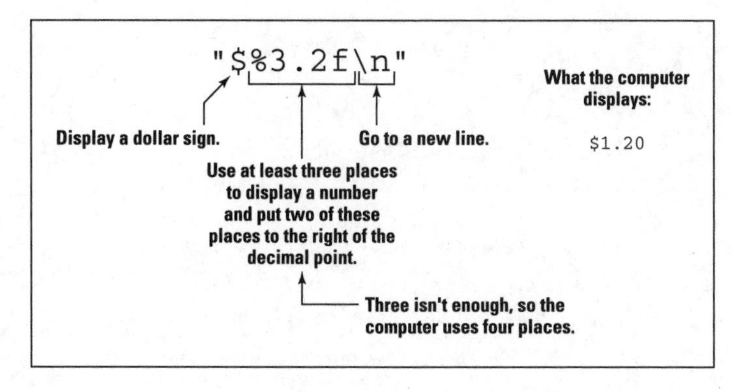

FIGURE 7-13: Specifying too few places to display a value.

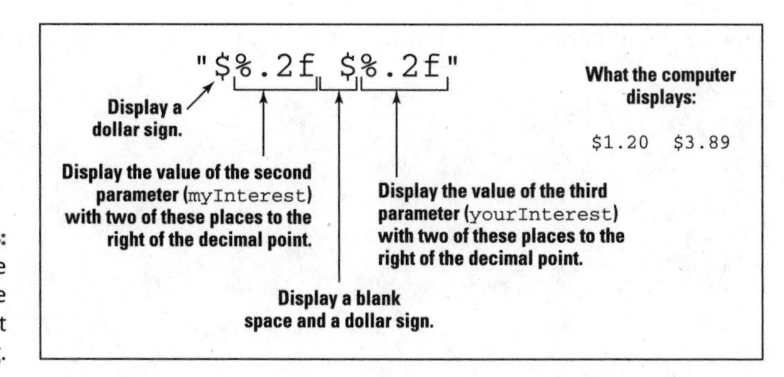

FIGURE 7-14: Displaying more than one value with a format string.

REMEMBER

The format string in a `printf` call doesn't change the way a number is stored internally for calculations. All the format string does is create a nice-looking bunch of digit characters that can be displayed on your screen.

CROSS REFERENCE

The `printf` method is good for formatting values of any kind: ordinary numbers, hexadecimal numbers, dates, strings of characters, and other strange values. That's why I show it to you in this section. But when you work with currency amounts, this section's `printf` tricks are fairly primitive. For some better ways to deal with currency amounts (such as the interest amounts in this section's example), see Chapter 11.

FILL IN THE BLANKS

TRY IT OUT

Here's a Java "unprogram." It's not a real Java program because I've masked some of the characters in the code. I replaced these characters with underscores (_):

```
import static java.lang.System.out;

public class Main {

    public static void main(String[] args) {
        out.printf("%s%_%s", ">>", 7, "<<\n");
        out.printf("%s%___%s", ">>", 7, "<<\n");
        out.printf("%s%____%s", ">>", 7, "<<\n");
        out.printf("%s%____%s", ">>", 7, "<<\n");
        out.printf("%s%__%s", ">>", 7, "<<\n");
        out.printf("%s%__%s", ">>", -7, "<<\n");
        out.printf("%s%__%s", ">>", -7, "<<\n");
        out.printf("%s%_____%s", ">>", 7.0, "<<\n");
        out.printf("%s%_%s", ">>", "Hello", "<<\n");
        out.printf("%s%_%s", ">>", 'x', "<<\n");
        out.printf("%s%_%s", ">>", 'x', "<<\n");
    }
}
```

Replace the underscores so that this program produces the following output:

```
>>7<<
>>         7<<
>>7         <<
>>0000000007<<
>>+7<<
>>-7<<
```

```
>>(7)<<
>>   7.00000<<
>>HELLO<<
>>x<<
>>X<<
```

To do this, look for clues in the `java.util.Formatter` page of Java's API documentation at `https://docs.oracle.com/en/java/javase/17/docs/api/java.base/java/util/Formatter.html`.

Hide-and-Seek

Put down this book and put on your hat. You've been such a loyal reader that I'm taking you out to lunch!

I have just one problem. I'm a bit short on cash. Would you mind if, on the way to lunch, we stopped at an automatic teller machine and picked up a few bucks? Also, we have to use your account. My account is a little low.

Fortunately, the teller machine is easy to use. Just step right up and enter your PIN. After you enter your PIN, the machine asks which of several variable names you want to use for your current balance. You have a choice of `balance324`, `myBal`, `currentBalance`, `b$`, `BALANCE`, `asj999`, or `conStanTinople`. Having selected a variable name, you're ready to select a memory location for the variable's value. You can select any number between 022FFF and 0555AA. (Those numbers are in hexadecimal format.) After you configure the teller machine's software, you can easily get your cash. You did bring a screwdriver, didn't you?

Good programming

When it comes to good computer programming practice, one word stands out above all others: *simplicity*. When you're writing complicated code, the last thing you want is to deal with somebody else's misnamed variables, convoluted solutions to problems, or clever, last-minute kludges. You want a clean interface that makes you solve your own problems and no one else's.

In the automatic teller machine scenario that I describe earlier, the big problem is that the machine's design forces you to worry about other people's concerns. When you should be thinking about getting money for lunch, you're thinking instead about variables and storage locations. Sure, someone has to work out the teller machine's engineering problems, but the banking customer isn't the person.

This section is about safety, not security. Safe code keeps you from making accidental programming errors. Secure code (a completely different story) keeps malicious hackers from doing intentional damage.

So, everything connected with every aspect of a computer program has to be simple, right? Well, no. That's not right. Sometimes, to make things simple in the long run, you have to do lots of preparatory work up front. The people who built the automated teller machine worked hard to make sure that the machine is consumer-proof. The machine's interface, with its screen messages and buttons, makes the machine a very complicated, but carefully designed, device.

The point is that making things look simple takes some planning. In the case of object-oriented programming, one of the ways to make things look simple is to prevent code outside a class from directly using fields defined inside the class. Take a peek at the code in Listing 7-1. You're working at a company that has just spent $10 million for the code in the `Account` class. (That's more than a million-and-a-half per line!) Now your job is to write the `UseAccount` class. You would like to write

```
myAccount.name = "Barry Burd";
```

but doing so would be getting you too far inside the guts of the `Account` class. After all, people who use an automatic teller machine aren't allowed to program the machine's variables. They can't use the machine's keypad to type the statement

```
balanceOnAccount29872865457 = balanceOnAccount29872865457 + 1000000.00;
```

Instead, they push buttons that do the job in an orderly manner. That's how a programmer achieves safety and simplicity.

To keep things nice and orderly, you need to change the `Account` class from Listing 7-1 by outlawing such statements as

```
myAccount.name = "Barry Burd";
```

and

```
out.print(yourAccount.balance);
```

Of course, this poses a problem. You're the person who's writing the code for the `UseAccount` class. If you can't write `myAccount.name` or `yourAccount.balance`, how will you accomplish anything at all? The answer lies in things called *accessor methods.* Listings 7-8 and 7-9 demonstrate these methods.

LISTING 7-8: **Hide Those Fields**

```
package com.example.accounts;

public class Account {
    private String name;
    private String address;
    private double balance;

    public void setName(String n) {
        name = n;
    }

    public String getName() {
        return name;
    }

    public void setAddress(String a) {
        address = a;
    }

    public String getAddress() {
        return address;
    }

    public void setBalance(double b) {
        balance = b;
    }

    public double getBalance() {
        return balance;
    }
}
```

LISTING 7-9: **Calling Accessor Methods**

```
package com.example.accounts;

import static java.lang.System.out;

public class UseAccount {

    public static void main(String[] args) {
        var myAccount = new Account();
        var yourAccount = new Account();
```

(continued)

LISTING 7-9: *(continued)*

```
        myAccount.setName("Barry Burd");
        myAccount.setAddress("222 Cyberspace Lane");
        myAccount.setBalance(24.02);

        yourAccount.setName("Jane Q. Public");
        yourAccount.setAddress("111 Consumer Street");
        yourAccount.setBalance(55.63);

        out.print(myAccount.getName());
        out.print(" (");
        out.print(myAccount.getAddress());
        out.print(") has $");
        out.print(myAccount.getBalance());
        out.println();

        out.print(yourAccount.getName());
        out.print(" (");
        out.print(yourAccount.getAddress());
        out.print(") has $");
        out.print(yourAccount.getBalance());
    }
}
```

A run of the code in Listings 7-8 and 7-9 looks no different from a run of Listings 7-1 and 7-2. Either program's run is shown earlier, in Figure 7-3. The big difference is that in Listing 7-8, the `Account` class enforces the carefully controlled use of its `name`, `address`, and `balance` fields.

Public lives and private dreams: Making a field inaccessible

Notice the addition of the word *private* in front of each of the `Account` class's field declarations. The word *private* is a Java keyword. When a field is declared private, no code outside of the class can make direct reference to that field. So if you put `myAccount.name = "Barry Burd"` in the `UseAccount` class of Listing 7-9, you get an error message such as `name has private access in Account`.

Rather than reference `myAccount.name`, the `UseAccount` programmer must call method `myAccount.setName` or method `myAccount.getName`. These methods, `setName` and `getName`, are called *accessor* methods because they provide access to the `Account` class's `name` field. (Actually, the term *accessor method* isn't formally a part of the Java programming language. It's just the term that people use for methods that do this sort of thing.) To zoom in even more, `setName` is called a *setter* method, and getName is called a *getter* method. (I bet you won't forget that terminology!)

TIP

With many IDEs, you don't have to type your own accessor methods. First, you type a field declaration like private String name. Then, on your IDE's menu bar, you choose Source⇨Generate Getters and Setters or choose Code⇨Insert Code⇨Setter or some mix of those commands. After you make all your choices, the IDE creates accessor methods and adds them to your code.

Notice that all the setter and getter methods in Listing 7-8 are declared to be public. This ensures that anyone from anywhere can call these two methods. The idea here is that manipulating the actual fields from outside the Account code is impossible, but you can easily reach the approved setter and getter methods for using those fields.

Think again about the automatic teller machine. Someone using the ATM can't type a command that directly changes the value in their account's balance field, but the procedure for depositing a million-dollar check is easy to follow. The people who build the teller machines know that if the check-depositing procedure is complicated, plenty of customers will mess it up royally. So that's the story — make impossible anything that people shouldn't do and make sure that the tasks people should be doing are easy.

TIP

Nothing about having setter and getter methods is sacred. You don't have to write any setter and getter methods that you won't use. For instance, in Listing 7-8, I can omit the declaration of method getAddress and everything still works. The only problem if I do this is that anyone else who wants to use my Account class and retrieve the address of an existing account is up a creek.

TIP

When you create a method to set the value in a balance field, you don't have to name your method setBalance. You can name it tunaFish or whatever you like. The trouble is that the set*Fieldname* convention (with lowercase letters in set and an uppercase letter to start the *Fieldname* part) is an established stylistic convention in the world of Java programming. If you don't follow the convention, you confuse the kumquats out of other Java programmers.

REMEMBER

When you call a setter method, you feed it a value of the type that's being set. That's why, in Listing 7-9, you call yourAccount.setBalance(55.63) with a parameter of type double. In contrast, when you call a getter method, you usually don't feed any values to the method. That's why, in Listing 7-9, you call yourAccount.getBalance() with an empty parameter list. Occasionally, you may want to get and set a value with a single statement. To add a dollar to your account's existing balance, you write yourAccount.setBalance(yourAccount.getBalance() + 1.00).

Enforcing rules with accessor methods

Go back to Listing 7-8 and take a quick look at the `setName` method. Imagine putting the method's assignment statement inside an `if` statement:

```
public void setName(String n) {
    if (!n.equals("")) {
        name = n;
    }
}
```

Now, if the programmer in charge of the `UseAccount` class writes `myAccount.setName("")`, the call to `setName` has no effect. Furthermore, because the name field is private, the following statement is illegal in the `UseAccount` class:

```
myAccount.name = "";
```

Of course, a call such as `myAccount.setName("Joe Schmoe")` still works because `"Joe Schmoe"` doesn't equal the empty string `""`.

That's cool. With a private field and an accessor method, you can prevent someone from assigning the empty string to an account's `name` field. With more elaborate `if` statements, you can enforce any rules you want.

PRIVATE EYE

TRY IT OUT

In previous sections, you create your own `Organization` and `FoodProduct` classes. In those classes, replace the default access fields with private fields. Create getter and setter methods for those fields. In the setter methods, add code to ensure that the `String` values aren't empty and that numeric values aren't negative.

Barry's Own GUI Class

You may be growing tired of the bland, text-based programs that litter this book's pages. You may want something a bit flashier — something with text fields and buttons. Well, I've got some examples for you!

I've created a class that I call `DummiesFrame`. When you import my `DummiesFrame` class, you can create a simple graphical user interface (GUI) application with very little effort.

Listing 7-10 uses my `DummiesFrame` class, and Figures 7-15, 7-16, and 7-17 show you the results.

LISTING 7-10: Your First DummiesFrame Example

```
package com.example.graphical;

import com.allmycode.dummiesframe.DummiesFrame;

public class Hello2U {

    public static void main(String[] args) {
        var frame = new DummiesFrame("Greet Me!");
        frame.addRow("Your first name");
        frame.go();
    }

    public static String calculate(String firstName) {
        return "Hello, " + firstName + "!";
    }
}
```

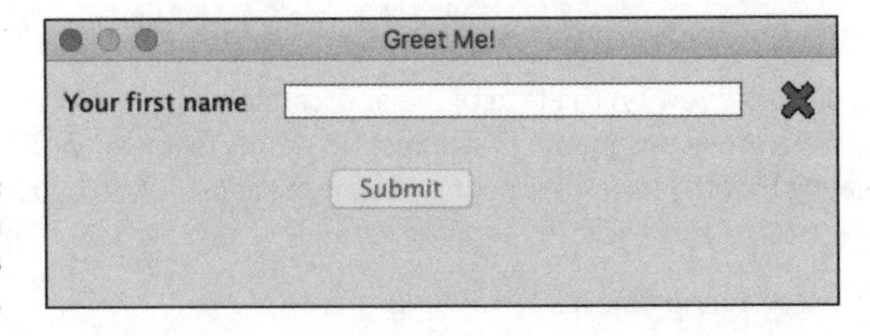

FIGURE 7-15: The code in Listing 7-10 starts running.

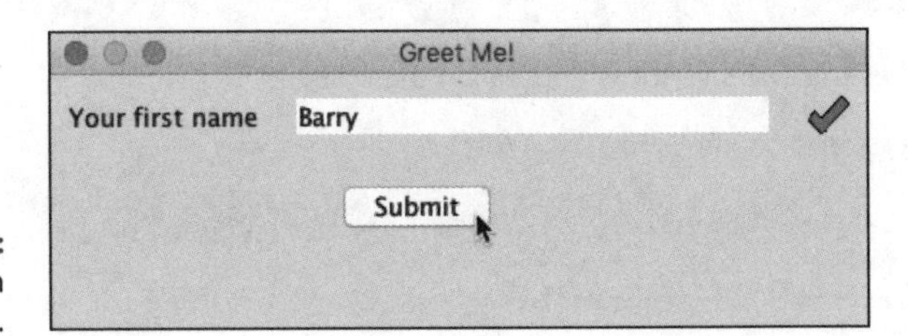

FIGURE 7-16: The user fills in the fields.

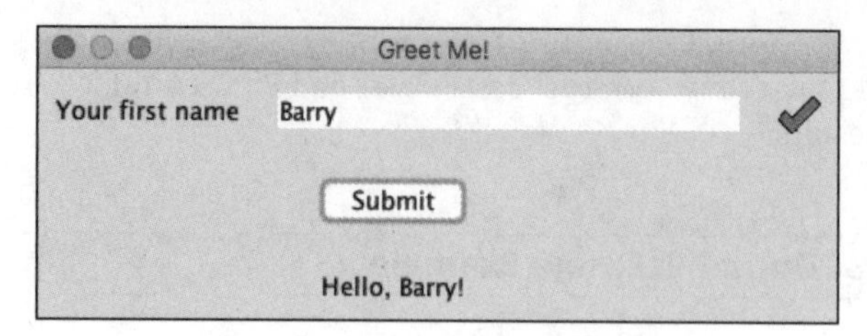

FIGURE 7-17: The user clicks the button.

Here's a blow-by-blow description of the lines in Listing 7-10:

- The first line

  ```
  import com.allmycode.dummiesframe.DummiesFrame;
  ```

  makes the name `DummiesFrame` available to the rest of the code in the listing.

- Inside the `main` method, the statement

  ```
  var frame = new DummiesFrame("Greet Me!");
  ```

  creates an instance of my `DummiesFrame` class and makes the variable name `frame` refer to that instance. A `DummiesFrame` object appears as a window on the user's screen. In this example, the text on the window's title bar is *Greet Me!*

- The next statement is a call to the `frame` object's `addRow` method:

  ```
  frame.addRow("Your first name");
  ```

  This call puts a row on the face of the application's window. The row consists of a label (whose text is *Your first name*), an empty text field, and a red X mark indicating that the user hasn't yet typed anything useful into the field. (Refer to Figure 7-15.)

- A call to the `frame` object's `go` method

  ```
  frame.go();
  ```

  makes the app's window appear on the screen.

- The header of the `calculate` method

  ```
  public static String calculate(String firstName) {
  ```

  tells Java two important things:

  - The `calculate` method returns a value of type `String`.
  - Java should expect the user to type a `String` value in the text field, and whatever the user types will become the `firstName` parameter's value.

  To use my `DummiesFrame` class, your code must have a method named `calculate`, and the `calculate` method must obey certain rules:

- The `calculate` method's header must start with the words `public static`.
- The method may return any Java type: `String`, `int`, `double`, or whatever. (That's actually not a rule; it's an opportunity!)
- The `calculate` method must have the same number of parameters as there are rows in the application's window.

  Listing 7-10 has only one addRow method call, so the window in Figure 7-10 has only one row (not including the *Submit* button), and so the `calculate` method has only one parameter.

When the user starts typing text into the window's text field, the red X mark turns into a green check mark. (Refer to Figure 7-16.) The green check mark indicates that the user has typed a value of the expected type (in this example, a `String` value) into the text field.

» When the user clicks the button, my `DummiesFrame` code works in the background and tells Java to execute the `calculate` method in Listing 7-10. The expression in the `calculate` method's `return` statement

```
return "Hello, " + firstName + "!";
```

tells Java what to display at the bottom of the window. (Refer to Figure 7-17.) In this example, the user types *Barry* in the one-and-only text field, so the value of `firstName` is `"Barry"`, and the calculate method returns the string `"Hello, Barry!"` (Ah! The perks of being a *For Dummies* author!)

Using my `DummiesFrame` class, you can build a simple GUI application with only ten lines of code.

CROSS REFERENCE

In this section's example, my `DummiesFrame` code, working in the background, calls the `calculate` method when the user clicks a button. How does `DummiesFrame` manage to do that? For some insight, see Chapter 16.

REMEMBER

The `DummiesFrame` class isn't built into the Java API, so, in order to run the code in Listing 7-10, my `DummiesFrame.java` file must be part of your project. When you download the code from this book's website (`http://javafordummies.allmycode.com`), you get a folder named `07-10` containing both the Listing 7-10 code and my `DummiesFrame.java` code. You can copy both these files to a project in your IDE, but the way you copy them depends on which IDE you use. One way or another, my `DummiesFrame` class is in a package named `com.allmycode.dummiesframe`, so the `DummiesFrame.java` file must be in a directory named `dummiesframe`, which is inside another directory named `all my code`, which is inside yet another directory named `com`. For some tips, refer to this chapter's "Package deal" sidebar.

TECHNICAL STUFF

To keep things simple, I include the `DummiesFrame.java` file in the `07-10` folder that you download from this book's website. But, really, is that the best way to add my own code to your project? In Chapter 1, I describe files with the `.class` extension, and the role that those files play in the running of a Java program. Instead of handing you my `DummiesFrame.java` file, I should be putting only a `DummiesFrame.class` file in the download. And, on some other occasion, if I have to give you hundreds of `.class` files, I should bundle them all into one big archive file. Java has a name for a big file that encodes many smaller `.class` files. It's called a *JAR file*, and it has the `.jar` extension. In a real-life application, if you're preparing your code for other people to use as part of their own applications, a JAR file is definitely the way to go.

My `DummiesFrame` class isn't exclusively for greetings and salutations. Listing 7-11 uses `DummiesFrame` to do arithmetic.

LISTING 7-11: A Really Simple Calculator

```
package com.example.graphical;

import com.allmycode.dummiesframe.DummiesFrame;

public class Addition {

    public static void main(String[] args) {
        var frame = new DummiesFrame("Adding Machine");
        frame.addRow("First number");
        frame.addRow("Second number");
        frame.setButtonText("Sum");
        frame.go();
    }

    public static int calculate(int firstNumber, int secondNumber) {
        return firstNumber + secondNumber;
    }
}
```

The window in Figure 7-18 has two rows because Listing 7-11 has two `addRow` calls and the listing's `calculate` method has two parameters. In addition, Listing 7-11 calls the `frame` object's `setButtonText` method. So, in Figure 7-18, the text on the face of the button isn't the default word *Submit*.

Listing 7-12 contains a GUI version of the Guessing Game application from Chapter 5, and Figure 7-19 shows the game in action.

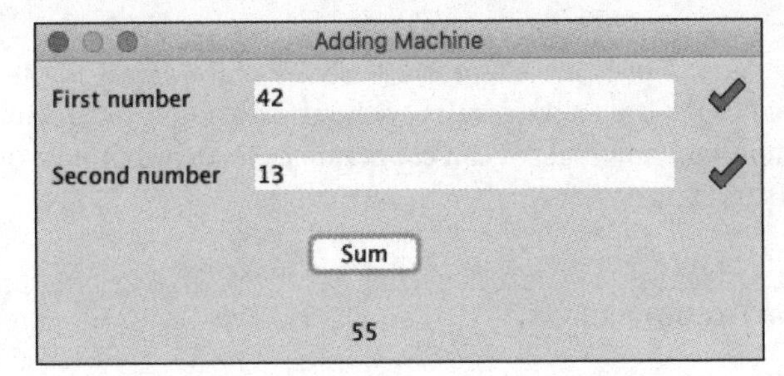

FIGURE 7-18: Look! The code in Listing 7-11 actually works!

LISTING 7-12: I'm Thinking of a Number

```
package com.example.graphical;

import java.util.Random;
import com.allmycode.dummiesframe.DummiesFrame;

public class GuessingGame {

    public static void main(String[] args) {
        DummiesFrame frame = new DummiesFrame("Guessing Game");
        frame.addRow("Enter an int from 1 to 10");
        frame.setButtonText("Submit your guess");
        frame.go();
    }

    public static String calculate(int inputNumber) {
        Random random = new Random();
        int randomNumber = random.nextInt(10) + 1;

        if (inputNumber == randomNumber) {
          return "You win.";
        } else {
          return "You lose. The random number was " + randomNumber + ".";
        }
    }
}
```

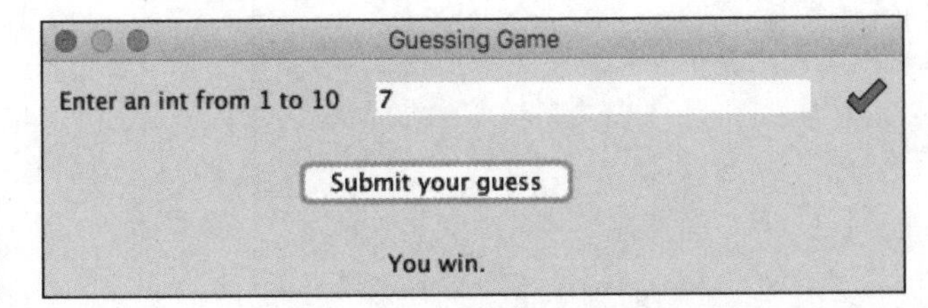

FIGURE 7-19: I win!

In Listing 7-13, I use this chapter's `Account` class alongside the `DummiesFrame` class. I could get the same results without creating an `Account` instance, but I want to show you how classes can cooperate to form a complete program. A run of the code is in Figure 7-20.

LISTING 7-13: Using the Account Class

```
package com.example.accounts;

import com.allmycode.dummiesframe.DummiesFrame;

public class UseAccount {

    public static void main(String[] args) {
        DummiesFrame frame = new DummiesFrame("Display an Account");
        frame.addRow("Full name");
        frame.addRow("Address");
        frame.addRow("Balance");
        frame.setButtonText("Display");
        frame.go();
    }

    public static String calculate(String name, String address,
                                                      double balance) {
        Account myAccount = new Account();

        myAccount.setName(name);
        myAccount.setAddress(address);
        myAccount.setBalance(balance);
        return myAccount.getName() + " (" + myAccount.getAddress() +
                 ") has $" + myAccount.getBalance();
    }
}
```

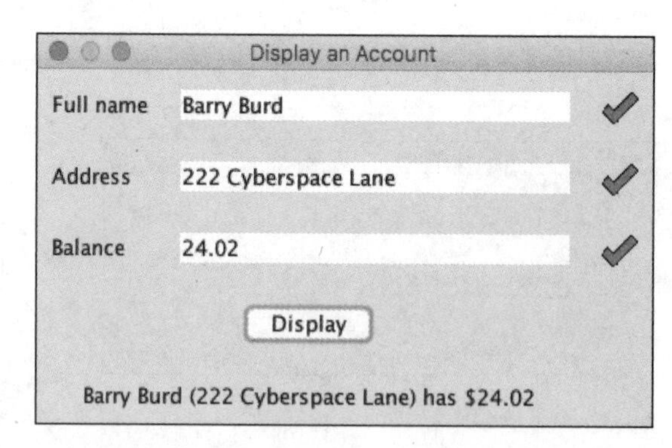

FIGURE 7-20: I'm rich.

TRY IT OUT

Use the `DummiesFrame` class to create two GUI programs.

WINDOW SHOPPING

A window has text fields for an organization's name, annual revenue, and status (profit-making or not profit-making). When the user clicks a button, the window displays the amount of tax the organization pays.

A profit-making organization pays 10 percent of its revenue in tax; a nonprofit organization pays 2 percent of its revenue in tax.

GUI CHOP SUEY

A window has text fields for a product's type of food, weight, cost, number of servings, and number of calories per serving. When the user clicks a button, the window displays the cost per 100 grams, the cost per serving, and the total number of calories in the product.

» Tweaking your code

» Making changes without spending a fortune

Chapter 8
Saving Time and Money: Reusing Existing Code

reuse /ree-YOOSS/ *noun* The act of using something for the *n*th time, where *n* is greater than 1. Example: "Reuse of the material in *Java For Dummies, 8th Edition* is strictly prohibited."

Reuse is good but, in many situations, reuse is a rarity. For example, in the United States, the Department of Agriculture estimates that 30 to 40 percent of the nation's food goes to waste. And, according to one source, only 9 percent of the world's plastics are recycled.

Information goes to waste too. As I revise this chapter for the 8th edition, I'm deciding not to reuse the 7th edition's chapter introduction. (That introduction's made-up story isn't amusing in the least.) In desperation, I looked at some of my other books to find an introduction that I could reuse for this chapter. No luck! I came up empty-handed.

I can't even reuse paragraphs to explain common concepts. My description of Java classes from another book wouldn't work well in this book. I've even experienced times when I had to scrap sections in several chapters because of a small update in Chapter 3.

No doubt about it! Reuse is a precious commodity, and it's in very short supply.

What It Means to Be an Employee

Wouldn't it be nice if every piece of software did just what you wanted it to do? In an ideal world, you could buy a program, make it work right away, plug it seamlessly into new situations, and update it easily whenever your needs change. Unfortunately, software of this kind doesn't exist. (*Nothing* of this kind exists.) The truth is that no matter what you want to do, you can find software that does some of it, but not all of it.

This is one of the reasons why object-oriented programming has been successful. For years, companies were buying prewritten code, only to discover that the code didn't do what they wanted it to do. So, what did the companies do about it? They started messing with the code. Their programmers dug deep into the program files, changed variable names, moved subprograms around, reworked formulas, and generally made the code worse. The reality was that if a program didn't already do what you wanted it to do (even if it did something ever so close to what you wanted), you could never improve the situation by mucking around inside the code. The best option was always to chuck the whole program (expensive as that was) and start all over again. What a sad state of affairs!

With object-oriented programming, a big change has come about. At its heart, an object-oriented program is made to be modified. With correctly written software, you can take advantage of features that are already built-in, add new features of your own, and override features that don't suit your needs. And the best part is that the changes you make are clean. No clawing and digging into other people's brittle program code. Instead, you make nice, orderly additions and modifications without touching the existing code's internal logic. It's the ideal solution.

The last word on employees

When you write an object-oriented program, you start by thinking about the data. You're writing about accounts. So, what's an account? You're writing code to handle button clicks. So, what's a button? You're writing a program to send payroll checks to employees. What's an employee?

In this chapter's first example, an employee is someone with a name and a job title. Sure, employees have other characteristics, but for now I stick to the basics. The code in Listing 8-1 defines what it means to be an employee.

LISTING 8-1: What Is an Employee?

```
package com.example.payroll;

import static java.lang.System.out;
```

```java
public class Employee {
    private String name;
    private String jobTitle;

    public void setName(String nameIn) {
        name = nameIn;
    }

    public String getName() {
        return name;
    }

    public void setJobTitle(String jobTitleIn) {
        jobTitle = jobTitleIn;
    }

    public String getJobTitle() {
        return jobTitle;
    }

    public void cutCheck(double amountPaid) {
        out.printf("Pay to the order of %s ", name);
        out.printf("(%s) ***$", jobTitle);
        out.printf("%,.2f\n", amountPaid);
    }
}
```

According to Listing 8-1, each employee has seven features. Two of these features are fairly simple: Each employee has a name and a job title. (In Listing 8-1, the `Employee` class has a `name` field and a `jobTitle` field.)

And what else does an employee have? Each employee has four methods to handle the values of the employee's name and job title. These methods are `setName`, `getName`, `setJobTitle`, and `getJobTitle`. I explain methods like these (*accessor* methods) in Chapter 7.

On top of all of that, each employee has a `cutCheck` method. The idea is that the method that writes payroll checks has to belong to one class or another. Because most of the information in the payroll check is customized for a particular employee, you may as well put the `cutCheck` method inside the `Employee` class.

For details about the `printf` calls in the `cutCheck` method, see the section "Cutting a check," later in this chapter.

Putting your class to good use

The Employee class in Listing 8-1 has no main method, so there's no starting point for executing code. To fix this deficiency, the programmer writes a separate program with a main method and uses that program to create Employee instances. Listing 8-2 shows a class with a main method — one that puts the code in Listing 8-1 to the test.

LISTING 8-2: Writing Payroll Checks

```
package com.example.payroll;

import java.io.File;
import java.io.IOException;
import java.util.Scanner;

public class DoPayroll {

    public static void main(String[] args) throws IOException {
        var diskScanner = new Scanner(new File("EmployeeInfo.txt"));

        for (int empNum = 1; empNum <= 3; empNum++) {
            payOneEmployee(diskScanner);
        }
        diskScanner.close();
    }

    static void payOneEmployee(Scanner aScanner) {
        var anEmployee = new Employee();

        anEmployee.setName(aScanner.nextLine());
        anEmployee.setJobTitle(aScanner.nextLine());
        anEmployee.cutCheck(aScanner.nextDouble());
        aScanner.nextLine();
    }
}
```

To run the code in Listing 8-2, your hard drive must contain a file named EmployeeInfo.txt. Fortunately, the stuff that you download from this book's website (http://javafordummies.allmycode.com) comes with an EmployeeInfo.txt file. Just copy the EmployeeInfo.txt file to a place where Listing 8-2 can find it. For example, imagine that you've created a project named *08-02*. Listing 8-2 lives somewhere inside the src subfolder of a folder named 08-02, Copy my EmployeeInfo.txt file directly inside the 08-02 folder — not inside the src subfolder. Java will normally look in only the 08-02 folder for files like EmployeeInfo.txt.

WHERE ON EARTH DO YOU LIVE?

Grouping separators vary from one country to another. This makes a big difference when you try to read `double` values using Java's `Scanner` class. To see what I mean, have a serious look at the following JShell session.

```
jshell> import java.util.Scanner

jshell> import java.util.Locale

jshell> var keyboard = new Scanner(System.in)
keyboard ==> java.util.Scanner[delimiters=\p{javaWhitespace}+] ... \E]
    [infinity string=\Q8\E]

jshell> keyboard.nextDouble()
1000.00
$4 ==> 1000.0

jshell> Locale.setDefault(Locale.FRANCE)

jshell> keyboard = new Scanner(System.in)
keyboard ==> java.util.Scanner[delimiters=\p{javaWhitespace}+] ... \E]
    [infinity string=\Q8\E]

jshell> keyboard.nextDouble()
1000,00
$7 ==> 1000.0

jshell> keyboard.nextDouble()
1000.00
| java.util.InputMismatchException thrown:
|       at Scanner.throwFor (Scanner.java:860)
|       at Scanner.next (Scanner.java:1497)
|       at Scanner.nextDouble (Scanner.java:2467)
|       at (#8:1)

jshell>
```

I conducted this session on a computer in the United States. The country of origin is relevant because, in response to the first `keyboard.nextDouble()` call, I type `1000.00` (with a period before the last two zeros) and Java accepts this as meaning "one thousand."

But then, in the JShell session, I call `Locale.setDefault(Locale.FRANCE)`, which tells Java to behave as if my computer is in France. When I create another `Scanner` instance

(continued)

and call `keyboard.nextDouble()` again, Java accepts `1000,00` (with a comma before the last two zeros) as an expression meaning *mille* (French for "one thousand"). What's more, Java no longer accepts the period in `1000.00`. When I type `1000.00` (with a period) I get an `InputMismatchException`.

By default, your computer's `Scanner` instance wants you to input `double` numbers the way you normally type them in your country. If you type numbers according to another country's convention, you get an `InputMismatchException`. So, when you run the code in Listing 8-2, the numbers in your `EmployeeInfo.txt` file must use your country's format.

This brings me to the running of the code in Listing 8-2. The `EmployeeInfo.txt` file that you download from this book's website starts with the following three lines:

- Barry Burd
- CEO
- 5000.00

That last number `5000.00` has a period in it, so if your country prefers a comma in place of my United States period, you get an `InputMismatchException`. In response to this, you have two choices:

- In the downloaded `EmployeeInfo.txt` file, change the periods to commas.
- In the code of Listing 8-2, add the statement `Locale.setDefault(Locale.US)` before the `diskScanner` declaration.

And finally, if you want your output to look like your own country's numbers, you can do it with Java's `Formatter` class. Add something like this to your code:

```
out.print(new java.util.Formatter().format(java.util.Locale.FRANCE, "%,.2f",
    1000.00));
```

For all the details, see the API (Application Programming Interface) documentation for Java's `Formatter` class (`https://docs.oracle.com/en/java/javase/17/docs/api/java.base/java/util/Formatter.html`) and `Locale` class (`https://docs.oracle.com/en/java/javase/17/docs/api/java.base/java/util/Locale.html`).

CROSS REFERENCE

For more words of wisdom about files on your hard drive, see the "Working with Disk Files (a Brief Detour)" section in this chapter.

The `DoPayroll` class in Listing 8-2 has two methods. One of the methods, `main`, calls the other method, `payOneEmployee`, three times. Each time around, the `payOneEmployee` method gets stuff from the `EmployeeInfo.txt` file and feeds this stuff to the `Employee` class's methods.

Here's how the variable name *anEmployee* is reused and recycled:

- The first time that `payOneEmployee` is called, the statement `anEmployee = new Employee()` makes `anEmployee` refer to a new object.
- The second time that `payOneEmployee` is called, the computer executes the same statement again. This second execution creates a new incarnation of the `anEmployee` variable that refers to a brand-new object.
- The third time around, all the same stuff happens again. A new `anEmployee` variable ends up referring to a third object.

The whole story is pictured in Figure 8-1.

There are always interesting things for you to try:

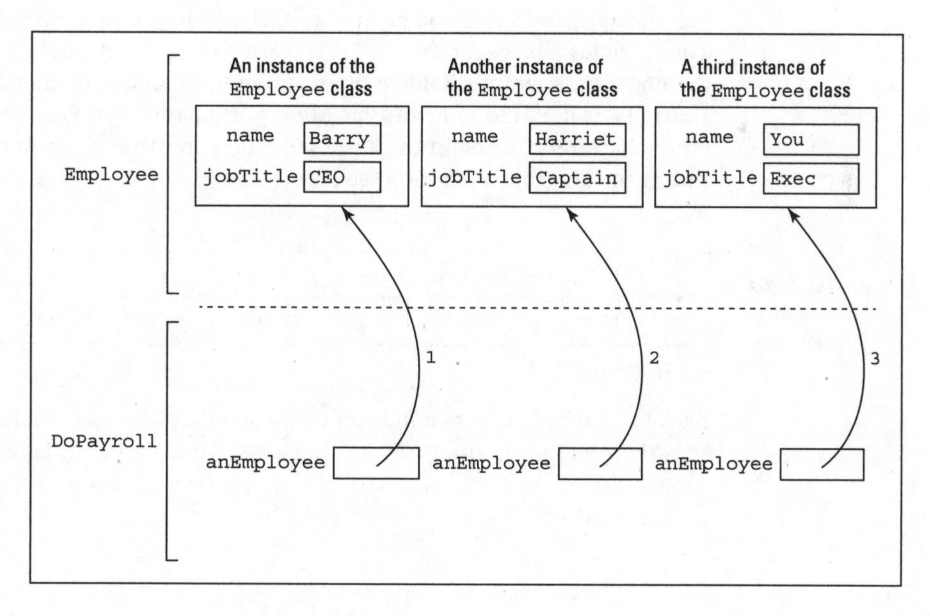

FIGURE 8-1: Three calls to the `payOneEmployee` method.

TRY IT OUT

LOCALE, LOCALE, LOCALE

A PlaceToLive has an address, a number of bedrooms, and an area (in square feet or square meters). Write the PlaceToLive class's code. Write code for a separate class named DisplayThePlaces. Your DisplayThePlaces class creates a few PlaceToLive instances by assigning values to their address, numberOfBedrooms, and area fields. The DisplayThePlaces class also reads (from the keyboard) the cost of living in each place. For each place, your code displays the cost per square foot (or square meter) and the cost per bedroom.

PAY PER CLICK

Use your new PlaceToLive class and my DummiesFrame class (from Chapter 7) to create a GUI application. The GUI application takes information about a place to live and displays the place's cost per square foot (or meter) and the cost per bedroom.

Cutting a check

Listing 8-1 has three printf calls. Each printf call has a format string (like "(%s) ***$") and a variable (like jobTitle). Each format string has a placeholder (like %s) that determines where and how the variable's value is displayed.

For example, in the second printf call, the format string has a %s placeholder. This %s holds a place for the jobTitle variable's value. According to Java's rules, the notation %s always holds a place for a string and, sure enough, the variable jobTitle is declared to be of type String in Listing 8-1. Parentheses and some other characters surround the %s placeholder, so parentheses surround each job title in the program's output. (See Figure 8-2.)

```
Pay to the order of Barry Burd (CEO) ***$5,000.00
Pay to the order of Harriet Ritter (Captain) ***$7,000.00
Pay to the order of Your Name Here (Honorary Exec of the Day) ***$10,000.00
```

FIGURE 8-2: Everybody gets paid.

Back in Listing 8-1, notice the comma inside the %,.2f placeholder. The comma tells the program to use *grouping separators*. That's why, in Figure 8-2, you see $5,000.00, $7,000.00, and $10,000.00 instead of $5000.00, $7000.00, and $10000.00.

Working with Disk Files (a Brief Detour)

In previous chapters, programs read characters from the computer's keyboard. But the code in Listing 8-2 reads characters from a specific file. The file (named `EmployeeInfo.txt`) lives on your computer's hard drive.

This `EmployeeInfo.txt` file is like a word processing document. The file can contain letters, digits, and other characters. But unlike a word processing document, the `EmployeeInfo.txt` file contains no formatting — no italics, no bold, no font sizes, nothing of that kind.

The `EmployeeInfo.txt` file contains only ordinary characters — the kinds of keystrokes that you type while you play a guessing game from Chapters 5 and 6. Of course, getting guesses from a user's keyboard and reading employee data from a disk file aren't exactly the same. In a guessing game, the program displays prompts, such as `Enter an int from 1 to 10`. The game program conducts a back-and-forth dialogue with the person sitting at the keyboard. In contrast, Listing 8-2 has no dialogue. This `DoPayroll` program reads characters from a hard drive and doesn't prompt or interact with anyone.

Most of this chapter is about code reuse. But Listing 8-2 stumbles upon an important idea — an idea that's not directly related to code reuse. Unlike the examples in previous chapters, Listing 8-2 reads data from a stored disk file. So, in the following sections, I take a short side trip to explore disk files.

Storing data in a file

The code in Listing 8-2 doesn't run unless you have some employee data sitting in a file. Listing 8-2 says that this file is `EmployeeInfo.txt`. So, before running the code of Listing 8-2, I created a small `EmployeeInfo.txt` file. The file is shown in Figure 8-3; refer to Figure 8-2 for the resulting output.

```
Barry Burd
CEO
5000.00
Harriet Ritter
Captain
7000.00
Your Name Here
Honorary Exec of the Day
10000.00
```

FIGURE 8-3: An EmployeeInfo.txt file.

When you visit this book's website (`http://javafordummies.allmycode.com`) and you download the book's code listings, you get a copy of the `EmployeeInfo.txt` file.

To keep Listing 8-2 simple, I insist that, when you type the characters in Figure 8-3, you finish up by typing `10000.00` and then pressing Enter. (Look again at Figure 8-3 and notice how the cursor is at the start of a brand-new line.) If you forget to finish by pressing Enter, the code in Listing 8-2 will crash when you try to run it.

Grouping separators vary from one country to another. The file shown in Figure 8-3 works on a computer configured in the United States where *5000.00* means "five thousand." But the file doesn't work on a computer that's configured in what I call a "comma country" — a country where *5000,00* means "five thousand." If you live in a comma country, be sure to read this chapter's "Where on Earth do you live?" sidebar.

Repeat after me

In almost any computer programming language, reading data from a file can be tricky. You add extra lines of code to tell the computer what to do. Sometimes you can copy and paste these lines from other peoples' code. For example, you can follow the pattern in Listing 8-2:

```
/*
* The pattern in Listing 8-2
*/
import java.io.File;
import java.io.IOException;
import java.util.Scanner;

class SomeClassName {

    public static void main(String[] args) throws IOException {

        var scannerName = new Scanner(new File("SomeFileName"));

        //Some code goes here

        scannerName.nextInt();
        scannerName.nextDouble();
        scannerName.next();
        scannerName.nextLine();
        //Some code goes here

        scannerName.close();
    }
}
```

You want to read data from a file. You start by imagining that you're reading from the keyboard. Put the usual `Scanner` and `next` codes into your program. Then add some extra items from the Listing 8-2 pattern:

- Add two new `import` declarations — one for `java.io.File` and another for `java.io.IOException`.
- Type **throws IOException** in your method's header.
- Type **new File(" ")** in your call to `new Scanner`.
- Take a file that's already on your hard drive. Type that filename inside the quotation marks.
- Take the word that you use for the name of your scanner. Reuse that word in calls to `next`, `nextInt`, `nextDouble`, and so on.
- Take the word that you use for the name of your scanner. Reuse that word in a call to `close`.

Occasionally, copying and pasting code can get you into trouble. Maybe you're writing a program that doesn't fit the simple Listing 8-2 pattern. You need to tweak the pattern a bit. But to tweak the pattern, you need to understand some of the ideas behind the pattern.

That's how the next section comes to your rescue. It covers some of these ideas.

TECHNICAL STUFF

This paragraph is actually a confession. In almost every computer programming language, input from a disk file is a nasty business. There's no such thing as a simple `INPUT` command. You normally have to set up a connection between the code and the disk device, prepare for possible trouble reading from the device, do your reading, convert the characters you read into the type of value that you want and, finally, break your connection with the disk device. It's a big mess. That's why, in this book, I rely on Java's `Scanner` class. The `Scanner` class makes input relatively painless. But, I admit, professional Java programmers hardly ever use the `Scanner` class to do input. Instead, they use something called a `BufferedReader` or classes in the `java.nio` package. If you're not content with my use of the `Scanner` class and you want to see Listing 8-2 translated into a `BufferedReader` program, visit this book's website (`http://javafordummies.allmycode.com/`).

Reading from a file

In previous chapters, programs read characters from the computer's keyboard. These programs use things like `Scanner`, `System.in`, and `nextDouble` — things defined in Java's API. The `DoPayroll` program in Listing 8-2 puts a new spin on this story. Rather than read characters from the keyboard, the program reads characters from the `EmployeeInfo.txt` file. The file lives on your computer's hard drive.

To read characters from a file, you use some of the same things that help you read characters from the keyboard. You use `Scanner`, `nextDouble`, and other goodies. But in addition to these goodies, you have a few extra hurdles to jump. Here's a list:

- **You need a new File object.** To be more precise, you need a new instance of the API's `File` class. You get this new instance with code like

  ```
  new File("EmployeeInfo.txt")
  ```

  The stuff in quotation marks is the name of a file — a file on your computer's hard drive. The file contains characters like those shown previously in Figure 8-3.

  At this point, the terminology makes mountains out of molehills. Sure, I use the phrases *new File object* and *new File instance,* but all you're doing is making `new File("EmployeeInfo.txt")` stand for a file on your hard drive. After you shove `new File("EmployeeInfo.txt")` into `new Scanner`,

  ```
  var diskScanner = new Scanner(new File("EmployeeInfo.txt"));
  ```

  you can forget all about the `new File` business. From that point on in the code, `diskScanner` stands for the `EmployeeInfo.txt` filename on your computer's hard drive. (The name `diskScanner` stands for a file on your hard drive just as, in previous examples, the name `keyboard` stands for those buttons you press day in and day out.)

  Creating a `new File` object in Listing 8-2 is like creating a `new Employee` object later in the same listing. It's also like creating a `new Account` object in the examples of Chapter 7. The only difference is that the `Employee` and `Account` classes are defined in this book's examples. The `File` class is defined in Java's API.

  When you connect to a disk file with `new Scanner`, don't forget the `new File` part. If you write `new Scanner("EmployeeInfo.txt")` without `new File`, the compiler won't mind. (You don't get any warnings or error messages before you run the code.) But when you run the code, you don't get anything like the results that you expect to get.

- **You must refer to the File class by its full name: `java.io.File`.** You can do this with an import declaration like the one in Listing 8-2. Alternatively, you can clutter your code with a statement like

  ```
  var diskScanner = new Scanner(new java.io.File("EmployeeInfo.txt"));
  ```

- **You need a throws IOException clause.** Lots of things can go wrong when your program connects to `EmployeeInfo.txt`. For one thing, your hard drive may not have a file named `EmployeeInfo.txt`. For another, the file `EmployeeInfo.txt` may be in the wrong directory. To brace for this kind of calamity, the Java programming language takes certain precautions. The language insists that when a disk file is involved, you acknowledge the possible dangers of calling `new Scanner`.

You can acknowledge the hazards in several possible ways, but the simplest way is to use a `throws` clause. In Listing 8-2, the `main` method's header ends with the words *throws IOException.* By adding these two words, you appease the Java compiler. It's as if you're saying "I know that calling `new Scanner` can lead to problems. You don't have to remind me." And, sure enough, adding `throws IOException` to your `main` method keeps the compiler from complaining. (Without this `throws` clause, you get an `unhandled exception` error message.)

CROSS REFERENCE

For the full story on Java exceptions, read Chapter 13. In the meantime, add `throws IOException` to the header of any method that calls `new Scanner (new File( ....`

- **You must refer to the `IOException` class by its full name: `java.io.IOException`.**

  You can do this with an `import` declaration like the one in Listing 8-2. Alternatively, you can enlarge the `main` method's `throws` clause:

```
public static void main(String[] args) throws java.io.IOException {
```

- **You must pass the file scanner's name to the `payOneEmployee` method.**

  In Listing 7-5 in Chapter 7, the `getInterest` method has a parameter named *percentageRate.* Whenever you call the `getInterest` method, you hand an extra, up-to-date piece of information to the method. (You hand a number — an interest rate — to the method. Figure 7-7 illustrates the idea.)

  The same thing happens in Listing 8-2. The `payOneEmployee` method has a parameter named *aScanner.* Whenever you call the `payOneEmployee` method, you hand an extra, up-to-date piece of information to the method. (You hand a scanner — a reference to a disk file — to the method.)

You may wonder why the `payOneEmployee` method needs a parameter. After all, in Listing 8-2, the `payOneEmployee` method always reads data from the same file. Why bother informing this method, each time you call it, that the disk file is still the `EmployeeInfo.txt` file?

Well, there are plenty of ways to shuffle the code in Listing 8-2. Some ways don't involve a parameter. But the way that this example has arranged things, you have two separate methods: a `main` method and a `payOneEmployee` method. You create a scanner once inside the `main` method and then use the scanner three times — once inside each call to the `payOneEmployee` method.

Anything you define inside a method is like a private joke that's known only to the code inside that method. So, the diskScanner that you define inside the main method isn't automatically known inside the payOneEmployee method. To make the payOneEmployee method aware of the disk file, you pass diskScanner from the main method to the payOneEmployee method.

CROSS REFERENCE

To read more about things that you declare inside (and outside) of methods, see Chapter 10.

Who moved my file?

When you download the code from this book's website (http://javafordummies.allmycode.com/), you'll find files named Employee.java and DoPayroll.java — the code in Listings 8-1 and 8-2. You'll also find a file named EmployeeInfo.txt. That's good because, if Java can't find the EmployeeInfo.txt file, the whole project doesn't run properly. Instead, you get a FileNotFoundException.

In general, when you get a FileNotFoundException, some file that your program needs isn't available to it. This is an easy mistake to make. It can be frustrating because, to you, a file such as EmployeeInfo.txt may look like it's available to your program. But remember: Computers are stupid. If you make a tiny mistake, the computer can't read between the lines for you. So, if your EmployeeInfo.txt file isn't in the right directory on your hard drive or the filename is spelled incorrectly, the computer chokes when it tries to run your code.

Sometimes you know darn well that an EmployeeInfo.txt (or *whatever.xyz*) file exists on your hard drive. But when you run your program, you still get a mean-looking FileNotFoundException. When this happens, the file is usually in the wrong directory on your hard drive. (Of course, it depends on your point of view. Maybe the file is in the right directory, but your Java program is looking for the file in the wrong directory.) To diagnose this problem, add the following code to Listing 8-2:

```
var employeeInfo = new File("EmployeeInfo.txt");
System.out.println("Looking for " + employeeInfo.getCanonicalPath());
```

When you run the code, Java tells you where, on your hard drive, the EmployeeInfo.txt file should be.

You moved your file!

Java normally looks in your project's top-level folder for a file like EmployeeInfo.txt. But you can override this behavior by naming a file's exact location in your

Java code. Code like `new File("C:\\Users\\bburd\\workspace\\08-01\\EmployeeInfo.txt")` looks really ugly, but it works.

In the preceding paragraph, did you notice the double backslashes in "`C:\\Users\\bburd\\workspace ...`"? If you're a Windows user, you'd be tempted to write `C:\Users\bburd\workspace ...` with single backslashes. But in Java, the single backslash has its own, special meaning. (For example, back in Listing 7-7, `\n` means to go to the next line.) So, in Java, to indicate a backslash inside a quoted string, you use a double backslash instead.

TECHNICAL STUFF

Macintosh and Linux users might find comfort in the fact that their path separator, `/`, has no special meaning in a Java string. On a Mac, the code `new File("/Users/bburd/workspace/08-01/EmployeeInfo.txt")` is as normal as breathing. (Well, it's almost that normal!) But Mac users and Linux wonks shouldn't claim superiority too quickly. Lines such as `new File("/Users/bburd/workspace ...` work in Windows as well. In Windows, you can use either a slash (`/`) or a backslash (`\`) as the path name separator. At the Windows command prompt, I can type `cd c:/users\bburd` to get to my home directory.

TIP

If you know where your Java program looks for files, you can worm your way from that place to the directory of your choice. Assume, for the moment, that the code in Listing 8-2 normally looks for the `EmployeeInfo.txt` file in a directory named `08-01`. As an experiment, go to the `08-01` directory and create a new subdirectory named *dataFiles*. Then move my `EmployeeInfo.txt` file to the new `dataFiles` directory. To read numbers and words from the file that you moved, modify Listing 8-2 with the code `new File("dataFiles\\EmployeeInfo.txt")` or `new File("dataFiles/EmployeeInfo.txt")`.

Reading a line at a time

In Listing 8-2, the `payOneEmployee` method illustrates some useful tricks for reading data. In particular, every scanner that you create has a `nextLine` method. (You might not use this `nextLine` method, but the method is available nonetheless.) When you call a scanner's `nextLine` method, the method grabs everything up to the end of the current line of text. In Listing 8-2, a call to `nextLine` can read a whole line from the `EmployeeInfo.txt` file. (In another program, a scanner's `nextLine` call may read everything the user types on the keyboard up to the pressing of the Enter key.)

Notice my careful choice of words: `nextLine` reads everything "up to the end of the current line." Unfortunately, what it means to read up to the end of the current line isn't always what you think it means. Intermingling `nextInt`, `nextDouble`, and `nextLine` calls can be messy. You have to watch what you're doing and check your program's output carefully.

To understand all of this, you need to be painfully aware of a data file's line breaks. Think of a line break as an extra character, stuck between one line of text and the next. Then imagine that calling `nextLine` means to read everything up to and including the next line break.

Now take a look at Figure 8-4:

- If one call to `nextLine` reads `Barry Burd[LineBreak]`, the subsequent call to `nextLine` reads `CEO[LineBreak]`.
- If one call to `nextDouble` reads the number 5000.00, the subsequent call to `nextLine` reads the `[LineBreak]` that comes immediately after the number 5000.00. (That's all the `nextLine` reads — a `[LineBreak]` and nothing more.)
- If a call to `nextLine` reads the `[LineBreak]` after the number 5000.00, the subsequent call to `nextLine` reads `Harriet Ritter[LineBreak]`.

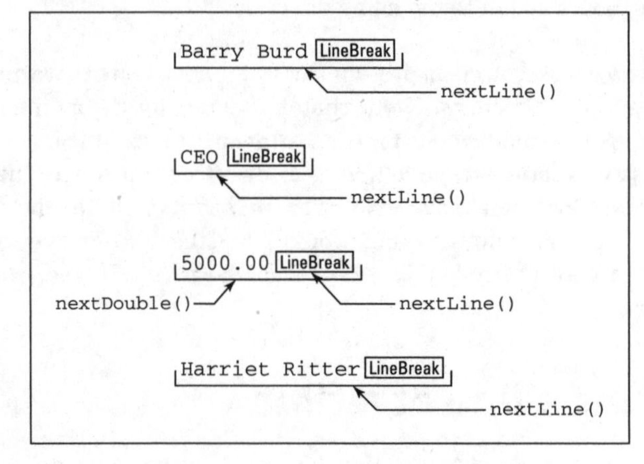

FIGURE 8-4: Calling `nextDouble` and `nextLine`.

So, after reading the number 5000.00, you need *two* calls to `nextLine` in order to scoop up the name *Harriet Ritter*. The mistake that I usually make is to forget the first of those two calls.

WARNING

Look again at the file in Figure 8-3. For this section's code to work correctly, you must have a line break after the last `10000.00`. If you don't, a final call to `nextLine` makes your program crash and burn. The error message reads `NoSuchElementException: No line found`.

TECHNICAL STUFF

I'm always surprised by the number of quirks that I find in each programming language's scanning methods. For example, the first `nextLine` that reads from the file in Figure 8-3 devours `Barry Burd[LineBreak]` from the file. But that `nextLine` call delivers `Barry Burd` (with no line break) to the running code. So `nextLine` looks for a line break, and then `nextLine` loses the line break. Yes, this is a subtle point. And no, this subtle point hardly ever causes problems for anyone.

TECHNICAL STUFF

If this business about `nextDouble` and `nextLine` confuses you, please don't put the blame on Java. Mixing input calls is delicate work in any computer programming language. And the really nasty thing is that each programming language approaches the problem a little differently. What you find out about `nextLine` in Java helps you understand the issues when you get to know C++ or Visual Basic, but it doesn't tell you all the details. Each language's details are unique to that language. (Yes, it's a big pain. But because all computer programmers become rich and famous, the pain eventually pays off.)

Clean up after yourself

To the average computer user, a keyboard doesn't feel anything like a file stored on a computer's hard drive. But disk files and keyboard input have a lot in common. In fact, a basic principle of computer operating systems dictates that any differences between two kinds of input be, for the programmer, as blurry as possible. As a Java programmer, you should treat disk files and keyboard input almost the same way. That's why Listing 8-2 contains a `diskScanner.close()` call.

When you run a Java program, you normally execute the `main` method's statements, starting with the first statement in the method body and ending with the last statement in the method body. You take detours along the way, skipping past `else` parts and diving into method bodies, but basically you finish executing statements at the end of the `main` method. That's why, in Listing 8-2, the call to `close` is at the end of the `main` method's body. When you run the code in Listing 8-2, the last thing you do is disconnect from the disk file. And, fortunately, that disconnection takes place after you've executed all the `nextLine` and `nextDouble` calls.

ON THE RECORD

TRY IT OUT

Previously in this chapter, you create instances of your own `PlaceToLive` class and display information about those instances. Modify the text-based version of your code so that it gets each instance's characteristics (address, number of bedrooms, and area) from a disk file.

Defining Subclasses (What It Means to Be a Full-Time or Part-Time Employee)

This time last year, your company paid $10 million for a piece of software. That software came in the `Employee.class` file. People at Burd Brain Consulting (the company that created the software) don't want you to know about the innards of the software. (Otherwise, you may steal their ideas.) So, you don't have the Java program file that the software came from. (In other words, you don't have `Employee.java`.) You can run the bytecode in the `Employee.class` file. You can also read the documentation in a web page named *Employee.html*. But you can't see the statements inside the `Employee.java` program, and you can't change any of the program's code.

Since this time last year, your company has grown. Unlike in the old days, your company now has two kinds of employees: full-time and part-time. Each full-time employee is on a fixed, weekly salary. (If the employee works nights and weekends, then in return for this monumental effort, the employee receives a hearty handshake.) In contrast, each part-time employee works for an hourly wage. Your company deducts an amount from each full-time employee's paycheck to pay for the company's benefits package. Part-time employees, however, don't get benefits.

The question is whether the software that your company bought last year can keep up with the company's growth. You invested in a great program to handle employees and their payroll, but the program doesn't differentiate between your full-time and part-time employees. You have several options:

» **Call your next-door neighbor, whose 12-year-old child knows more about computer programming than anyone in your company.** Get this uppity little brat to take the employee software apart, rewrite it, and hand it back to you with all the changes and additions your company requires.

On second thought, you can't do that. No matter how smart that kid is, the complexities of the employee software will probably confuse the kid. By the time you get the software back, it'll be filled with bugs and inconsistencies. Besides, you don't even have the `Employee.java` file to hand to the kid. All you have is the `Employee.class` file, which can't be read or modified with a text editor. (See Chapter 2.) Besides, your kid just beat up the neighbor's kid. You don't want to give your neighbor the satisfaction of seeing you beg for the whiz kid's help.

- **Scrap the $10 million employee software.** Get someone in your company to rewrite the software from scratch.

  In other words, say goodbye to your time and money.

- **Write a new front end for the employee software.** That is, build a piece of code that does some preliminary processing on full-time employees and then hands the preliminary results to your $10 million software. Do the same for part-time employees.

  This idea could be decent or spell disaster. Are you sure that the existing employee software has convenient *hooks* in it? (That is, does the employee software contain entry points that allow your front-end software to easily send preliminary data to the expensive employee software?) Remember: This plan treats the existing software as one big, monolithic lump, which can become cumbersome. Dividing the labor between your front-end code and the existing employee program is difficult. And if you add layer upon layer to existing black box code, you'll probably end up with a fairly inefficient system.

- **Call Burd Brain Consulting, the company that sold you the employee software.** Tell Dr. Burd that you want the next version of his software to differentiate between full-time and part-time employees.

  "No problem," says Dr. Burd. "It'll be ready by the start of the next fiscal quarter." That evening, Dr. Burd makes a discreet phone call to his next-door neighbor. . . .

- **Create two new Java classes named *FullTimeEmployee* and *PartTimeEmployee*.** Have each new class extend the existing functionality of the expensive `Employee` class, but have each new class define its own, specialized functionality for certain kinds of employees.

  Way to go! Figure 8-5 shows the structure that you want to create.

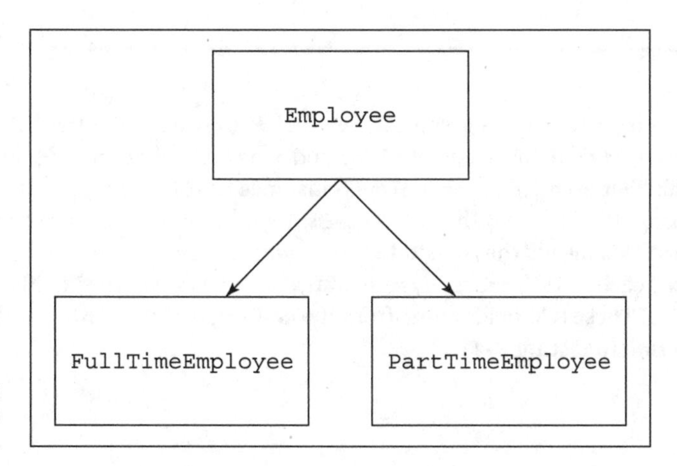

FIGURE 8-5: The Employee class family tree.

Creating a subclass

In Listing 8-1, I define an Employee class. I can use what I define in Listing 8-1 and extend the definition to create new, more specialized classes. So, in Listing 8-3, I define a new class: a FullTimeEmployee class.

LISTING 8-3: What Is a FullTimeEmployee?

```
package com.example.payroll;

public class FullTimeEmployee extends Employee {
    private double weeklySalary;
    private double benefitDeduction;

    public void setWeeklySalary(double weeklySalaryIn) {
        weeklySalary = weeklySalaryIn;
    }

    public double getWeeklySalary() {
        return weeklySalary;
    }

    public void setBenefitDeduction(double benefitDedIn) {
        benefitDeduction = benefitDedIn;
    }

    public double getBenefitDeduction() {
        return benefitDeduction;
    }

    public double findPaymentAmount() {
        return weeklySalary - benefitDeduction;
    }
}
```

Looking at Listing 8-3, you can see that each instance of the FullTimeEmployee class has two fields: weeklySalary and benefitDeduction. But are those the only fields that each FullTimeEmployee instance has? No, they're not. The first line of Listing 8-3 says that the FullTimeEmployee class extends the existing Employee class. This means that in addition to having a weeklySalary and a benefitDeduction, each FullTimeEmployee instance also has two other fields: name and jobTitle. These two fields come from the definition of the Employee class, which you can find in Listing 8-1.

In Listing 8-3, the magic word is *extends*. When one class extends an existing class, the extending class automatically inherits functionality that's defined in the existing class. So, the `FullTimeEmployee` class *inherits* the `name` and `jobTitle` fields. The `FullTimeEmployee` class also inherits all the methods that are declared in the `Employee` class: `setName`, `getName`, `setJobTitle`, `getJobTitle`, and `cutCheck`. The `FullTimeEmployee` class is a *subclass* of the `Employee` class. That means the `Employee` class is the *superclass* of the `FullTimeEmployee` class. You can also talk in terms of blood relatives: The `FullTimeEmployee` class is the *child* of the `Employee` class, and the `Employee` class is the *parent* of the `FullTimeEmployee` class.

It's almost (but not quite) as if the `FullTimeEmployee` class were defined by the code in Listing 8-4.

LISTING 8-4: Fake (But Informative) Code

```
package com.example.payroll;

import static java.lang.System.out;

public class FullTimeEmployee {
    private String name;
    private String jobTitle;
    private double weeklySalary;
    private double benefitDeduction;

    public void setName(String nameIn) {
        name = nameIn;
    }

    public String getName() {
        return name;
    }

    public void setJobTitle(String jobTitleIn) {
        jobTitle = jobTitleIn;
    }
    public String getJobTitle() {
        return jobTitle;
    }

    public void setWeeklySalary(double weeklySalaryIn) {
        weeklySalary = weeklySalaryIn;
    }
```

(continued)

LISTING 8-4: ***(continued)***

```
    public double getWeeklySalary() {
        return weeklySalary;
    }

    public void setBenefitDeduction(double benefitDedIn) {
        benefitDeduction = benefitDedIn;
    }

    public double getBenefitDeduction() {
        return benefitDeduction;
    }

    public double findPaymentAmount() {
        return weeklySalary - benefitDeduction;
    }

    public void cutCheck(double amountPaid) {
        out.printf("Pay to the order of %s ", name);
        out.printf("(%s) ***$", jobTitle);
        out.printf("%,.2f\n", amountPaid);
    }
}
```

TECHNICAL STUFF

Why does the title for Listing 8-4 call that code fake? (Should the code feel insulted?) Well, the main difference between Listing 8-4 and the inheritance situation in Listings 8-1 and 8-3 is this: A child class can't directly reference the private fields of its parent class. To do anything with the parent class's private fields, the child class has to call the parent class's accessor methods. Back in Listing 8-3, calling `setName("Rufus")` would be legal, but the code `name="Rufus"` wouldn't be. If you believe everything you read in Listing 8-4, you'd think that code in the `FullTimeEmployee` class can do `name="Rufus"`. Well, it can't. (My, what a subtle point this is!)

REMEMBER

You don't need the `Employee.java` file on your hard drive to write code that extends the `Employee` class. All you need is the file `Employee.class`.

Creating subclasses is habit-forming

After you're accustomed to extending classes, you can get extend-happy. If you created a `FullTimeEmployee` class, you might as well create a `PartTimeEmployee` class, as shown in Listing 8-5.

LISTING 8-5: What Is a PartTimeEmployee?

```
package com.example.payroll;

public class PartTimeEmployee extends Employee {
    private double hourlyRate;

    public void setHourlyRate(double rateIn) {
        hourlyRate = rateIn;
    }

    public double getHourlyRate() {
        return hourlyRate;
    }

    public double findPaymentAmount(int hours) {
        return hourlyRate * hours;
    }
}
```

Unlike the `FullTimeEmployee` class, `PartTimeEmployee` has no salary or deduction. Instead `PartTimeEmployee` has an `hourlyRate` field. (Adding a `numberOfHoursWorked` field would also be a possibility. I chose not to do this, figuring that the number of hours a part-time employee works will change drastically from week to week.)

Using Subclasses

The preceding section tells a story about creating subclasses. It's a good story, but it's incomplete. Creating subclasses is fine, but you gain nothing from these subclasses unless you write code to use them.

Listing 8-6 contains the simplest possible example of a program that uses the subclasses `FullTimeEmployee` and `PartTimeEmployee`. (For a look as some more interesting examples, visit this book's website).

LISTING 8-6: Putting Subclasses to Good Use

```
package com.example.payroll;

public class PayrollForTwo {

    public static void main(String[] args) {

        var ftEmployee = new FullTimeEmployee();

        ftEmployee.setName("Barry Burd");
        ftEmployee.setJobTitle("CEO");
        ftEmployee.setWeeklySalary(5000.00);
        ftEmployee.setBenefitDeduction(500.00);
        ftEmployee.cutCheck(ftEmployee.findPaymentAmount());
        System.out.println();

        var ptEmployee = new PartTimeEmployee();

        ptEmployee.setName("Steve Surace");
        ptEmployee.setJobTitle("Driver");
        ptEmployee.setHourlyRate(7.53);
        ptEmployee.cutCheck(ptEmployee.findPaymentAmount(10));
    }
}
```

Figure 8-6 shows the output of the code in Listing 8-6.

```
Pay to the order of Barry Burd (CEO) ***$4,500.00

Pay to the order of Steve Surace (Driver) ***$75.30
```

FIGURE 8-6: I earn a lot more than Steve.

To understand Listing 8-6, you need to keep an eye on three classes: `Employee`, `FullTimeEmployee`, and `PartTimeEmployee`. (For a look at the code that defines these classes, see Listings 8-1, 8-3, and 8-5.)

The first half of Listing 8-6 deals with a full-time employee. Notice how many methods are available for use with the `ftEmployee` variable? For instance, you can call `ftEmployee.setWeeklySalary` because `ftEmployee` has type `FullTimeEmployee`. You can also call `ftEmployee.setName` because the `FullTimeEmployee` class extends the `Employee` class.

Because cutCheck is declared in the Employee class, you can call ftEmployee.cutCheck. But you can also call ftEmployee.findPaymentAmount because a findPaymentAmount method is in the FullTimeEmployee class.

Making types match

Look again at the first half of Listing 8-6. Take special notice of that last statement — the one in which the full-time employee is actually cut a check. The statement forms a nice, long chain of values and their types. You can see this by reading the statement from the inside out:

- Method ftEmployee.findPaymentAmount is called with an empty parameter list. (Refer to Listing 8-6.) That's good because the findPaymentAmount method takes no parameters. (Refer to Listing 8-3.)
- The findPaymentAmount method returns a value of type double. (Again, refer to Listing 8-3.)
- The double value that ftEmployee.findPaymentAmount returns is passed to method ftEmployee.cutCheck. (Refer to Listing 8-6.) That's good because the cutCheck method takes one parameter of type double. (Refer to Listing 8-1.)

For a fanciful graphical illustration, see Figure 8-7.

Always feed a method the value types that it wants in its parameter list.

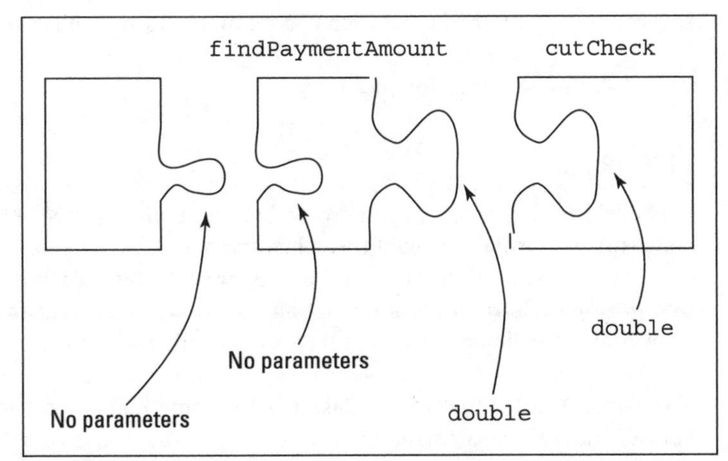

FIGURE 8-7: Matching parameters.

The second half of the story

In the second half of Listing 8-6, the code creates an object of type PartTimeEmployee. A variable of type PartTimeEmployee can do some of the same things a FullTimeEmployee variable can do. But the PartTimeEmployee class doesn't have the setWeeklySalary and setBenefitDeduction methods. Instead, the PartTimeEmployee class has the setHourlyRate method. (See Listing 8-5.) So, in Listing 8-6 the next-to-last line is a call to the setHourlyRate method.

The last line of Listing 8-6 is by far the most interesting. On that line, the code hands the number 10 (the number of hours worked) to the findPaymentAmount method. Compare this with the earlier call to findPaymentAmount — the call for the full-time employee in the first half of Listing 8-6. Between the two subclasses, FullTimeEmployee and PartTimeEmployee, are two different findPaymentAmount methods. The two methods have two different kinds of parameter lists:

- The FullTimeEmployee class's findPaymentAmount method takes no parameters (refer to Listing 8-3).
- The PartTimeEmployee class's findPaymentAmount method takes one int parameter (refer to Listing 8-5).

This is par for the course. Finding the payment amount for a part-time employee isn't the same as finding the payment amount for a full-time employee. A part-time employee's pay changes each week, depending on the number of hours the employee works in a week. The full-time employee's pay stays the same each week. So, the FullTimeEmployee and PartTimeEmployee classes both have findPaymentAmount methods, but each class's method works quite differently.

TRY IT OUT

Yes, I have some things for you to try:

BUY OR RENT

Previously in this chapter, you create instances of your own PlaceToLive class and display information about those instances. Create two subclasses of your PlaceToLive class: a House class and an Apartment class. Each House object has a mortgage cost (a monthly amount) and a property tax cost (a yearly amount). Each Apartment object has a rental cost (a monthly amount).

A separate DisplayThePlaces class creates some houses and some apartments. For each house or apartment, your DisplayThePlaces class displays the total cost per square foot (or square meter) and the total cost per bedroom, both calculated monthly.

TAX BREAKS

In Chapter 7, you create an `Organization` class. Each instance of your `Organization` class has a name, an annual revenue amount, and a `boolean` value indicating whether the organization is or is not a profit-making organization.

Create a new `Organization_2` class. Each instance of this new class has only a name and an annual revenue amount. Create two subclasses: a `ProfitMakingOrganization` class and a `NonProfitOrganization` class. A profit-making organization pays 10 percent of its revenue in tax, but a nonprofit organization pays only 2 percent of its revenue in tax.

Make a separate class that creates `ProfitMakingOrganization` instances and `NonProfitOrganization` instances while also displaying information about each instance, including the amount of tax the organization pays.

Changing the Payments for Only Some of the Employees

Wouldn't you know it! Some knucklehead in the human resources department offered double pay for overtime to one of your part-time employees. Now word is getting around, and some of the other part-timers want double pay for their overtime work. If this keeps up, you'll end up in the poorhouse, so you need to send out a memo to all the part-time employees, explaining why earning more money is not to their benefit.

In the meantime, you have two kinds of part-time employees — the ones who receive double pay for overtime hours and the ones who don't — so you need to modify your payroll software. What are your options?

- Well, you can dig right into the `PartTimeEmployee` class code, make a few changes, and hope for the best. (Not a good idea!)
- You can follow the previous section's advice and create a subclass of the existing `PartTimeEmployee` class. "But wait," you say. "The existing `PartTimeEmployee` class already has a `findPaymentAmount` method. Do I need some tricky way of bypassing this existing `findPaymentAmount` method for each double-pay-for-overtime employee?"

At this point, you can thank your lucky stars that you're doing object-oriented programming in Java. With object-oriented programming, you can create a subclass that overrides the functionality of its parent class. Listing 8-7 has just such a subclass.

LISTING 8-7: Yet Another Subclass

```
package com.example.payroll;

public class PartTimeWithOver extends PartTimeEmployee {

    @Override
    public double findPaymentAmount(int hours) {

        if (hours <= 40) {
            return getHourlyRate() * hours;
        } else {
            return getHourlyRate() * 40 + getHourlyRate() * 2 * (hours - 40);
        }
    }
}
```

Figure 8-8 shows the relationship between the code in Listing 8-7 and other pieces of code in this chapter. In particular, PartTimeWithOver is a subclass of a subclass. In object-oriented programming, a chain of this kind is not the least bit unusual. In fact, as subclasses go, this chain is rather short.

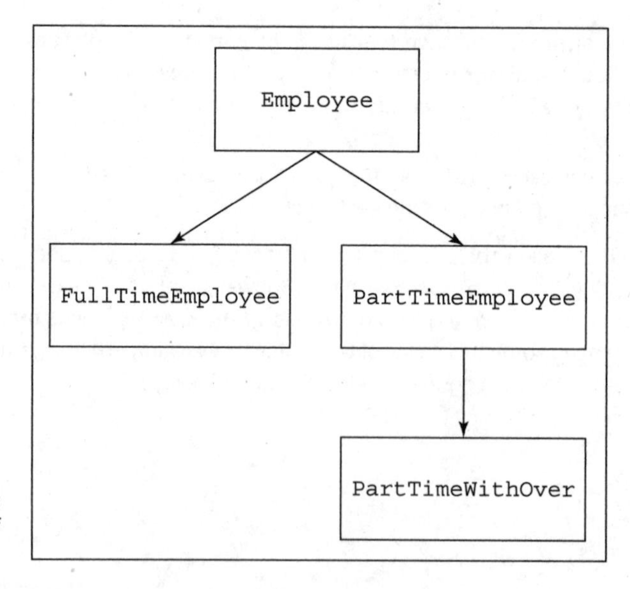

FIGURE 8-8: A tree of classes.

The PartTimeWithOver class extends the PartTimeEmployee class, but PartTimeWithOver picks and chooses what it wants to inherit from the PartTimeEmployee class. Because PartTimeWithOver has its own declaration for the findPaymentAmount method, the PartTimeWithOver class doesn't inherit a findPaymentAmount method from its parent. (See Figure 8-9.)

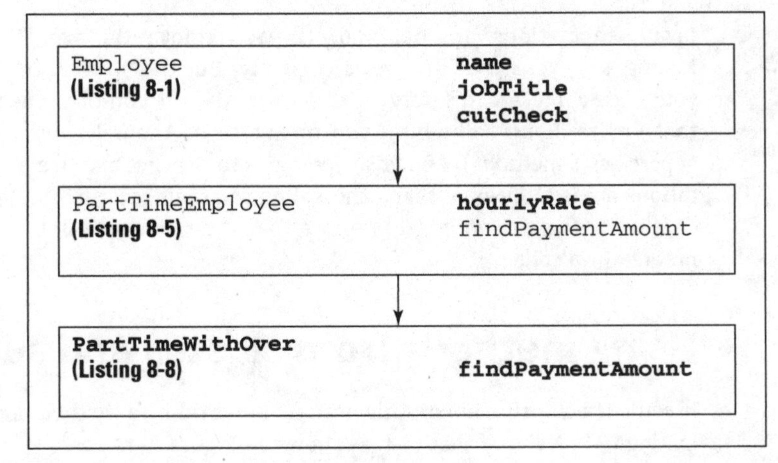

FIGURE 8-9: Method findPaymentAmount isn't inherited.

According to the official terminology, the PartTimeWithOver class *overrides* its parent class's findPaymentAmount method. If you create an object from the PartTimeWithOver class, that object has the name, jobTitle, hourlyRate, and cutCheck of the PartTimeEmployee class, but the object has the findPaymentAmount method that's defined in Listing 8-7.

A Java annotation

The word @Override in Listing 8-7 is an example of an *annotation*. A Java annotation tells your computer something about your code. In particular, the @Override annotation in Listing 8-7 tells the Java compiler to be on the lookout for a common coding error. The annotation says, "Make sure that the method immediately following this annotation has the same stuff (the same name, the same parameters, and so on) as one of the methods in the superclass. If not, then display an error message."

So if I accidentally type

```
public double findPaymentAmount(double hours) {
```

instead of `int hours` as in Listings 8-5 and 8-7, the compiler reminds me that my new `findPaymentAmount` method doesn't really override anything that's in Listing 8-5.

Java has other kinds of annotations (such as `@Deprecated` and `@SuppressWarnings`). You can read a bit about the `@SuppressWarnings` annotation in Chapter 9.

TECHNICAL STUFF

Java's annotations are optional. If you remove the word `@Override` from Listing 8-7, your code still runs correctly. But the `@Override` annotation gives your code some added safety. With `@Override`, the compiler checks to make sure that you're doing something you intend to do (namely, overriding one of the superclass's methods). And with apologies to George Orwell, some types of annotations are less optional than others. You can omit certain annotations from your code only if you're willing to replace the annotation with lots and lots of unannotated Java code.

Using methods from classes and subclasses

If you need clarification on this notion of overriding a method, look at the code in Listing 8-8. A run of that code is shown in Figure 8-10.

```
Pay to the order of Barry Burd (CEO) ***$4,500.00
Pay to the order of Chris Apelian (Computer Book Author) ***$376.50
Pay to the order of Steve Surace (Driver) ***$451.80
```

FIGURE 8-10: Running the code of Listing 8-8.

LISTING 8-8: Testing the Code from Listing 8-7

```
package com.example.payroll;

public class PayrollForThree {

    public static void main(String[] args) {

        var ftEmployee = new FullTimeEmployee();

        ftEmployee.setName("Barry Burd");
        ftEmployee.setJobTitle("CEO");
        ftEmployee.setWeeklySalary(5000.00);
        ftEmployee.setBenefitDeduction(500.00);
        ftEmployee.cutCheck(ftEmployee.findPaymentAmount());
```

```
        var ptEmployee = new PartTimeEmployee();

        ptEmployee.setName("Chris Apelian");
        ptEmployee.setJobTitle("Computer Book Author");
        ptEmployee.setHourlyRate(7.53);
        ptEmployee.cutCheck(ptEmployee.findPaymentAmount(50));

        PartTimeWithOver ptoEmployee = new PartTimeWithOver();

        ptoEmployee.setName("Steve Surace");
        ptoEmployee.setJobTitle("Driver");
        ptoEmployee.setHourlyRate(7.53);
        ptoEmployee.cutCheck(ptoEmployee.findPaymentAmount(50));
    }
}
```

The code in Listing 8-8 writes checks to three employees. The first employee is a full-timer. The second is a part-time employee who hasn't yet gotten wind of the overtime payment scheme. The third employee knows about the overtime payment scheme and demands a fair wage.

With the subclasses, all three of these employees coexist in Listing 8-8. Sure, one subclass comes from the old `PartTimeEmployee` class, but that doesn't mean you can't create an object from the `PartTimeEmployee` class. In fact, Java is smart about this. Listing 8-8 has three calls to the `findPaymentAmount` method, and each call reaches out to a different version of the method:

- In the first call, `ftEmployee.findPaymentAmount`, the `ftEmployee` variable is an instance of the `FullTimeEmployee` class. So the method that's called is the one in Listing 8-3.
- In the second call, `ptEmployee.findPaymentAmount`, the `ptEmployee` variable is an instance of the `PartTimeEmployee` class. So, the method that's called is the one in Listing 8-5.
- In the third call, `ptoEmployee.findPaymentAmount`, the `ptoEmployee` variable is an instance of the `PartTimeWithOver` class. So, the method that's called is the one in Listing 8-7.

This code is fantastic. It's clean, elegant, and efficient. With all the money that you save on software, you can afford to pay everyone double for overtime hours. (Whether you do that or keep the money for yourself is another story.)

TRY IT OUT

Here are some things for you to try.

PAY MORE AND MORE

In previous sections, you create House and Apartment subclasses of your PlaceTo-Live class. Create an ApartmentWithFees subclass of your Apartment class. In addition to the monthly rental price, someone living in an ApartmentWithFees pays a fixed amount every quarter (every three months). Create a separate class that displays the monthly cost of living in a House instance, an Apartment instance, and an ApartmentWithFees instance.

VIRTUAL METHODS

Create a project with four Java source files. Each file defines a particular Java class.

```
public class Main {

    public static void main(String[] args) {
        MyThing myThing, myThing2;

        myThing = new MySubThing();
        myThing2 = new MyOtherThing();

        myThing.value = 7;
        myThing2.value = 44;
        myThing.display();
        myThing2.display();
    }
}

public class MyThing {
    int value;

    public void display() {
        System.out.println("In MyThing, value is " + value);
    }
}

public class MySubThing extends MyThing {

    @Override
    public void display() {
        System.out.println("In MySUBThing, value is " + value);
    }
```

```
}

public class MyOtherThing extends MyThing {

    @Override
    public void display() {
        System.out.println("In MyOTHERThing, value is " + value);
    }
}
```

What output do you see when you run the `Main.java` file's code? What does this output tell you about variable declarations and method calling in Java?

» **Inheriting constructors from a parent class**

» **Using Java's default constructor features**

» **Constructing a simple GUI from scratch**

Chapter 9

Constructing New Objects

Ms. Jennie Burd

121 Schoolhouse Lane

Anywhere, Kansas

Dear Ms. Burd,

In response to your letter of June 21, I believe I can say with complete assurance that objects are not created spontaneously from nothing. Although I've never actually seen an object being created (and no one else in this office can claim to have seen an object in its moment of creation), I have every confidence that some process or another is responsible for the building of these interesting and useful thingamajigs. We here at ObjectsAndClasses.com support the unanimous opinions of both the scientific community and the private sector in matters of this nature. Furthermore, we agree with the recent finding of a blue ribbon panel that concludes, beyond any doubt, that spontaneous object creation would impede the present economic outlook.

Please be assured that I have taken all steps necessary to ensure the safety and well-being of you, our loyal customer. If you have any further questions, please do not hesitate to contact our complaint department. The department's manager is Mr. Blake Wholl. You can contact him by visiting our company's website.

Once again, let me thank you for your concern, and I hope you continue to patronize ObjectsAndClasses.com.

Yours truly,

Mr. Scott Brickenchicker

The one who couldn't get on the elevator in Chapter 4

Defining Constructors (What It Means to Be a Temperature)

Here's a statement that creates an object:

```
Account myAccount = new Account();
```

I know it works — I got it from one of my own examples in Chapter 7. Anyway, in Chapter 7 I say, "[W]hen Java executes `new Account()`, you're creating an object by calling the `Account` class's constructor." What does this pithy sentence mean?

Well, the keyword `new` tells Java to create an object — an instance of a class. Java responds by performing certain actions. For starters, Java finds a place in its memory to store information about the new object. If the object has fields, the fields should eventually have meaningful values.

CROSS REFERENCE

To find out about fields, see Chapter 7.

When you ask Java to create a new object, you may want to specify what's placed in the object's fields. And what if you're interested in doing more than filling fields? Perhaps, when Java creates a new object, you have a whole list of jobs for Java to carry out. For instance, when Java creates a new window object, you want Java to realign the sizes of all buttons in that window.

Creating a new object can involve all kinds of tasks, so in this chapter you create constructors. A *constructor* tells Java to perform a new object's start-up tasks.

What is a temperature?

"Good morning, and welcome to Object News. The local temperature in your area is a pleasant 73 degrees Fahrenheit."

Each temperature consists of two parts: a number and a temperature scale. A number is just a `double` value, such as 32.0 or 70.52. But what's a temperature scale? Is it a string of characters, like `"Fahrenheit"` or `"Celsius"`? Not really, because some strings aren't temperature scales. There's no `"Quelploof"` temperature scale, and a program that can display the temperature `"73 degrees Quelploof"` is a bad program. So how can you limit the temperature scales to the small number of scales that people use? One way to do it is with Java's `enum` type.

What is a temperature scale? (Java's enum type)

Java provides lots of ways for you to group things together. In Chapter 11, you group things to form an array. And in Chapter 12, you group things to form a collection. In this chapter, you group things into an `enum` type. (Of course, you can't group anything unless you can pronounce `enum`. The word *enum* is pronounced "ee-NOOM," like the first two syllables of the word *enumeration.*)

Creating a complicated `enum` type isn't easy, but to create a simple `enum` type, just write a bunch of words inside a pair of curly braces. Listing 9-1 defines an `enum` type. The name of the `enum` type is `TempScale`.

LISTING 9-1: The TempScale Type (an enum Type)

```
package com.example.weather;

public enum TempScale {
    CELSIUS, FAHRENHEIT, KELVIN, RANKINE,
    NEWTON, DELISLE, RÉAUMUR, RØMER, LEIDEN
}
```

WARNING

In Listing 9-1, I'm showing off my physics prowess by naming not two, not four, but *nine* different temperature scales. Some readers' computers have trouble with the special characters in the words `RÉAUMUR` and `RØMER`. If you're one of those readers, simply delete the words `RÉAUMUR` and `RØMER` from the code. I promise: It won't mess up the example.

When you define an enum type, two important things happen:

- **You create values.**

  Just as 13 and 151 are int values, CELSIUS and FAHRENHEIT are TempScale values.

- **You can create variables to refer to those values.**

  In Listing 9-2, I declare the fields number and scale. Just as

  ```
  double number;
  ```

  declares that a number variable is of type double,

  ```
  TempScale scale;
  ```

  declares variable scale to be of type TempScale.

  "To be of type TempScale" means that you can have values CELSIUS, FAHRENHEIT, KELVIN, and so on. So, in Listing 9-2, I can give the scale variable the value FAHRENHEIT (or TempScale.FAHRENHEIT, to be more precise).

TECHNICAL STUFF

An enum type is a Java class in disguise. That's why Listing 9-1 contains an entire file devoted to one thing — namely, the declaration of an enum type (the TempScale type). Like the declaration of a class, an enum type declaration belongs in a file all its own. The code in Listing 9-1 belongs in a file named TempScale.java.

Okay, so then what is a temperature?

Each temperature consists of two things: a number and a temperature scale. The code in Listing 9-2 makes this fact abundantly clear.

LISTING 9-2: The Temperature Class

```
package com.example.weather;

public class Temperature {
    private double number;

    private TempScale scale;

    public Temperature() {
        number = 0.0;
        scale = TempScale.FAHRENHEIT;
    }
```

```
    public Temperature(double number) {
        this.number = number;
        scale = TempScale.FAHRENHEIT;
    }

    public Temperature(TempScale scale) {
        number = 0.0;
        this.scale = scale;
    }

    public Temperature(double number, TempScale scale) {
        this.number = number;
        this.scale = scale;
    }

    public void setNumber(double number) {
        this.number = number;
    }

    public double getNumber() {
        return number;
    }

    public void setScale(TempScale scale) {
        this.scale = scale;
    }

    public TempScale getScale() {
        return scale;
    }
}
```

The code in Listing 9-2 has the usual setter and getter methods (accessor methods for the number and scale fields).

For some good reading on setter and getter methods (also known as accessor methods), see Chapter 7.

On top of all of that, Listing 9-2 has four other method-like-looking things. Each of these method-like things has the name Temperature, which happens to be the same as the name of the class. None of these Temperature method-like things has a return type of any kind — not even void, which is the cop-out return type.

Each of these method-like things is called a constructor. A *constructor* is like a method, except that a constructor has a special purpose: to create new objects.

REMEMBER

Whenever the computer creates a new object, the computer executes the statements inside a constructor.

You can omit the word `public` in the first lines of Listings 9-1 and 9-2. If you omit `public`, other Java programs might not be able to use the features defined in the `TempScale` type and in the `Temperature` class. (Don't worry about the programs in this chapter: With or without the word `public`, all programs in this chapter can use the code in Listings 9-1 and 9-2. To find out which Java programs can use classes that aren't public, see Chapter 14.) If you *do* use the word `public` in the first line of Listing 9-1, Listing 9-1 *must* be in a file named `TempScale.java`, starting with a capital letter `T`. And if you *do* use the word `public` in the first line of Listing 9-2, Listing 9-2 *must* be in a file named `Temperature.java`, starting with a capital letter `T`. (For an introduction to public classes, see Chapter 7.)

What you can do with a temperature

Listing 9-3 gives form to some of the ideas that I describe in the preceding section. In Listing 9-3, you call the constructors that are declared earlier, in Listing 9-2. Figure 9-1 shows what happens when you run all this code.

```
70.00 degrees FAHRENHEIT
32.00 degrees FAHRENHEIT
 0.00 degrees CELSIUS
 2.73 degrees KELVIN
```

FIGURE 9-1: Running the code from Listing 9-3.

LISTING 9-3: Using the Temperature Class

```
package com.example.weather;

import static java.lang.System.out;

public class UseTemperature {

    public static void main(String[] args) {
        final String format = "%5.2f degrees %s\n";

        var temp = new Temperature();
        temp.setNumber(70.0);
        temp.setScale(TempScale.FAHRENHEIT);
        out.printf(format, temp.getNumber(), temp.getScale());

        temp = new Temperature(32.0);
        out.printf(format, temp.getNumber(), temp.getScale());
```

```
        temp = new Temperature(TempScale.CELSIUS);
        out.printf(format, temp.getNumber(), temp.getScale());

        temp = new Temperature(2.73, TempScale.KELVIN);
        out.printf(format, temp.getNumber(), temp.getScale());
    }
}
```

In Listing 9-3, each statement of the kind

```
temp = new Temperature(blah,blah,blah);
```

calls one of the constructors from Listing 9-2. So, by the time the code in Listing 9-3 finishes running, it creates four instances of the `Temperature` class. Each instance is created by calling a different constructor from Listing 9-2.

In Listing 9-3, the last of the four constructor calls has two parameters: `2.73` and `TempScale.KELVIN`. This isn't particular to constructor calls. A method call or a constructor call can have a bunch of parameters. You separate one parameter from another with a comma. Another name for "a bunch of parameters" is a *parameter list*.

The only rule you must follow is to match the parameters in the call with the parameters in the declaration. For example, in Listing 9-3, the fourth and last constructor call

```
new Temperature(2.73, TempScale.KELVIN)
```

has two parameters: the first of type `double` and the second of type `TempScale`. Java approves of this constructor call because Listing 9-2 contains a matching declaration. That is, the header

```
public Temperature(double number, TempScale scale)
```

has two parameters: the first of type `double` and the second of type `TempScale`. If a `Temperature` constructor call in Listing 9-3 had no matching declaration in Listing 9-2, Listing 9-3 would crash and burn. (To state things more politely, Java would display errors when you tried to compile the code in Listing 9-3.)

By the way, this business about multiple parameters isn't new. Over in Chapter 6, I write `keyboard.findWithinHorizon(".",0).charAt(0)`. In that line, the method call `findWithinHorizon(".",0)` has two parameters: a string and an `int` value. Luckily for me, the Java API has a method declaration for `findWithinHorizon` — a declaration whose first parameter is a string and whose second parameter is an `int` value.

HOW TO CHEAT: ENUM TYPES AND SWITCHES

Listings 9-2 and 9-3 contain long-winded names such as TempScale.FAHRENHEIT and TempScale.CELSIUS. Names such as FAHRENHEIT and CELSIUS belong to my TempScale type (the type defined in Listing 9-1). These names have no meaning outside of my TempScale context. (If you think I'm being egotistical with this "no meaning outside of my context" remark, try deleting the TempScale. part of TempScale.FAHRENHEIT in Listing 9-2. Suddenly, Java tells you that your code contains an error.)

Java is normally fussy about type names and dots. But when they created enum types, the makers of Java decided that enum types in switch statements and expressions deserved special treatment. You can use an enum value to decide which case to execute in a switch statement or switch expression. When you do this, you don't use the enum type name in the case expressions. For example, the following Java code is correct:

```
TempScale scale = TempScale.RANKINE;
char letter =
        switch (scale) {
            case CELSIUS -> 'C';
            case KELVIN -> 'K';
            case RANKINE, RÉAUMUR, RØMER -> 'R';
            default -> 'X';
        };
```

In the first line of code, I write TempScale.RANKINE because this first line isn't inside a switch. But in the next several lines of code, I write case CELSIUS, case KELVIN, and case RANKINE without the word TempScale. In fact, if I create a case clause by writing case TempScale.RANKINE, Java complains with a loud, obnoxious error message.

Constructing a temperature; a slow-motion replay

When the computer executes one of the new Temperature statements in Listing 9-3, the computer has to decide which of the constructors in Listing 9-2 to use. The computer decides by looking at the parameter list — the stuff in parentheses after the words new Temperature. For instance, when the computer executes

```
temp = new Temperature(32.0);
```

from Listing 9-3, the computer says to itself, "The number 32.0 in parentheses is a double value. One of the Temperature constructors in Listing 9-2 has just one parameter with type double. The constructor's header looks like this:

```
public Temperature(double number)
```

"So I guess I'll execute the statements inside that particular constructor." The computer goes on to execute the following statements:

```
this.number = number;
scale = TempScale.FAHRENHEIT;
```

As a result, you get a brand-new object whose `number` field has the value `32.0` and whose `scale` field has the value `TempScale.FAHRENHEIT`.

In the two lines of code, you have two statements that set values for the fields `number` and `scale`. Take a look at the second of these statements, which is a bit easier to understand. The second statement sets the new object's `scale` field to `TempScale.FAHRENHEIT`. You see, the constructor's parameter list is `(double number)`, and that list doesn't include a `scale` value. So whoever programmed this code had to make a decision about what value to use for the `scale` field. The programmer could have chosen `FAHRENHEIT` or `CELSIUS`, but they could also have chosen `KELVIN`, `RANKINE`, or any of the other obscure scales named in Listing 9-1. (This programmer happens to live in New Jersey, in the United States, where people commonly use the old Fahrenheit temperature scale.)

Marching back to the first of the two statements, this first statement assigns a value to the new object's `number` field. The statement uses a cute trick that you can see in many constructors (and in other methods that assign values to objects' fields). To understand the trick, take a look at Listing 9-4. The listing shows you two ways that I could have written the same constructor code.

LISTING 9-4: Two Ways to Accomplish the Same Thing

```
//Use this constructor...

    public Temperature(double whatever) {
        number = whatever;
        scale = TempScale.FAHRENHEIT;
    }

//... or use this constructor...

    public Temperature(double number) {
        this.number = number;
        scale = TempScale.FAHRENHEIT;
    }

//... but don't put both constructors in your code.
```

Listing 9-4 has two constructors in it. In the first constructor, I use two different names: `number` and `whatever`. In the second constructor, I don't need two names. Rather than make up a new name for the constructor's parameter, I reuse an existing name by writing `this.number`.

Here's what's going on in Listing 9-2:

» In the statement `this.number = number`, the name *this.number* refers to the new object's `number` field — the field that's declared near the top of Listing 9-2. (See Figure 9-2.)

  In the statement `this.number = number`, *number* (on its own, without `this`) refers to the constructor's parameter. (Again, see Figure 9-2.)

In general, `this.`*`someName`* refers to a field belonging to the object that contains the code. In contrast, plain old *`someName`* refers to the closest place where *`some-Name`* happens to be declared. In the statement `this.number = number` (refer to Listing 9-2), that closest place happens to be the `Temperature` constructor's parameter list.

WHAT'S THIS ALL ABOUT?

Suppose that your code contains a constructor — the first of the two constructors in Listing 9-4. The `whatever` parameter is passed a number like 32.0, for instance. Then the first statement in the constructor's body assigns that value, `32.0`, to the new object's `number` field. The code works. But in writing this code, you had to make up a new name for a parameter — the name *whatever.* And the only purpose for this new name is to hand a value to the object's `number` field. What a waste! To distinguish between the parameter and the `number` field, you gave a name to something that was just momentary storage for the `number` value.

Making up names is an art, not a science. I've gone through plenty of naming phases. Years ago, whenever I needed a new name for a parameter, I picked a confusing misspelling of the original variable name. (I'd name the parameter something like `numbr` or `nuhmber`.) I've also tried changing a variable name's capitalization to come up with a parameter name. (I'd use parameter names like `Number` or `nUMBER`.) In Chapter 8, I name all my parameters by adding the suffix *In* to their corresponding variable names. (The `jobTitle` variable matched up with the `jobTitleIn` parameter.) None of these naming schemes works well — I can never remember the quirky new names I've created. The good news is that this parameter-naming effort isn't necessary. You can give the parameter the same name as the variable. To distinguish between the two, you use the Java keyword `this`.

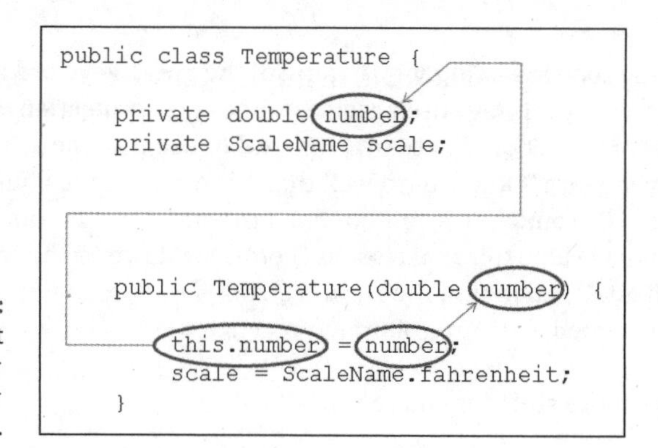

public class Temperature {
private double number;
private ScaleName scale;
public Temperature(double number) {
this.number = number;
scale = ScaleName.fahrenheit;
}

FIGURE 9-2: What this.number and number mean.

Some things never change

Chapter 7 introduces the printf method and explains that each printf call starts with a format string. The format string describes the way the other parameters are to be displayed.

In previous examples, this format string is always a quoted literal. For instance, the first printf call in Listing 7-7 (see Chapter 7) is

```
out.printf("$%4.2f\n", myInterest);
```

In Listing 9-3, I break with tradition and begin the printf call with a variable that I name *format:*

```
out.printf(format, temp.getNumber(), temp.getScale());
```

That's okay as long as my format variable is of type String. And indeed, in Listing 9-3, the first variable declaration is

```
final String format = "%5.2f degrees %s\n";
```

In this declaration of the format variable, take special note of the word final. This Java keyword indicates that the value of format can't be changed. If I add another assignment statement to Listing 9-3:

```
format = "%6.2f (%s)\n";
```

the compiler barks back at me with the message cannot assign a value to final variable.

When I write the code in Listing 9-3, the use of the `final` keyword isn't absolutely necessary. But the `final` keyword provides some extra protection. When I initialize `format` to `"%5.2f degrees %s\n"`, I intend to use this same format just as it is, over and over again. I know darn well that I don't intend to change the `format` variable's value. Of course, in a 10,000-line program, I can become confused and try to assign a new value to `format` somewhere deep down in the code. To prevent me from accidentally changing the `format` string, I declare the `format` variable to be final. It's just good, safe programming practice.

TRY IT OUT

There's always more stuff for you to try.

SCHOOL DAYS

Create a `Student` class with a name, an ID number, a grade point average (GPA), and a major area of study. The student's name is a `String`. The student's ID number is an `int` value. The GPA is a `double` value between 0.0 and 4.0. The `Major` is an `enum` type, with values such as `COMPUTER_SCIENCE`, `MATHEMATICS`, `LITERATURE`, `PHYSICS`, and `HISTORY`.

Every student has a name and an ID number, but a brand-new student might not have a GPA or a major. Create constructors with and without GPA and `Major` parameters.

As usual, create a separate class that makes use of your new `Student` class.

FLIGHT OF FANCY

Create an `AirplaneFlight` class with a flight number, a departure airport, the time of departure, an arrival airport, and a time of arrival. The flight number is an `int` value. The departure and arrival airport fields belong to an `Airport` enum type, with values corresponding to some of the official IATA airport codes. (For example, London Heathrow Airport's code is LHR; Los Angeles International Airport's code is LAX; check out `www.iata.org/publications/Pages/code-search.aspx` for a searchable database of airline codes.)

For the times of arrival and departure, use Java's `LocalTime` class. (For more on `LocalTime`, check out the `LocalTime` documents page at `https://docs.oracle.com/en/java/javase/17/docs/api/java.base/java/time/LocalTime.html`.) To create a `LocalTime` object that's set to 2:15 P.M. (also known as 14:15), execute

```
LocalTime twoFifteen = LocalTime.of(14, 15);
```

To create a `LocalTime` object that's set to the current time (according to the computer's system clock), execute

```
LocalTime currentTime = LocalTime.now();
```

Every flight has a number, a departure airport, and an arrival airport. But some flights might not have departure and arrival times. Create constructors with and without departure and arrival time parameters.

Create a separate class that makes use of your new `AirplaneFlight` class.

MAKE A HIT RECORD

Newer versions of Java (from Java 16 onward) have a fancy feature called record classes. For an introduction to these beauties, name two files `TemperatureRecord.java` and `UseTemperatureRecord.java`. Put the following code in these files and then give the code a spin:

```
//TemperatureRecord.java

package com.example.weather;

public record TemperatureRecord(double number, TempScale scale) {
}

//UseTemperatureRecord.java

package com.example.weather;

import static java.lang.System.out;

public class UseTemperatureRecord {

    public static void main(String[] args) {
        final String format = "%5.2f degrees %s\n";

        TemperatureRecord temp = new TemperatureRecord(2.73, TempScale.KELVIN);
        out.printf(format, temp.number(), temp.scale());
    }
}
```

Doing Something about the Weather

In Chapter 8, I make a big fuss over the notion of subclasses. That's the right thing to do. Subclasses make code reusable, and reusable code is good code. With that in mind, it's time to create a subclass of the `Temperature` class (which I develop in this chapter's first section).

Building better temperatures

After perusing the code in Listing 9-3, you decide that the responsibility for displaying temperatures has been seriously misplaced. Listing 9-3 has several tedious repetitions of the lines to print temperature values. A 1970s programmer would tell you to collect those lines into one place and turn them into a method. (The 1970s programmer wouldn't have used the word *method*, but that's not important right now.) Collecting lines into methods is fine, but with today's object-oriented programming methodology, you think in broader terms. Why not get each `temperature` object to take responsibility for displaying itself? After all, if you develop a `display` method, you probably want to share the method with other people who use temperatures. So put the method right inside the declaration of a `temperature` object. That way, anyone who uses the code for temperatures has easy access to your `display` method.

Now replay the tape from Chapter 8. "Blah, blah, blah . . . don't want to modify existing code . . . blah, blah, blah . . . too costly to start again from scratch . . . blah, blah, blah . . . extend existing functionality." It all adds up to one thing:

> Don't abuse it. Instead, reuse it.

So you decide to create a subclass of the `Temperature` class — the class defined in Listing 9-2. Your new subclass complements the `Temperature` class's functionality by having methods to display values in a nice, uniform fashion. The new class, `TemperatureNice`, is shown in Listing 9-5.

LISTING 9-5: The TemperatureNice Class

```
package com.example.weather;

import static java.lang.System.out;

public class TemperatureNice extends Temperature {

    public TemperatureNice() {
        super();
    }
```

```
    public TemperatureNice(double number) {
        super(number);
    }

    public TemperatureNice(TempScale scale) {
        super(scale);
    }

    public TemperatureNice(double number, TempScale scale) {
        super(number, scale);
    }

    public void display() {
        out.printf("%5.2f degrees %s\n", getNumber(), getScale());
    }
}
```

In the `display` method of Listing 9-5, notice the calls to the `Temperature` class's `getNumber` and `getScale` methods. Why do I do this? Well, inside the `TemperatureNice` class's code, any direct references to the `number` and `scale` fields would generate error messages. It's true that every `TemperatureNice` object has its own `number` and `scale` fields. (After all, `TemperatureNice` is a subclass of the `Temperature` class, and the code for the `Temperature` class defines the `number` and `scale` fields.) But because number and scale are declared to be private inside the `Temperature` class, only code that's right inside the `Temperature` class can directly use these fields.

Don't put additional declarations of the `number` and `scale` fields inside the `TemperatureNice` class's code. If you do, you inadvertently create four different variables (two called `number` and another two called `scale`). You'll assign values to one pair of variables. Then you'll be shocked that when you display the other pair of variables, those values seem to have disappeared.

When an object's code contains a call to one of the object's own methods, you don't need to preface the call with a dot. For instance, in the last statement of Listing 9-5, the object calls its own methods with `getNumber()` and `getScale()`, not with *`someObject`*`.getNumber()` and `somethingOrOther.getScale()`. If going dotless makes you queasy, you can compensate by taking advantage of yet another use for the `this` keyword: Just write `this.getNumber()` and `this.getScale()` in the last line of Listing 9-5.

Constructors for subclasses

By far, the biggest news in Listing 9-5 is the way the code declares constructors. The `TemperatureNice` class has four of its own constructors. If you've gotten in gear thinking about subclass inheritance, you may wonder why these constructor declarations are necessary. Doesn't `TemperatureNice` inherit the parent `Temperature` class's constructors? No, subclasses don't inherit constructors.

REMEMBER

Subclasses don't inherit constructors.

That's right. Subclasses don't inherit constructors. In one oddball case, a constructor may look like it's being inherited, but that oddball situation is a fluke, not the norm. In general, when you define a subclass, you declare new constructors to go with the subclass.

I describe the oddball case (in which a constructor looks like it's being inherited) later in this chapter, in the section "The default constructor."

So the code in Listing 9-5 has four constructors. Each constructor has the name `TemperatureNice`, and each constructor has its own uniquely identifiable parameter list. That's the boring part. The interesting part is that each constructor makes a call to something named `super`, which is a Java keyword.

In Listing 9-5, `super` stands for a constructor in the parent class:

- The statement `super()` in Listing 9-5 calls the parameterless `Temperature()` constructor that's in Listing 9-2. That parameterless constructor assigns `0.0` to the `number` field and `TempScale.FAHRENHEIT` to the `scale` field.
- The statement `super(number, scale)` in Listing 9-5 calls the constructor `Temperature(double number, TempScale scale)` that's in Listing 9-2. In turn, the constructor assigns values to the `number` and `scale` fields.
- In a similar way, the statements `super(number)` and `super(scale)` in Listing 9-5 call constructors from Listing 9-2.

The computer decides which of the `Temperature` class's constructors is being called by looking at the parameter list after the word `super`. For instance, when the computer executes

```
super(number, scale);
```

from Listing 9-5, the computer says to itself, "The `number` and `scale` fields in parentheses have types `double` and `TempScale`. But only one of the `Temperature` constructors in Listing 9-2 has two parameters with types `double` and `TempScale`. The constructor's header looks like this:

```
public Temperature(double number, TempScale scale)
```

"So I guess I'll execute the statements inside that particular constructor."

Using all this stuff

In Listing 9-5, I define what it means to be in the `TemperatureNice` class. Now it's time to put this `TemperatureNice` class to good use. Listing 9-6 has code that uses `TemperatureNice`.

LISTING 9-6: Using the TemperatureNice Class

```
package com.example.weather;

public class UseTemperatureNice {

    public static void main(String[] args) {

        var temp = new TemperatureNice();
        temp.setNumber(70.0);
        temp.setScale(TempScale.FAHRENHEIT);
        temp.display();

        temp = new TemperatureNice(32.0);
        temp.display();

        temp = new TemperatureNice(TempScale.CELSIUS);
        temp.display();

        temp = new TemperatureNice(2.73, TempScale.KELVIN);
        temp.display();
    }
}
```

The code in Listing 9-6 is much like its cousin code in Listing 9-3. The big differences are described here:

- Listing 9-6 creates instances of the `TemperatureNice` class. That is, Listing 9-6 calls constructors from the `TemperatureNice` class, not the `Temperature` class.

» Listing 9-6 takes advantage of the `display` method in the `TemperatureNice` class. So the code in Listing 9-6 is much tidier than its counterpart in Listing 9-3.

A run of Listing 9-6 looks exactly like a run of the code in Listing 9-3 — it just reaches the finish line in a far more elegant fashion. (The run is shown previously, in Figure 9-1.)

The default constructor

The main message in the previous section is that subclasses don't inherit constructors. So, what gives with all the listings over in Chapter 8? In Listing 8-6, a statement says

```
FullTimeEmployee ftEmployee = new FullTimeEmployee();
```

But here's the problem: The code defining `FullTimeEmployee` (refer to Listing 8-3) doesn't seem to have any constructors declared inside it. So, in Listing 8-6, how can you possibly call the `FullTimeEmployee` constructor?

Here's what's going on. When you create a subclass and don't put any explicit constructor declarations in your code, Java creates one constructor for you. It's called a *default constructor.* If you're creating the `public FullTimeEmployee` subclass, the default constructor looks like the one in Listing 9-7.

LISTING 9-7: A Default Constructor

```
public FullTimeEmployee() {
    super();
}
```

The constructor in Listing 9-7 takes no parameters, and its single statement calls the constructor of whatever class you're extending. (Woe be to you if the class you're extending has no parameterless constructor.)

You've just read about default constructors, but watch out! Notice one thing that this talk about default constructors *doesn't* say: It doesn't say that you always get a default constructor. In particular, if you create a subclass and define any constructors yourself, Java doesn't add a default constructor for the subclass (and the subclass doesn't inherit any constructors, either).

How can this trip you up? Listing 9-8 has a copy of the code from Listing 8-3, but with one constructor added to it. Take a look at this modified version of the `FullTimeEmployee` code.

LISTING 9-8: Look, I Have a Constructor!

```
package com.example.payroll;

public class FullTimeEmployee extends Employee {
    private double weeklySalary;
    private double benefitDeduction;

    public FullTimeEmployee(double weeklySalary) {
        this.weeklySalary = weeklySalary;
    }

    public void setWeeklySalary(double weeklySalaryIn) {
        weeklySalary = weeklySalaryIn;
    }

    public double getWeeklySalary() {
        return weeklySalary;
    }

    public void setBenefitDeduction(double benefitDedIn) {
        benefitDeduction = benefitDedIn;
    }

    public double getBenefitDeduction() {
        return benefitDeduction;
    }

    public double findPaymentAmount() {
        return weeklySalary - benefitDeduction;
    }
}
```

If you use the `FullTimeEmployee` code in Listing 9-8, a line like the following doesn't work:

```
FullTimeEmployee ftEmployee = new FullTimeEmployee();
```

It doesn't work because, having declared a `FullTimeEmployee` constructor that takes one `double` parameter, you no longer get a default parameterless constructor for free.

What do you do about this? If you declare any constructors, declare all constructors that you'll possibly need. Take the constructor in Listing 9-7 and add it to the code in Listing 9-8. Then the call `new FullTimeEmployee()` starts working again.

Under certain circumstances, Java automatically adds an invisible call to a parent class's constructor at the top of a constructor body. This automatic addition of a `super` call is a tricky bit of business that doesn't appear often, so when it does appear, it may seem quite mysterious. For more information, see this book's website (`http://javafordummies.allmycode.com`).

TRY IT OUT

In this section, I have four (count 'em — *four*) experiments for you to try:

STUDENT SHOWCASE

In a previous section, you create your own `Student` class. Create a subclass that has a method named `getString`.

Like the `display` method in this chapter's `TemperatureNice` class, the `getString` method creates a nice-looking `String` representation of its object. But unlike the `TemperatureNice` class's `display` method, the `getString` method doesn't print that `String` representation on the screen. Instead, the `getString` method simply returns that `String` representation as its result.

In a way, a `getString` method is much more versatile than a `display` method. With a `display` method, all you can do is show a `String` representation on the screen. But with a `getString` method, you can create a `String` representation and then do whatever you want with it.

Create a separate class that creates some instances of your new subclass and puts their `getString` methods to good use.

THE WAITING GAME

In a previous section, you create your own `AirplaneFlight` class. Create a subclass that has a method named `duration`. The `duration` method, which has no parameters, returns the amount of time between the flight's departure time and arrival time.

To find the number of hours between two `LocalTime` objects (such as `twoFifteen` and `currentTime`), execute

```
long hours = ChronoUnit.HOURS.between(twoFifteen, currentTime);
```

To find the number of minutes between two LocalTime objects (such as twoFifteen and currentTime), execute

```
long minutes = ChronoUnit.MINUTES.between(twoFifteen, currentTime);
```

A CONVERSION DIVERSION

Create a new TemperatureEvenNicer class — a subclass of this section's TemperatureNice class. The TemperatureEvenNicer class has a convertTo method. If the variable temp refers to a Fahrenheit temperature and Java executes

```
temp.convertTo(TempScale.CELSIUS);
```

then the temp object changes to a Celsius temperature, with the number converted appropriately. The same kind of thing happens if Java executes

```
temp.convertTo(TempScale.FAHRENHEIT);
```

with temp already referring to a Celsius temperature.

SET A NEW RECORD

Follow up on the "Make a Hit Record" experiment from earlier in this chapter. Name two files TemperatureNiceRecord.java and UseTemperatureNiceRecord.java. Put the following code in these files and see how they run.

```
//TemperatureNiceRecord.java

package com.example.weather;

import static java.lang.System.out;

public record TemperatureNiceRecord(double number, TempScale scale) {

    public TemperatureNiceRecord() {
        this(0, TempScale.CELSIUS);
    }

    public void display() {
        out.printf("%5.2f degrees %s\n", number, scale);
    }
}

//UseTemperatureNiceRecord.java
```

```
package com.example.weather;

public class UseTemperatureNiceRecord {

    public static void main(String[] args) {
        var temp = new TemperatureNiceRecord();
        temp.display();

        temp = new TemperatureNiceRecord(2.73, TempScale.KELVIN);
        temp.display();
    }
}
```

A Constructor That Does More

Here's a quote from somewhere near the start of this chapter: "And what if you're interested in doing more than filling fields? Perhaps, when the computer creates a new object, you have a whole list of jobs for the computer to carry out." Okay, what-if?

This section's example has a constructor that does more than just assign values to fields. The example is in Listings 9-9 and 9-10. The result of running the example's code is shown in Figure 9-3.

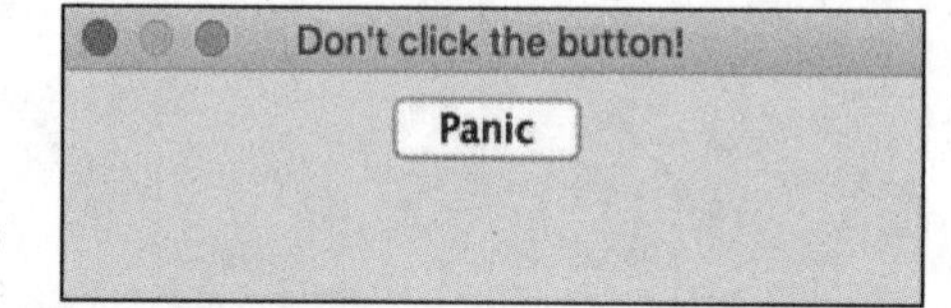

FIGURE 9-3: Don't panic.

LISTING 9-9: Defining a Frame

```
package com.example.graphical;

import javax.swing.JButton;
import javax.swing.JFrame;
import java.awt.FlowLayout;

public class SimpleFrame extends JFrame {
```

```
    public SimpleFrame() {
        setTitle("Don't click the button!");
        setLayout(new FlowLayout());
        setDefaultCloseOperation(EXIT_ON_CLOSE);
        add(new JButton("Panic"));
        setSize(300, 100);
        setVisible(true);
    }
}
```

LISTING 9-10: Displaying a Frame

```
package com.example.graphical;

public class ShowAFrame {

    public static void main(String[] args) {
        new SimpleFrame();
    }
}
```

Like my DummiesFrame examples, the code in Listings 9-9 and 9-10 displays a window on the computer screen. But unlike my DummiesFrame examples, all the method calls in Listings 9-9 and 9-10 refer to methods in Java's standard API (application programming interface).

CROSS REFERENCE

To find my DummiesFrame examples, refer to Chapter 7.

The code in Listing 9-9 contains lots of names that are probably unfamiliar to you — names from Java's API. When I was first becoming acquainted with Java, I foolishly believed that knowing Java meant remembering all these names. Quite the contrary: These names are just carry-on baggage. The real Java is the way the language implements object-oriented concepts.

Anyway, Listing 9-10's main method has only one statement: a call to the constructor in the SimpleFrame class. Notice how the object that this call creates isn't even assigned to a variable. That's okay because the code doesn't need to refer to the object anywhere else.

Up in the SimpleFrame class, there's only one constructor declaration. Far from just setting variables' values, this constructor calls method after method from the Java API.

All the methods called in the SimpleFrame class's constructor come from the parent class, JFrame. The JFrame class lives in the javax.swing package. This package and another package, java.awt, have classes that help you put windows, images, drawings, and other gizmos on a computer screen. (In the java.awt package, the letters *awt* stand for *a*bstract *w*indowing *t*oolkit.)

For a little gossip about the notion of a Java package, see Chapters 7 and 14.

In the Java API, what people normally call a *window* is an instance of the javax.swing.JFrame class.

Classes and methods from the Java API

Looking at Figure 9-3, you can probably tell that an instance of the SimpleFrame class doesn't do much. The frame has only one button, and when you click the button, nothing happens. I made the frame this way to keep the example from becoming too complicated. Even so, the code in Listing 9-9 uses several API classes and methods. The setTitle, setLayout, setDefaultCloseOperation, add, setSize, and setVisible methods all belong to the javax.swing.JFrame class. Here's a list of names used in the code:

- setTitle: Calling setTitle puts words on the frame's title bar. (The new SimpleFrame object is calling its own setTitle method.)
- FlowLayout: An instance of the FlowLayout class positions objects on the frame in a centered, typewriter fashion. Because the frame in Figure 9-3 has only one button on it, that button is centered near the top of the frame. If the frame had eight buttons, five of them may be lined up in a row across the top of the frame and the remaining three would be centered along a second row.
- setLayout: Calling setLayout puts the new FlowLayout object in charge of arranging components, such as buttons, on the frame. (The new SimpleFrame object is calling its own setLayout method.)
- setDefaultCloseOperation: Calling setDefaultCloseOperation tells Java what to do when you click the little × in the frame's upper right corner. (On a Mac, you click the little red circle in the frame's upper left corner.) Without this method call, the frame itself disappears, but the Java virtual machine (JVM) keeps running. To stop your program's run, you have to perform one more step. (You may have to look for a Terminate option in Eclipse, IntelliJ IDEA, or NetBeans.)

  Calling setDefaultCloseOperation(EXIT_ON_CLOSE) tells Java to shut itself down when you click the × in the frame's upper right corner. The alternatives to EXIT_ON_CLOSE are HIDE_ON_CLOSE, DISPOSE_ON_CLOSE, and, my personal favorite, DO_NOTHING_ON_CLOSE. Use one of these alternatives when your program has more work to do after the user closes your frame.

- » JButton: The JButton class lives in the javax.swing package. One of the class's constructors takes a String instance (such as "Panic") for its parameter. Calling this constructor makes that String instance into the label on the face of the new button.
- » add: The new SimpleFrame object calls its add method. Calling the add method places the button on the object's surface (in this case, the surface of the frame).
- » setSize: The frame becomes 300 pixels wide and 100 pixels tall. (In the javax.swing package, whenever you specify two dimension numbers, the width number always comes before the height number.)
- » setVisible: When it's first created, a new frame is invisible. But when the new frame calls setVisible(true), the frame appears on your computer screen.

Live dangerously

Your IDE may warn you that the SimpleFrame in Listing 9-9 has no serialVersionUID field. "And what," you ask, "is a serialVersionUID field?" It's something having to do with storing a JFrame object — something you don't care about. Not having a serialVersionUID field generates a warning, not an error. So throw caution to the wind and ignore the warning.

If, for some reason, you can't ignore the warning, suppress the warning by adding the line @SuppressWarnings("serial") with no semicolon immediately above the public class SimpleFrame line. (Like @Override from Chapter 8, @SuppressWarnings is a Java annotation.)

If, for some other reason, you don't want to suppress the warning, add the statement private static final long serialVersionUID = 1L; immediately below the public class SimpleFrame line.

THE NULL HYPOTHESIS

TRY IT OUT

In JShell, type the following sequence of declarations and statements. What happens? Why?

```
jshell> import javax.swing.JFrame

jshell> JFrame frame
```

```
jshell> frame.setSize(100, 100)

jshell> frame = new JFrame()

jshell> frame.setSize(100, 100)

jshell> frame.setVisible(true)
```

WIDESPREAD PANIC

In Listing 9-9, change the statement

```
setLayout(new FlowLayout());
```

to

```
setLayout(new BorderLayout));
```

What difference does this change make when you run the program?

4 Smart Java Techniques

IN THIS PART . . .

Decide where declarations belong in your Java program.

Deal with bunches of things (bunches of rooms, bunches of sales, and even bunches of bunches).

Fully embrace Java's object-oriented features.

Create a windowed app and respond to mouse clicks.

Work with your favorite database.

entire class

» Putting variables inside and outside methods

» Improving your batting average

Chapter 10

Putting Variables and Methods Where They Belong

Hello, again. You're listening to radio station WWW, and I'm your host, Sam Burd. It's the start again of the big baseball season, and today station WWW brought you live coverage of the Hankees-versus-Socks game. At this moment, I'm awaiting news of the game's final score.

If you remember from earlier this afternoon, the Socks looked like they were going to take those Hankees to the cleaners. Then the Hankees belted ball after ball, giving the Socks a run for their money. Those Socks! I'm glad I wasn't in their shoes.

Anyway, as the game went on, the Socks pulled themselves up. Now the Socks are nose-to-nose with the Hankees. We'll get the final score in a minute, but first, a few reminders. Stay tuned after this broadcast for the big Jerseys game. And don't forget to tune in next week when the Cleveland Gowns play the Bermuda Shorts.

Okay, here's the final score. Which team has the upper hand? Which team will come out a head? And the winner is . . . oh, no — it's a tie!

Defining a Class (What It Means to Be a Baseball Player)

As far as I'm concerned, a baseball player has a name and a batting average. Listing 10-1 puts my belief about baseball players into Java program form.

LISTING 10-1: The Player Class

```
package com.example.baseball;

import java.text.DecimalFormat;

public class Player {
    private String name;
    private double average;

    public Player(String name, double average) {
        this.name = name;
        this.average = average;
    }

    public String getName() {
        return name;
    }

    public double getAverage() {
        return average;
    }

    public String getAverageString() {
        var decFormat = new DecimalFormat();
        decFormat.setMaximumIntegerDigits(0);
        decFormat.setMaximumFractionDigits(3);
        decFormat.setMinimumFractionDigits(3);
        return decFormat.format(average);
    }
}
```

Here I go, picking apart the code in Listing 10-1. Luckily, earlier chapters cover lots of stuff in this code. The code defines what it means to be an instance of the `Player` class. Here's what's in the code:

- **Declarations of the fields** `name` **and** `average`: For bedtime reading about field declarations, see Chapter 7.

» **A constructor to make new instances of the `Player` class:** For the lowdown on constructors, see Chapter 9.

» **Getter methods for the fields `name` and `average`:** For chitchat about accessor methods (that is, setter and getter methods), see Chapter 7.

» **A method that returns the player's batting average in `String` form:** For the good word about methods, see Chapter 7. (I put a lot of good stuff in Chapter 7, didn't I?)

Another way to beautify your numbers

The `getAverageString` method in Listing 10-1 takes the value from the `average` field (a player's batting average), converts that value (normally of type `double`) into a `String`, and then sends that `String` value right back to the method caller. The use of `DecimalFormat`, which comes directly from the Java application programming interface (API), ensures that the `String` value looks like a baseball player's batting average. According to the `decFormat.setMaximum...` and `decFormat.setMinimum...` method calls, the `String` value has no digits to the left of the decimal point and has exactly three digits to the right of the decimal point.

Java's `DecimalFormat` class can be quite handy. For example, to display the values 345 and -345 in an accounting-friendly format, you can use the following code:

```
DecimalFormat decFormat = new DecimalFormat();
decFormat.setMinimumFractionDigits(2);
decFormat.setNegativePrefix("(");
decFormat.setNegativeSuffix(")");
System.out.println(decFormat.format(345));
System.out.println(decFormat.format(-345));
```

In this little example's format string, everything before the semicolon dictates the way positive numbers are displayed, and everything after the semicolon determines the way negative numbers are displayed. So, with this format, the numbers 345 and −345 appear as follows:

```
345.00
(345.00)
```

To discover some other tricks with numbers, visit the `DecimalFormat` page of Java's API documentation (`https://docs.oracle.com/en/java/javase/17/docs/api/java.base/java/text/DecimalFormat.html`).

Using the Player class

Listings 10-2 and 10-3 have code that uses the `Player` class — the class that's defined earlier, in Listing 10-1.

LISTING 10-2: Using the Player Class

```
package com.example.baseball;

import javax.swing.JFrame;
import javax.swing.JLabel;
import java.awt.GridLayout;
import java.io.File;
import java.io.IOException;
import java.util.Scanner;

public class TeamFrame extends JFrame {

  public TeamFrame() throws IOException {
    Player;
    var hankeesData = new Scanner(new File("Hankees.txt"));

    for (int num = 1; num <= 9; num++) {
      player = new Player(hankeesData.nextLine(), hankeesData.nextDouble());
      hankeesData.nextLine();
      addPlayerInfo(player);
    }

    setTitle("The Hankees");
    setLayout(new GridLayout(10, 2, 20, 6));
    setDefaultCloseOperation(EXIT_ON_CLOSE);
    pack();
    setVisible(true);

    hankeesData.close();
  }

  void addPlayerInfo(Player player) {
    add(new JLabel("    " + player.getName()));
    add(new JLabel(player.getAverageString()));
  }
}
```

LISTING 10-3: Displaying a Frame

```
package com.example.baseball;

import java.io.IOException;

public class ShowTeamFrame {

  public static void main(String[] args) throws IOException {
    new TeamFrame();
  }
}
```

For a run of the code in Listings 10-1, 10-2, and 10-3, see Figure 10-1.

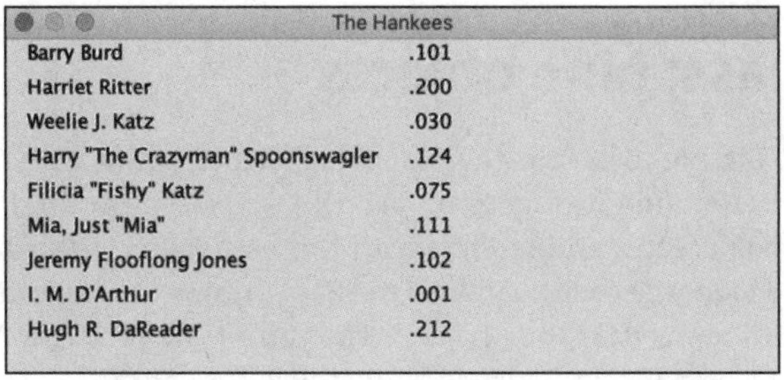

The Hankees

Barry Burd	.101
Harriet Ritter	.200
Weelie J. Katz	.030
Harry "The Crazyman" Spoonswagler	.124
Filicia "Fishy" Katz	.075
Mia, Just "Mia"	.111
Jeremy Flooflong Jones	.102
I. M. D'Arthur	.001
Hugh R. DaReader	.212

FIGURE 10-1: Would you bet money on these people?

To run this program, you need a file containing data on your favorite baseball players. Fortunately, the stuff you download from this book's website comes with a `Hankees.txt` file. (See Figure 10-2.)

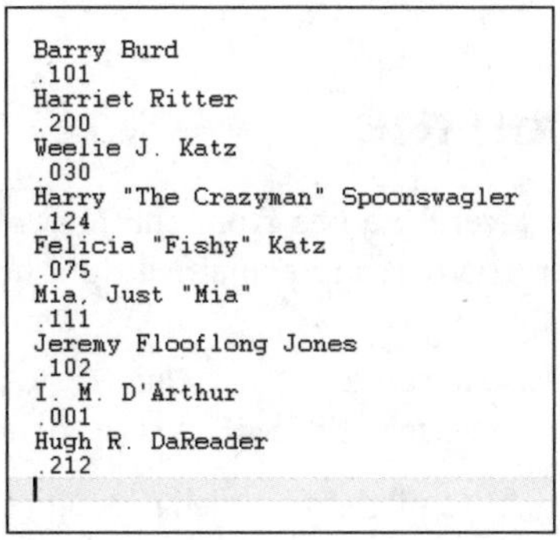

```
Barry Burd
.101
Harriet Ritter
.200
Weelie J. Katz
.030
Harry "The Crazyman" Spoonswagler
.124
Felicia "Fishy" Katz
.075
Mia, Just "Mia"
.111
Jeremy Flooflong Jones
.102
I. M. D'Arthur
.001
Hugh R. DaReader
.212
```

FIGURE 10-2: What a team!

The Hankees.txt file can't be in just any old folder on your computer's hard drive. The file has to be in a place where Java can find it, which is most likely your project's root folder. If Java can't find the file, then a run of this section's example gives you an unpleasant FileNotFoundException message. For advice on choosing locations for files like Hankees.txt, refer to Chapter 8.

WARNING

You may live in a country where the value of π is approximately 3,14159 (with a comma) instead of 3.14159 (with a period). If you do, the file shown in Figure 10-2 won't work for you. The program will crash with an InputMismatchException. To run this section's example, you have to change the periods in the Hankees.txt file into commas. Alternatively, you can add a statement such as Locale.setDefault(Locale.US) to your code. For details, see Chapter 8.

REMEMBER

For this section's code to work correctly, you must have a line break after the last .212 in Figure 10-2. For details about line breaks, see Chapter 8.

One class; nine objects

The code in Listing 10-2 calls the Player constructor nine times. The result is that the code creates nine instances of the Player class. The first time through the loop, the code creates an instance with the name Barry Burd. The second time through the loop, the code abandons the Barry Burd instance and creates another instance with name Harriet Ritter. The third time through, the code abandons poor Harriet Ritter and creates an instance for Weelie J. Katz. The code has only one instance at a time but, all in all, the code creates nine instances.

Each Player instance has its own name and average fields. Each instance also has its own Player constructor and its own getName, getAverage, and getAverageString methods. Look at Figure 10-3 and think of the Player class with its nine incarnations.

Don't get all GUI on me

The code in Listing 10-2 uses several names from the Java API. Some of these names are explained in Chapter 9. Others are explained right here:

- JLabel: A JLabel is an object with some text in it. One way to display text inside the frame is to add an instance of the JLabel class to the frame.

  In Listing 10-2, the addPlayerInfo method is called nine times, once for each player on the team. Each time addPlayerInfo is called, the method adds two new JLabel objects to the frame. The text for each JLabel object comes from a player object's getter method.

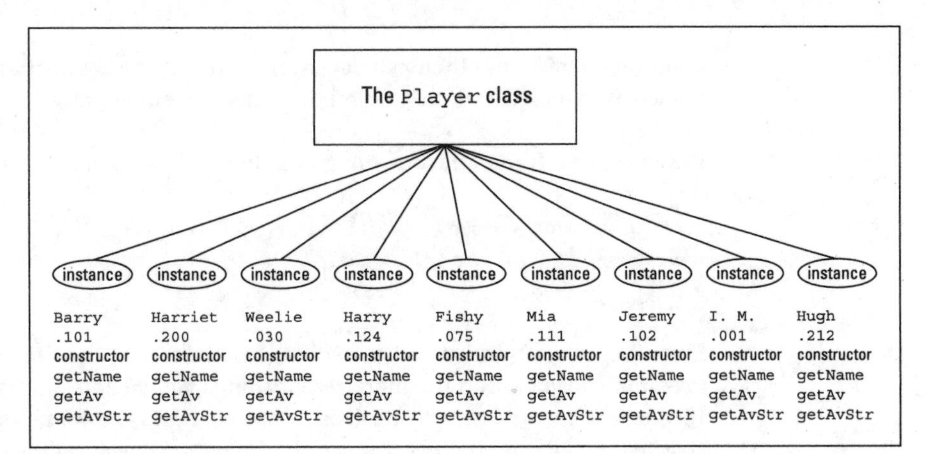

FIGURE 10-3: A class and its objects.

- `GridLayout`: A `GridLayout` arranges things in evenly spaced rows and columns. This constructor for the `GridLayout` class takes two parameters: the number of rows and the number of columns.

  In Listing 10-2, the call to the `GridLayout` constructor takes parameters `(10, 2, 20, 6)`. So, in Figure 10-1,

  - The display has ten rows — one for each of the nine players and an empty one at the bottom for spacing.
  - The display also has two columns: one for a player's name and another for the player's average.
  - The horizontal gap between the two columns is 20 pixels wide.
  - The vertical gap between any two rows is 6 pixels tall.

- `pack`: When you pack a frame, you set the frame's size. That's the size the frame has when it appears on your computer screen. Packing a frame shrink-wraps the frame around whatever objects you've added inside the frame.

  In Listing 10-2, by the time you've reached the call to `pack`, you've already called `addPlayerInfo` nine times and added 18 labels to the frame. In executing the `pack` method, the computer picks a nice size for each label, given whatever text you've placed inside the label. Then the computer picks a nice size for the whole frame, given that the frame has these 18 labels inside it.

When you plop stuff onto frames, you have quite a bit of leeway with the order in which you do things. For instance, you can set the layout before or after you've added labels and other stuff to the frame. If you call `setLayout` and then add labels, the labels appear in nice, orderly positions on the frame. If you reverse this order (add labels and then call `setLayout`), the calling of `setLayout` rearranges the labels in a nice, orderly fashion. It works fine either way.

In setting up a frame, the one thing you shouldn't do is violate this sequence:

```
Add things to the frame, then
pack();
setVisible(true);
```

If you call `pack` and then add more things to the frame, the `pack` method doesn't take into consideration the more recent things you've added. If you call `setVisible` before you add things or call `pack`, the user sees the frame as it's being constructed. Finally, if you forget to set the frame's size (by calling `pack` or another sizing method), the frame you see looks like the one in Figure 10-4. (Normally, I wouldn't show you an anomalous run like the one in Figure 10-4, but I've made the mistake so many times that I feel as though this puny frame is an old friend of mine.)

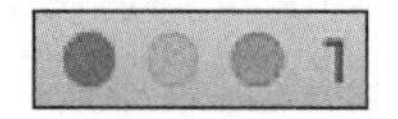

FIGURE 10-4: A shrunken frame.

Some facts about potatoes

Chapter 8 introduces input from a disk file, and along with that topic comes the notion of an exception. When you tinker with a disk file, you need to acknowledge the possibility of raising an `IOException`. That's the lesson from Chapter 8, and that's why the constructor in Listing 10-2 has a `throws IOException` clause.

What about the `main` method in Listing 10-3? With no apparent reference to disk files in this `main` method, why does the method need its own `throws IOException` clause? Well, an exception is a hot potato. If you have one, you either have to eat it (as you can see in Chapter 13) or use a `throws` clause to toss it to someone else. If you toss an exception with a `throws` clause, someone else is stuck with the exception just the way you were.

The constructor in Listing 10-2 throws an `IOException`, but to whom is this exception thrown? Who in this chain of code becomes the bearer of responsibility for the problematic `IOException`? Well, who called the constructor in Listing 10-2? It was the `main` method in Listing 10-3 — that's who called the `TeamFrame` constructor. Because the `TeamFrame` constructor throws its hot potato to the `main` method in Listing 10-3, the `main` method has to deal with it. As shown in Listing 10-3, the `main` method deals with it by tossing the `IOException` again (by having a `throws IOException` clause of its own). That's how the `throws` clause works in Java programs.

REMEMBER

If a method calls another method and the called method has a `throws` clause, the calling method must contain code that deals with the exception. To find out more about dealing with exceptions, read Chapter 13.

TECHNICAL STUFF

At this point in the book, the astute *For Dummies* reader may pose a follow-up question or two: "When a `main` method has a `throws` clause, someone else has to deal with the exception in that `throws` clause. But who called the `main` method? Who deals with the `IOException` in the `throws` clause of Listing 10-3?" The answer is that the Java virtual machine (or JVM, the thing that runs all your Java code) called the `main` method. So the JVM takes care of the `IOException` in Listing 10-3. If the program has any trouble reading the `Hankees.txt` file, the responsibility ultimately falls on the JVM. The JVM takes care of the situation by displaying an error message and then ending the run of your program. How convenient!

TRY IT OUT

Would you like some practice with the material in this section? If so, try this:

COURTING DISASTER

The code in Listing 10-2 reads from a file named `Hankees.txt`. Delete that `Hankees.txt` file from your computer's hard drive, or temporarily move the file to a different directory. Then try to run the program in Listings 10-1 to 10-3. What horrible things happen when you do this?

MEN'S CLOTHING

A line of men's clothing features shirts, pants, jackets, overcoats, neckties, and shoes. Create an `enum` to represent the six kinds of items. Then create a `MensClothingItem` class. Each instance of the class has a kind (one of the six `enum` values) and a name (such as *Casual Summer Design #7*).

Write code to display a frame (like the frame in Figure 10-1). The frame has six rows to describe one complete men's wardrobe.

DEAL ME IN

Create an `enum` to represent the suits in a deck of playing cards (`CLUBS`, `DIAMONDS`, `HEARTS`, and `SPADES`). Create a `PlayingCard` class. Each playing card has a number (from 1 to 13) and a suit. In the numbering scheme, 11 stands for a Jack, 12 stands for a Queen, and 13 stands for a King. Write code that creates several cards and displays them on the screen (in either text-only format or as a frame, like the one in Figure 10-1).

RECORD KEEPING

Some challenges in the Try It Out paragraphs in Chapter 9 introduce Java's `record` feature. Modify Listing 10-1 so that `Player` is a record. Modify Listing 10-2 to use your new `Player` record.

Making Static (Finding the Team Average)

Thinking about the code in Listings 10-1 through 10-3, you decide that you want to find the team's overall batting average. Not a bad idea! The Hankees in Figure 10-1 have an average of about .106, so the team needs some intensive training. While the players are out practicing on the ball field, you have a philosophical hurdle to overcome.

In Listings 10-1 through 10-3, you have three classes: a `Player` class and two other classes that help display data from the `Player` class. So, in this class morass, where should the variables storing your overall, team-average tally go?

- **It makes no sense to put tally variables in either of the displaying classes (`TeamFrame` or `ShowTeamFrame`).** After all, the tally has something-or-other to do with players, teams, and baseball. The displaying classes are about creating windows, not about playing baseball.
- **You're uncomfortable putting an overall team average in an instance of the `Player` class because an instance of the `Player` class represents just one player on the team.** What business does a single player have storing overall team data? Sure, you could make the code work, but it wouldn't be an elegant solution to the problem.

Lucky for you, Java has its `static` keyword. Anything that's declared to be `static` belongs to the whole class, not to any particular instance of the class. When you create the `static` field, `totalOfAverages`, you create just one copy of the field. This copy stays with the entire `Player` class. No matter how many instances of the `Player` class you create — one, nine, or none — you have just one `totalOfAverages` field. And, while you're at it, you create other `static` fields (`playerCount` and `decFormat`) and `static` methods (`findTeamAverage` and `findTeamAverageString`). To see what I mean, look at Figure 10-5.

Going along with your passion for subclasses, you put code for teamwide tallies in a subclass of the `Player` class. The `PlayerPlus` subclass code is shown in Listing 10-4.

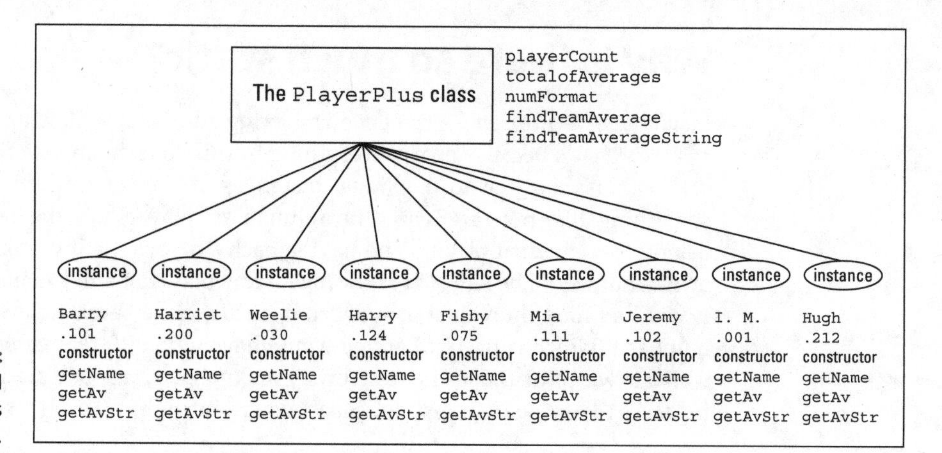

FIGURE 10-5: Some static and nonstatic fields and methods.

LISTING 10-4: Creating a Team Batting Average

```
package com.example.baseball;

import java.text.DecimalFormat;

public class PlayerPlus extends Player {
    private static int playerCount = 0;
    private static double totalOfAverages = .000;
    private static DecimalFormat decFormat = new DecimalFormat();

    static {
        decFormat.setMaximumIntegerDigits(0);
        decFormat.setMaximumFractionDigits(3);
        decFormat.setMinimumFractionDigits(3);
    }

    public PlayerPlus(String name, double average) {
        super(name, average);
        playerCount++;
        totalOfAverages += average;
    }

    public static double findTeamAverage() {
        return totalOfAverages / playerCount;
    }

    public static String findTeamAverageString() {
        return decFormat.format(totalOfAverages / playerCount);
    }
}
```

Why is there so much static?

Maybe you've noticed — the code in Listing 10-4 is overflowing with the word `static`. That's because nearly everything in this code belongs to the entire `PlayerPlus` class and not to individual instances of the class. That's good, because something like `playerCount` (the number of players on the team) shouldn't belong to individual players and having each `PlayerPlus` object keep track of its own count would be silly. ("I know how many players I am. I'm just one player!") If you had nine individual `playerCount` fields, either each field would store the number 1 (which is useless) or you would have nine different copies of the count, which is wasteful and prone to error. By making `playerCount` static, you're keeping the `playerCount` in just one place, where it belongs.

The same kind of reasoning holds for the `totalOfAverages`. Eventually, the `totalOfAverages` field will store the sum of the players' batting averages. For all nine members of the Hankees, this adds up to .956. It's not until someone calls the `findTeamAverage` or `findTeamAverageString` method that the computer actually finds the overall Hankees team batting average.

You also want the methods `findTeamAverage` and `findTeamAverageString` to be static. Without the word `static`, there would be nine `findTeamAverage` methods — one for each instance of the `PlayerPlus` class. This wouldn't make much sense. Each instance would have the code to calculate `totalOfAverages / playerCount` on its own, and each of the nine calculations would yield the same answer.

In general, any task that all instances have in common (and that yields the same result for each instance) should be coded as a `static` method.

REMEMBER

Constructors are never static.

Meet the static initializer

In Listing 10-4, the `decFormat` field is `static`. This makes sense because `decFormat` makes `totalOfAverages / playerCount` look nice, and both fields in the expression `totalOfAverages / playerCount` are `static`. Thinking more directly, the code needs only one thing for formatting numbers. If you have several numbers to format, the same `decFormat` thing that belongs to the entire class can format each number. Creating a `decFormat` for each player is not only inelegant but also wasteful.

But declaring `decFormat` to be `static` presents a little problem. To set up the formatting, you have to call methods like `decFormat.setMaximumIntegerDigits(0)`.

You can't just plop these method calls anywhere in the PlayerPlus class. For example, the following code is bad, invalid, illegal, and otherwise un-Java-like:

```
// THIS IS BAD CODE:
public class PlayerPlus extends Player {
    private static DecimalFormat decFormat = new DecimalFormat();

    decFormat.setMaximumIntegerDigits(0); // Bad!
    decFormat.setMaximumFractionDigits(3); // Bad!
    decFormat.setfsMinimumFractionDigits(3); // Bad!
```

Look at the examples from previous chapters. In those examples, I never let a method call just dangle on its own, the way I do in the bad, bad code. In this chapter, in Listing 10-1, I don't call setMaximumIntegerDigits without putting the method call inside the getAverageString method's body. This no-dangling-method-calls business isn't an accident. Java's rules restrict the places in the code where you can issue calls to methods, and putting a lonely method call on its own immediately inside a class definition is a big no-no.

In Listing 10-4, where can you put the necessary setMax and setMin calls? You can put them inside the body of the findTeamAverageString method, much the way I put them inside the getAverageString method in Listing 10-1. But putting those method calls inside the findTeamAverageString method's body might defeat the purpose of having decFormat be static. After all, a programmer might call findTeamAverageString several times, calling decFormat.setMaximumIntegerDigits(0) each time. But that would be quite wasteful. The entire PlayerPlus class has only one decFormat field, and that decFormat field's MaximumIntegerDigits value is always 0. Don't keep setting MaximumIntegerDigits(0) over and over again.

The best alternative is to take the bad lines in this section's bad code and put them inside a *static initializer*. Then they become good lines inside good code. (See Listing 10-4.) A static initializer is a block that's preceded by the word static. Java executes the static initializer's statements once for the entire class. That's exactly what you want for something called *static*.

Displaying the overall team average

You may be noticing a pattern. When you create code for a class, you generally write two pieces of code: One piece of code defines the class, and the other piece of code uses the class. (The ways to use a class include calling the class's constructor, referencing the class's nonprivate fields, and calling the class's methods.) Listing 10-4, shown previously, contains code that defines the PlayerPlus class, and Listing 10-5 contains code that uses this PlayerPlus class.

LISTING 10-5: Using the Code from Listing 10-4

```
package com.example.baseball;

import javax.swing.JFrame;
import javax.swing.JLabel;
import java.awt.GridLayout;
import java.io.File;
import java.io.IOException;
import java.util.Scanner;

public class TeamFrame extends JFrame {

  public TeamFrame() throws IOException {
    PlayerPlus player;
    var hankeesData = new Scanner(new File("Hankees.txt"));

    for (int num = 1; num <= 9; num++) {
      player =
          new PlayerPlus(hankeesData.nextLine(), hankeesData.nextDouble());
      hankeesData.nextLine();

      addPlayerInfo(player);
    }

    add(new JLabel());
    add(new JLabel("----"));
    add(new JLabel("    Team Batting Average:"));
    add(new JLabel(PlayerPlus.findTeamAverageString()));

    setTitle("The Hankees");
    setLayout(new GridLayout(12, 2, 20, 6));
    setDefaultCloseOperation(EXIT_ON_CLOSE);
    pack();
    setVisible(true);

    hankeesData.close();
  }

  void addPlayerInfo(PlayerPlus player) {
    add(new JLabel("    " + player.getName()));
    add(new JLabel(player.getAverageString()));
  }
}
```

To run the code in Listing 10-5, you need a class with a `main` method. The `Show-TeamFrame` class in Listing 10-3 works just fine.

Figure 10-6 shows a run of the code from Listing 10-5. This run depends on the availability of the `Hankees.txt` file from Figure 10-2. The code in Listing 10-5 is almost an exact copy of the code from Listing 10-2. (So close is the copy that, if I could afford it, I'd sue myself for theft of intellectual property.) The only thing new in Listing 10-5 is the stuff shown in bold.

The Hankees

Barry Burd	.101
Harriet Ritter	.200
Weelie J. Katz	.030
Harry "The Crazyman" Spoonswagler	.124
Filicia "Fishy" Katz	.075
Mia, Just "Mia"	.111
Jeremy Flooflong Jones	.102
I. M. D'Arthur	.001
Hugh R. DaReader	.212

Team Batting Average:	.106

FIGURE 10-6: A run of the code in Listing 10-5.

In Listing 10-5, the `GridLayout` has two extra rows: one row for spacing and another row for the Hankees team average. Each of these rows has two `Label` objects in it:

- **The spacing row has a blank label and a label with a dashed line.** The blank label is a placeholder. When you add components to a `GridLayout`, the components are added row by row, starting at the left end of a row and working toward the right end. Without this blank label, the dashed-line label would appear at the left end of the row, under Hugh R. DaReader's name.
- **The other row has a label displaying the words *Team Batting Average,* and another label displaying the number .106.** The method call that gets the number .106 is interesting. The call looks like this:

  ```
  PlayerPlus.findTeamAverageString()
  ```

  Take a look at that method call. That call has the following form:

  ```
  ClassName.methodName()
  ```

  That's new and different. In earlier chapters, I say that you normally preface a method call with an object's name, not a class's name. So why does this dot notation use a class name? The answer: When you call a `static` method, you preface the method's name with the name of the class that contains the method. The same holds true whenever you reference another class's `static` field. This makes sense. ***Remember:*** The whole class that defines a `static` field or method owns that field or method. So, to refer to a `static` field or method, you preface the field or method's name with the class's name.

TIP

With dot notation that refers to a `static` field or method, you can cheat and use an object's name in place of the class name. For instance, in Listing 10-5, with judicious rearranging of some other statements, you can use the expression `player.findTeamAverageString()`.

The static keyword is yesterday's news

This section makes a big noise about `static` fields and methods, but `static` things have been part of the picture since early in this book. For example, Chapter 3 introduces `System.out.println`. The name `System` refers to a class, and `out` is a `static` field in that class. That's why, in Chapter 4 and beyond, I use the `static` keyword to import the `out` field:

```
import static java.lang.System.out;
```

In Java, `static` fields and methods show up all over the place. When they're declared in someone else's code and you're making use of them in your code, you hardly ever have to worry about them. But when you're declaring your own fields and methods and must decide whether to make them `static`, you have to think a little harder.

TECHNICAL STUFF

In this book, my first serious use of the word `static` is way back in Listing 3-1. I use the `static` keyword as part of every main method (and lots of `main` methods are in this book's listings). So why does `main` have to be static? Well, remember that nonstatic things belong to objects, not classes. If the `main` method isn't static, you can't have a `main` method until you create an object. But, when you start up a Java program, no objects have been created yet. The statements that are executed in the `main` method start creating objects. So if the `main` method isn't static, you have a big chicken-and-egg problem.

Could cause static; handle with care

When I first started writing Java programs, I had recurring dreams about seeing a certain error message. The message was `Nonstatic field or method cannot be referenced from a static context`. So often did I see this message, so thoroughly was I perplexed, that the memory of this message became burned into my subconscious existence.

These days, I know why I saw that error message so often. I can even make the message occur, if I want. But I still feel a little shiver whenever I see this message on my screen.

Before you can understand why the message occurs and how to fix the problem, you need to get some terminology under your belt. If a field or method isn't static, it's called *nonstatic.* (Real surprising, hey?) Given that terminology, you have at least two ways to make the dreaded message appear:

» Put `Class.nonstaticThing` somewhere in your program.

» Put `nonstaticThing` somewhere inside a `static` method.

In either case, you're getting yourself into trouble. You're taking something that belongs to an object (the nonstatic thing) and putting it in a place where no objects are in sight.

Take, for instance, the first of the two situations I just described. To see this calamity in action, go back to Listing 10-5. Toward the end of the listing, change `player.getName()` to `Player.getName()`. That does the trick. What could `Player.getName` possibly mean? If anything, the expression `Player.getName` means "Call the `getName` method that belongs to the entire `Player` class." But look back at Listing 10-1. The `getName` method isn't static. Each instance of the `Player` (or `PlayerPlus`) class has a `getName` method. None of the `getName` methods belongs to the entire class. So the call `Player.getName` makes no sense. (Maybe the computer is pulling punches when it displays the inoffensive `cannot be referenced...` message. Perhaps a harsh, nonsensical expression message would be more fitting.)

For a taste of the second situation (in the bullet list earlier in this section), go back to Listing 10-4. While no one's looking, quietly remove the word `static` from the declaration of the `decFormat` field (near the top of the listing). This removal turns `decFormat` into a nonstatic field. Suddenly, each player on the team has a separate `decFormat` field.

Well, things are just hunky-dory until the computer reaches the `findTeamAverageString` method. That static method has four `decFormat.SuchAndSuch` statements in it. Once again, you're forced to ask what a statement of this kind could possibly mean. Method `findTeamAverageString` belongs to no instance in particular. (The method is static, so the entire `PlayerPlus` class has one `findTeamAverageString` method.) But with the way you've just butchered the code, plain old `decFormat` without reference to a particular object has no meaning. So again, you're referencing the nonstatic field, `decFormat`, from inside a `static` method's context. For shame, for shame, for shame!

TRY IT OUT

I don't know about you, but I can always use some practice with static variables and methods:

DESIGNER DUMMIES

In a previous section, you create a class to represent items in a line of men's clothing. Create a subclass that includes the name of the designer *(Dummies House of Fashion)*, the color of the item, and the cost of the item.

The designer's name will be static because all items in the line have the same designer. The color can be a static field from Java's own `Color` class. (See `https://docs.oracle.com/en/java/javase/17/docs/api/java.desktop/java/awt/Color.html`.)

Write code to display a frame (like the frame in Figure 10-1). The frame has eight rows. The first row displays the name of the designer. The next six rows describe one complete men's wardrobe. The last row shows the wardrobe's total cost.

COUNT ME IN

In a previous section, you create a class to represent a playing card. Add a static field to your `PlayingCard` class. The field keeps track of the number of times the class's constructor has been called, and thus has a count of the number of playing cards.

ATTACK OF THE MUTANTS

What's the output of the following code? Make some predictions, and then run the code to see whether your predictions are correct:

```
import static java.lang.System.out;

public class Main {

  public static void main(String[] args) {

    out.println("bigValue: " + MutableInteger.bigValue);
    // out.println("bigValue: " + IntegerHolder.value); ILLEGAL

    var holder1 = new MutableInteger(42);
    var holder2 = new MutableInteger(7);

    out.println("holder1: " + holder1.value);
    out.println("holder2: " + holder2.value);
```

```
    out.println();
    holder1.value++;
    holder2.value++;
    MutableInteger.bigValue++;

    out.println("bigValue: " + MutableInteger.bigValue);
    out.println("holder1: " + holder1.value);
    out.println("holder2: " + holder2.value);

    out.println();
    holder1.bigValue++;
    out.println("bigValue according to holder1: " + holder1.bigValue);
    out.println("bigValue according to holder2: " + holder2.bigValue);
  }
}

class MutableInteger {
  int value;
  static int bigValue = 1_000_000;

  public MutableInteger(int value) {
    this.value = value;
  }
}
```

Experiments with Variables

One summer during my college days, I was sitting on the front porch, loafing around and talking with someone I'd just met. I think her name was Janine. "Where are you from?" I asked. "Mars," she answered. She paused to see whether I'd ask a follow-up question.

As it turned out, Janine was from Mars, Pennsylvania, a small town about 20 miles north of Pittsburgh. Okay, what's my point? The point is that the meaning of a name depends on the context. If you're just north of Pittsburgh and ask, "How do I get to Mars from here?" you may receive a sensible, nonchalant answer. But if you ask the same question standing on a street corner in Manhattan, you'll probably arouse some suspicion. (Okay, knowing Manhattan, people would probably just ignore you.)

Of course, the people who live in Mars, Pennsylvania, are very much aware that their town has an oddball name. Fond memories of teenage years at Mars High School don't prevent a person from knowing about the big red planet. On a clear evening in August, you can still have the following conversation with one of the local residents:

> *You:* How do I get to Mars?
>
> *Local resident:* You're in Mars, pal. What particular part of Mars are you looking for?
>
> *You:* No, I don't mean Mars, Pennsylvania. I mean the planet Mars.
>
> *Local resident:* Oh, the planet! Well, then, catch the 8:19 train leaving for Cape Canaveral. No, wait — that's the local train. That'd take you through West Virginia. . . .

The meaning of a name depends on where you're using the name. Although most English-speaking people think of Mars as a place with a carbon dioxide atmosphere, some folks in Pennsylvania think about all the shopping they can do in Mars. And those folks in Pennsylvania really have two meanings for the name *Mars.* In Java, those names may look like this: `Mars` and `planets.Mars`.

Putting a variable in its place

Your first experiment is shown in Listings 10-6 and 10-7. The listings' code highlights the difference between variables that are declared inside and outside methods.

LISTING 10-6: Two Meanings for Mars

```
import static java.lang.System.out;

class EnglishSpeakingWorld {
    String mars = " red planet";

    void visitPennsylvania() {
        out.println("visitPA is running:");

        String mars = " Janine's home town";

        out.println(mars);
        out.println(this.mars);
    }
}
```

LISTING 10-7: Calling the Code of Listing 10-6

```
import static java.lang.System.out;

public class GetGoing {

    public static void main(String[] args) {
        out.println("main is running:");
        var e = new EnglishSpeakingWorld();

        //out.println(mars); cannot resolve symbol
        out.println(e.mars);
        e.visitPennsylvania();
    }
}
```

Figure 10-7 shows a run of the code in Listings 10-6 and 10-7. Figure 10-8 shows a diagram of the code's structure. In the `GetGoing` class, the `main` method creates an instance of the `EnglishSpeakingWorld` class. The variable `e` refers to the new instance. The new instance is an object with a variable named `mars` inside it. That `mars` variable has the value `"red planet"`. This `mars` (`"red planet"`) variable is a field.

```
main is running:
   red planet
visitPA is running:
   Janine's home town
   red planet
```

FIGURE 10-7: A run of the code in Listings 10-6 and 10-7.

TECHNICAL STUFF

Another way to describe that `mars` field is to call it an *instance variable* because that `mars` variable (the variable whose value is `"red planet"`) belongs to an *instance* of the `EnglishSpeakingWorld` class. In contrast, you can refer to `static` fields (like the `playerCount`, `totalOfAverages`, and `decFormat` fields in Listing 10-4) as *class variables*. For example, `playerCount` in Listing 10-4 is a class variable because one copy of `playerCount` belongs to the entire `PlayerPlus` class.

Now look at the `main` method in Listing 10-7. Inside the `GetGoing` class's `main` method, you aren't permitted to write `out.println(mars)`. In other words, a bare-faced reference to any `mars` variable is a definite no-no. The `mars` variable that I mention in the preceding paragraph belongs to the `EnglishSpeakingWorld` object, not the `GetGoing` class.

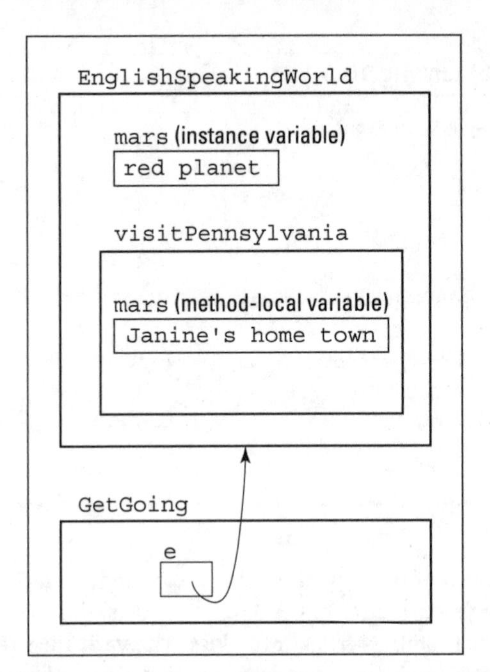

FIGURE 10-8: The structure of the code in Listings 10-6 and 10-7.

However, inside the `GetGoing` class's `main` method, you can certainly write `e.mars` because the `e` variable refers to your `EnglishSpeakingWorld` object. That's nice.

Near the bottom of the code, the `visitPennsylvania` method is called. When you're inside `visitPennsylvania`, you have another declaration of a `mars` variable, whose value is `"Janine's home town"`. This particular `mars` variable is called a *method-local variable* because it belongs to just one method: the `visitPennsylvania` method.

Now you have two variables, both with the name *mars.* One `mars` variable, a field, has the value `"red planet"`. The other `mars` variable, a method-local variable, has the value `"Janine's home town"`. In the code, when you use the word *mars,* to which of the two variables are you referring?

The answer is, when you're visiting Pennsylvania, the variable with value `"Janine's home town"` wins. When in Pennsylvania, think the way the Pennsylvanians think. When you're executing code inside the `visitPennsylvania` method, resolve any variable name conflicts by going with method-local variables — variables declared right inside the `visitPennsylvania` method.

What if you're in Pennsylvania and need to refer to that 2-mooned celestial object? More precisely, how does code inside the `visitPennsylvania` method refer to the field with value `"red planet"`? The answer is, use `this.mars`. The word *this* points to whatever object contains all this code (and not to any methods inside the code). That object, an instance of the `EnglishSpeakingWorld` class, has a big, fat `mars` field, and that field's value is `"red planet"`. So that's how you can force code to see outside the method it's in — you use the Java keyword `this`.

For more information on the keyword `this`, see Chapter 9.

Telling a variable where to go

Years ago, when I lived in Milwaukee, Wisconsin, I made frequent use of the local bank's automatic teller machines. Machines of this kind were just beginning to become standardized. The local teller machine system was named TYME, which stood for Take Your Money Everywhere.

I remember traveling by car out to California. At one point, I got hungry and stopped for a meal, but I was out of cash. So I asked a gas station attendant, "Do you know where there's a TYME machine around here?"

So you see, a name that works well in one place could work terribly, or not at all, in another place. In Listings 10-8 and 10-9, I illustrate this point (with more than just an anecdote about teller machines).

LISTING 10-8: Tale of Atomic City

```
import static java.lang.System.out;

class EnglishSpeakingWorld2 {
    String mars;

    void visitIdaho() {
        out.println("visitID is running:");

        mars = " red planet";
        String atomicCity = " Population: 25";

        out.println(mars);
        out.println(atomicCity);
    }

    void visitNewJersey() {
        out.println("visitNJ is running:");
        out.println(mars);
        //out.println(atomicCity); cannot resolve symbol
    }
}
```

LISTING 10-9: Calling the Code of Listing 10-8

```
public class GetGoing2 {

    public static void main(String[] args) {
        var e = new EnglishSpeakingWorld2();

        e.visitIdaho();
        e.visitNewJersey();
    }
}
```

Figure 10-9 shows a run of the code in Listings 10-8 and 10-9. Figure 10-10 shows a diagram of the code's structure. The code for `EnglishSpeakingWorld2` has two variables: The `mars` variable, which isn't declared inside a method, is a field; the other variable, `atomicCity`, is a method-local variable and is declared inside the `visitIdaho` method.

```
visitID is running:
   red planet
   Population: 25
visitNJ is running:
   red planet
```

FIGURE 10-9: A run of the code in Listings 10-8 and 10-9.

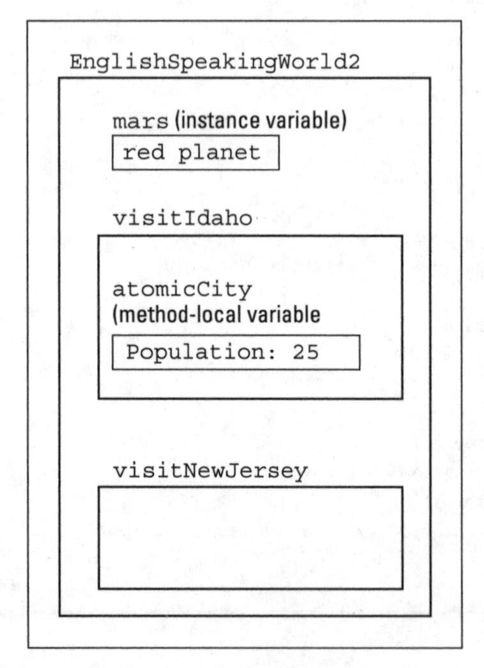

FIGURE 10-10: The structure of the code in Listings 10-8 and 10-9.

In Listing 10-8, notice where each variable can and can't be used. When you try to use the `atomicCity` variable inside the `visitNewJersey` method, you see an error message. Literally, the message says `cannot resolve symbol`. Figuratively, the message says, "Hey, buddy, Atomic City is in Idaho, not New Jersey." Technically, the message says that the method-local variable `atomicCity` is available only in the `visitIdaho` method because that's where the variable was declared.

Back inside the `visitIdaho` method, you're free to use the `atomicCity` variable as much as you want. After all, the `atomicCity` variable is declared inside the `visitIdaho` method.

And what about Mars? Have you forgotten about your old friend, that lovely 80-degrees-below-0 planet? Well, both the `visitIdaho` and `visitNewJersey` methods can access the `mars` variable. That's because the `mars` variable is a field. That is, the `mars` variable is declared in the code for the `EnglishSpeakingWorld2` class but not inside any particular method. (In my stories about the names for things, remember that people who live in both states, Idaho and New Jersey, have heard of the planet Mars.)

The life cycle of the `mars` field has three separate steps:

1. When the `EnglishSpeakingWorld2` class first flashes into existence, the computer sees `String mars` and creates space for the `mars` field.
2. When the `visitIdaho` method is executed, the method assigns the value `"red planet"` to the `mars` field. (The `visitIdaho` method also prints the value of the `mars` field.)
3. When the `visitNewJersey` method is executed, the method prints the `mars` value once again.

In this way, the `mars` field's value is passed from one method to another.

TRY IT OUT

Try out these programs. See what you think.

WHO'S WHO?

What's the output of the following code? Why?

```
public class Main1 {
    static String name = "Nancy";

    public static void main(String[] args) {

        System.out.println(name);
```

```
        String name = "Barry";
        System.out.println(name);
    }
}
```

WHO MOVED MY NAME?

What's the output of the following code? Why?

```
public class Main2 {
    String name = "George";

    public static void main(String[] args) {
        new Main2();
    }

    Main2() {
        System.out.println(name);

        String name = "Barry";
        System.out.println(name);

        System.out.println(this.name);
    }
}
```

MISTER WHO

What's the output of the following code? Why?

```
public class Main3 {
    static String name = "George";

    public static void main(String[] args) {
        String name = "Barry";
        new OtherClass();
    }
}

class OtherClass {
```

```
    OtherClass() {
        String name = "Leonard";
        System.out.println(name);
        System.out.println(Main3.name);
    }
}
```

WHO ELSE IS WHO?

What's the output of the following code? Why?

```
public class Main4 {
    String name = "Betty";

    public static void main(String[] args) {
        new Main4();
    }

    Main4() {
        String name = "Barry";
        new YetAnotherClass(this);
    }
}

class YetAnotherClass {

    YetAnotherClass(Main4 whoCreatedMe) {
        String name = "Leonard";
        System.out.println(name);
        // System.out.println(Main4.name); ILLEGAL
        System.out.println(whoCreatedMe.name);
    }
}
```

Passing Parameters

A method can communicate with another part of your Java program in several ways. One way is by way of the method's parameter list. Using a parameter list, you pass on-the-fly information to a method as the method is being called.

Imagine that the information you pass to the method is stored in one of your program's variables. What, if anything, does the method actually do with that variable? The following sections present a few interesting case studies.

Pass by value

According to my web research, the town of Smackover, Arkansas, has 2,232 people in it. But my research isn't current. Just yesterday, Dora Kermongoos celebrated a joyous occasion over at Smackover General Hospital — the birth of her healthy, blue-eyed baby girl. (The girl weighs 7 pounds, 4 ounces, and is 21 inches tall.) Now the town's population has risen to 2,233.

Listing 10-10 has a very bad program in it. The program is supposed to add 1 to a variable that stores Smackover's population, but the program doesn't work. Take a look at Listing 10-10 to see why.

LISTING 10-10: This Program Doesn't Work

```
public class TrackPopulation {

    public static void main(String[] args) {
        int smackoverARpop = 2232;

        birth(smackoverARpop);
        System.out.println(smackoverARpop);
    }

    static void birth(int cityPop) {
        cityPop++;
    }
}
```

When you run the program in Listing 10-10, the program displays the number 2,232 onscreen. After nine months of planning and anticipation and Dora's whopping seven hours in labor, the Kermongoos family's baby girl wasn't registered in the system. What a shame!

The improper use of parameter passing caused the problem. In Java, when you pass a parameter that has one of the eight primitive types, that parameter is *passed by value*.

CROSS REFERENCE

For a review of Java's eight primitive types, see Chapter 4.

Here's what this means in plain English: Any changes that the method makes to the value of its parameter don't affect the values of variables back in the calling code. In Listing 10-10, the `birth` method can apply the ++ operator to `cityPop` all it wants — the application of ++ to the `cityPop` parameter has absolutely no effect on the value of the `smackoverARpop` variable back in the main method.

Technically, what's happening is the copying of a value. (See Figure 10-11.) When the `main` method calls the `birth` method, the value stored in `smackoverARpop` is copied to another memory location — a location reserved for the `cityPop` parameter's value. During the `birth` method's execution, 1 is added to the `city-Pop` parameter. But the place where the original 2232 value was stored — the memory location for the `smackoverARpop` variable — remains unaffected.

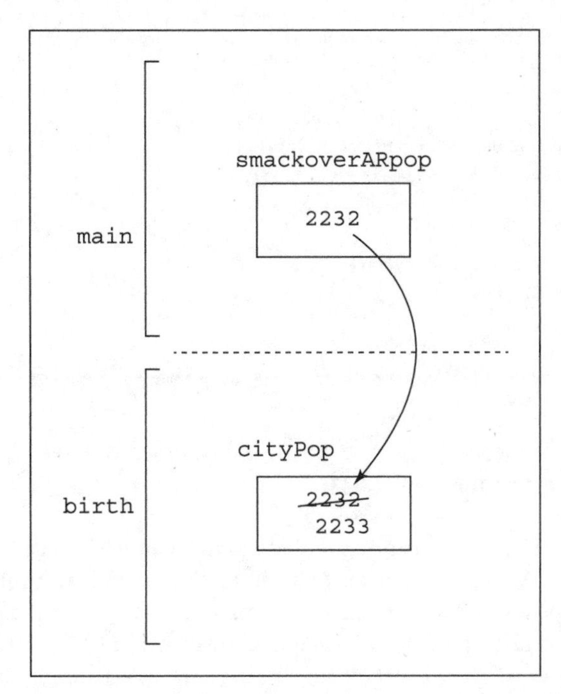

FIGURE 10-11: Pass by value, under the hood.

REMEMBER

When you do parameter passing with any of the eight primitive types, the computer uses *pass by value.* The value stored in the calling code's variable remains unchanged. This happens even if the calling code's variable and the called method's parameter happen to have exactly the same name.

Returning a result

You must fix the problem that the code in Listing 10-10 poses. After all, a young baby Kermongoos can't go through life untracked. To record this baby's existence, you have to add 1 to the value of the `smackoverARpop` variable. You can do this in plenty of ways, and the way presented in Listing 10-11 isn't the simplest. Even so, the way shown in Listing 10-11 illustrates a point: Returning a value from a method call can be an acceptable alternative to parameter passing. Look at Listing 10-11 to see what I mean.

LISTING 10-11: This Program Works

```
public class TrackPopulation2 {

    public static void main(String[] args) {
        int smackoverARpop = 2232;

        smackoverARpop = birth(smackoverARpop);
        System.out.println(smackoverARpop);
    }

    static int birth(int cityPop) {
        return cityPop + 1;
    }
}
```

After running the code in Listing 10-11, the number you see on your computer screen is the correct number: 2,233.

The code in Listing 10-11 has no new features in it (unless you call *working correctly* a new feature). The most important idea in Listing 10-11 is the `return` statement, which also appears in Chapter 7. Even so, Listing 10-11 presents a nice contrast to the approach in Listing 10-10, which had to be discarded.

Pass by reference

In the previous section or two, I take great pains to emphasize a certain point — that when a parameter has one of the eight primitive types, the parameter is passed by value. If you read this, you probably missed the emphasis on the parameter's having one of the eight primitive types. The emphasis is needed because passing objects (reference types) doesn't quite work the same way.

When you pass an object to a method, the object is *passed by reference.* What this means to you is that statements in the called method *can* change any values that are stored in the object's variables. Those changes *do* affect the values that are seen by whatever code called the method. Listings 10-12 and 10-13 illustrate the point.

LISTING 10-12: What Is a City?

```
class City {
    int population;
}
```

LISTING 10-13: Passing an Object to a Method

```
public class TrackPopulation3 {

    public static void main(String[] args) {
        var smackoverAR = new City();
        smackoverAR.population = 2232;
        birth(smackoverAR);
        System.out.println(smackoverAR.population);
    }

    static void birth(City aCity) {
        aCity.population++;
    }
}
```

When you run the code in Listings 10-12 and 10-13, the output you see is the number 2,233. That's good because the code has things like ++ and the word *birth* in it. The deal is, adding 1 to `aCity.population` inside the `birth` method actually changes the value of `smackoverAR.population`, as it's known in the `main` method.

To see how the `birth` method changes the value of `smackoverAR.population`, look at Figure 10-12. When you pass an object to a method, the computer doesn't make a copy of the entire object. Instead, the computer makes a copy of a *reference to* that object. (Think of it the way it's shown in Figure 10-12. The computer makes a copy of an arrow that points to the object.)

In Figure 10-12, you see just one instance of the `City` class, with a `population` variable inside it. Now keep your eye on that object as you read the following steps:

1. Just before the `birth` method is called, the `smackoverAR` variable refers to that object — the instance of the `City` class.

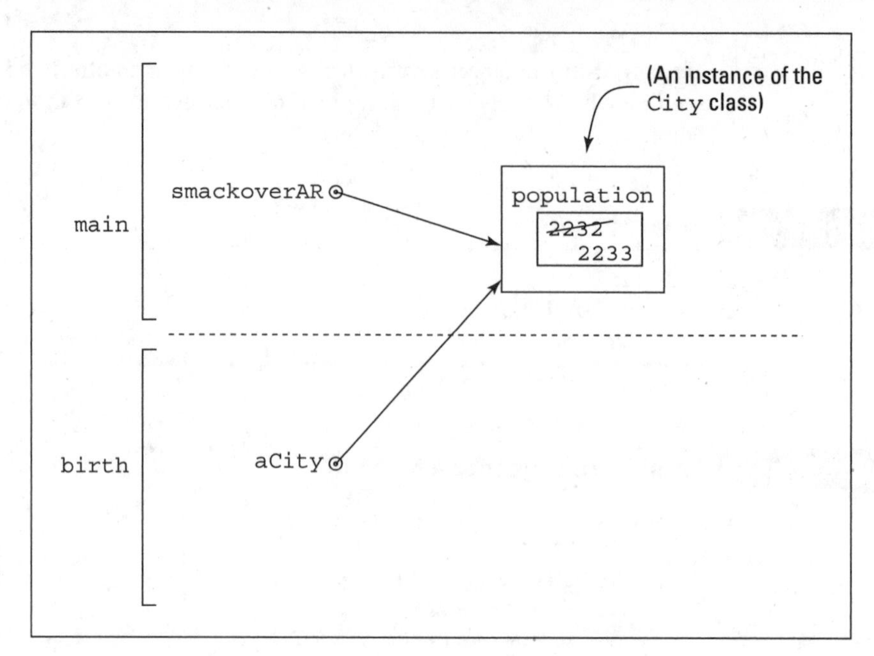

FIGURE 10-12: Pass by reference, under the hood.

2. When the `birth` method is called and `smackoverAR` is passed to the `birth` method's `aCity` parameter, the computer copies the reference from `smackoverAR` to `aCity`. Now `aCity` refers to that same object — the instance of the `City` class.

3. When the statement `aCity.population++` is executed inside the `birth` method, the computer adds 1 to the object's `population` field. Now the program's one and only `City` instance has 2233 stored in its `population` field.

4. The flow of execution goes back to the `main` method. The value of `smack-overAR.population` is printed. But `smackoverAR` refers to that one instance of the `City` class. So `smackoverAR.population` has the value 2233. The Kermongoos family is so proud.

Returning an object from a method

Believe it or not, the previous sections on parameter passing left unexplored one nook-and-cranny of Java methods: When you call a method, the method can return something right back to the calling code. In previous chapters and sections, I return primitive values, such as `int` values, or nothing (otherwise known as *void*). In this section, I return a whole object. It's an object of type `City` from Listing 10-12. The code that makes this happen is in Listing 10-14.

LISTING 10-14: Here, Have a City

```
public class TrackPopulation4 {

    public static void main(String[] args) {
        var smackoverAR = new City();
        smackoverAR.population = 2232;
        smackoverAR = doBirth(smackoverAR);
        System.out.println(smackoverAR.population);
    }

    static City doBirth(City aCity) {
        var myCity = new City();
        myCity.population = aCity.population + 1;
        return myCity;
    }
}
```

If you run the code in Listing 10-14, you get the number 2,233. That's good. The code works by telling the `doBirth` method to create another `City` instance. In the new instance, the value of `population` is `2233`. (See Figure 10-13.)

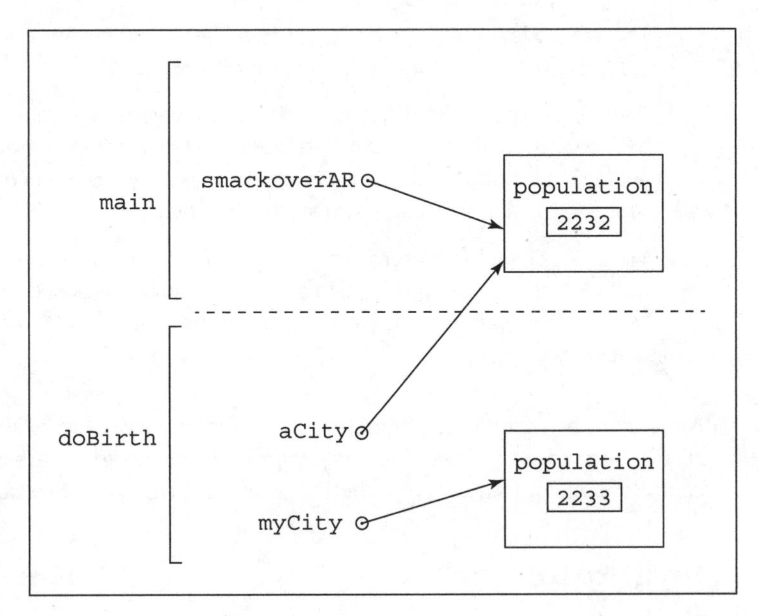

FIGURE 10-13: The doBirth method creates a City instance.

After the `doBirth` method is executed, that `City` instance is returned to the `main` method. Then, back in the `main` method, that instance (the one that `doBirth` returns) is assigned to the `smackoverAR` variable. (See Figure 10-14.) Now `smackoverAR` refers to a brand-new `City` instance — an instance whose population is 2,233.

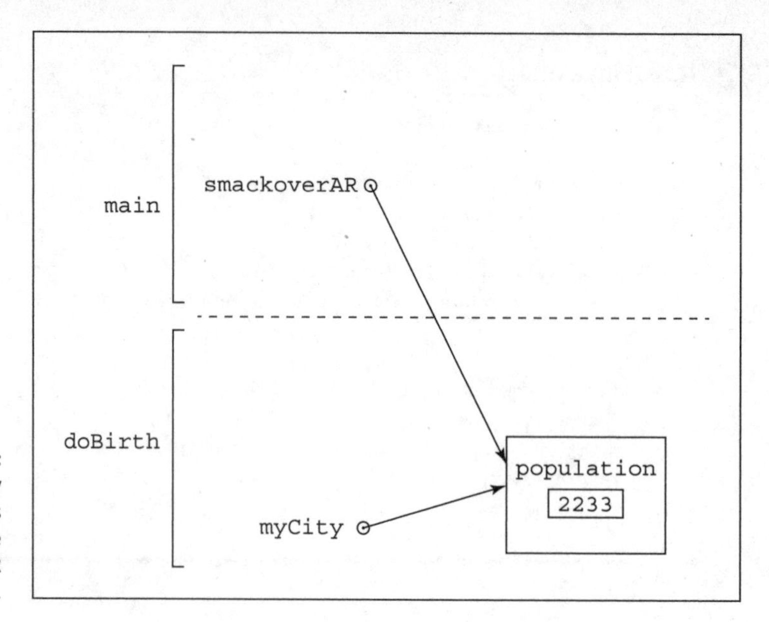

FIGURE 10-14: The new City instance is assigned to the smackoverAR variable.

In Listing 10-14, notice the type consistency in the calling and returning of the `doBirth` method:

- The `smackoverAR` variable has type `City`. The `smackoverAR` variable is passed to the `aCity` parameter, which is also of type `City`.
- The `myCity` variable is of type `City`. The `myCity` variable is sent back in the `doBirth` method's `return` statement. That's consistent, because the `doBirth` method's header begins with `static City doBirth(`*`blah, blah, blah`* `...` — a promise to return an object of type `City`.
- The `doBirth` method returns an object of type `City`. Back in the `main` method, the object that the call to `doBirth` returns is assigned to the `smackoverAR` variable, and (you guessed it) the `smackoverAR` variable is of type `City`.

Aside from being quite harmonious, all this type agreement is absolutely necessary. If you write a program in which your types don't agree with one another, the compiler spits out an unsympathetic `incompatible types` message.

Epilogue

Dora Kermongoos and her newborn baby daughter are safe, healthy, and resting happily in their Smackover, Arkansas, home.

» Searching for things

» Creating values as you get a program running

Chapter 11

Using Arrays to Juggle Values

Welcome to the Java Motel! No haughty bellhops, no overpriced room service, none of the usual silly puns. Just a clean double room that's a darn good value!

Getting Your Ducks All in a Row

The Java Motel, with its ten comfortable rooms, sits in a quiet place off the main highway. Aside from a small, separate office, the motel is just one long row of ground floor rooms. Each room is easily accessible from the spacious front parking lot.

Oddly enough, the motel's rooms are numbered 0 through 9. I could say that the numbering is a fluke — something to do with the builder's original design plan. But the truth is that starting with 0 makes the examples in this chapter easier to write.

Anyway, you're trying to keep track of the number of guests in each room. Because you have ten rooms, you may think about declaring ten variables:

```
int guestsInRoomNum0, guestsInRoomNum1, guestsInRoomNum2,
    guestsInRoomNum3, guestsInRoomNum4, guestsInRoomNum5,
    guestsInRoomNum6, guestsInRoomNum7, guestsInRoomNum8,
    guestsInRoomNum9;
```

Doing it this way may seem a bit inefficient — but inefficiency isn't the only thing wrong with this code. Even more problematic is the fact that you can't loop through these variables. To read a value for each variable, you have to copy the `nextInt` method ten times:

```
guestsInRoomNum0 = diskScanner.nextInt();
guestsInRoomNum1 = diskScanner.nextInt();
guestsInRoomNum2 = diskScanner.nextInt();
//... and so on.
```

Surely a better way exists.

That better way involves an array. An *array* is a row of values, like the row of rooms in a 1-floor motel. To picture the array, just picture the Java Motel:

- » First, picture the rooms, lined up next to one another.
- » Next, picture the same rooms with their front walls missing. Inside each room you can see a certain number of guests.
- » If you can, forget that the two guests in Room 9 are putting piles of bills into a big briefcase. Ignore the fact that the guests in Room 6 haven't moved away from the TV set in a day-and-a-half. Instead of all these details, see only numbers. In each room, see a number representing the count of guests in that room. (If free-form visualization isn't your strong point, look at Figure 11-1.)

In the lingo of this chapter, the entire row of rooms is called an *array*. Each room in the array is called a *component* of the array (also known as an array *element*). Each component has two numbers associated with it:

- » The room number (a number from 0 to 9), which is called an *index* of the array
- » A number of guests, which is a *value* stored in a component of the array

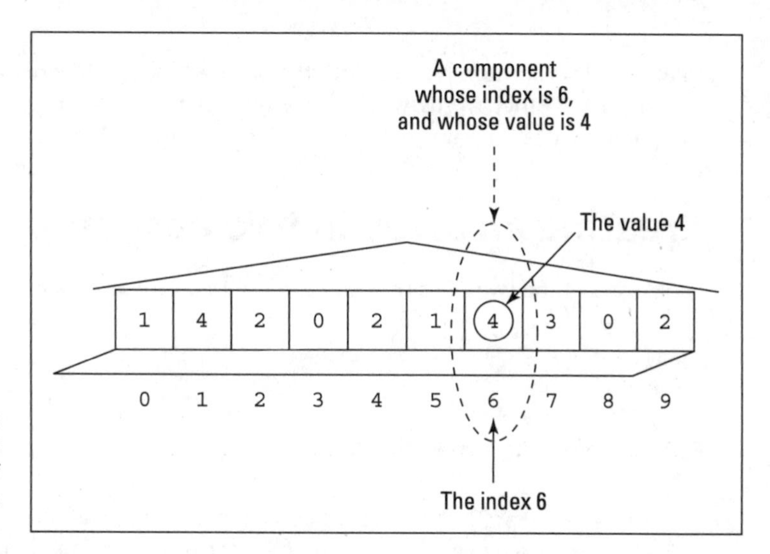

FIGURE 11-1: An abstract snapshot of rooms in the Java Motel.

Using an array saves you from all the repetitive nonsense in the sample code shown at the beginning of this section. For instance, to declare an array with ten values in it, you can write one fairly short statement:

```
int[] guests = new int[10];
```

If you're especially verbose, you can expand this statement so that it becomes two separate statements:

```
int[] guests;
guests = new int[10];
```

In either of these code snippets, notice the use of the number 10. This number tells the computer to make the `guests` array have ten components. Each component of the array has a name of its own. The starting component is named *guests[0]*, the next is named *guests[1]*, and so on. The last of the ten components is named *guests[9]*.

REMEMBER

In creating an array, you always specify the number of components. The array's indices start with 0 and end with the number that's one less than the total number of components.

The snippets that I show you give you two ways to create an array. The first way uses one line. The second way uses two lines. If you take the single-line route, you can put that line inside or outside a method. The choice is yours. On the other hand, if you use two separate lines, the second line, `guests = new int[10]`, should be inside a method.

TIP

In an array declaration, you can put the square brackets before or after the variable name. In other words, you can write `int[] guests` or `int guests[]`. The computer creates the same `guests` variable no matter which form you use.

Creating an array in two easy steps

Look again at the two lines that you can use to create an array:

```
int[] guests;
guests = new int[10];
```

Each line serves its own distinct purpose:

- `int[] guests`: This first line is a declaration. The declaration reserves the array name (a name like *guests*) for use in the rest of the program. In the Java Motel metaphor, this line says, "I plan to build a motel here and assign a certain number of guests to each room." (See Figure 11-2.)

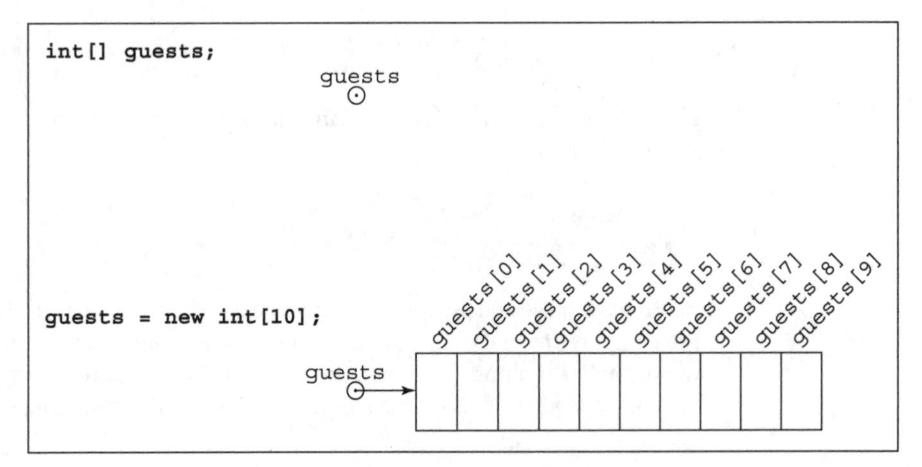

FIGURE 11-2: Two steps in creating an array.

Never mind what the declaration `int[] guests` actually does. It's more important to notice what the declaration `int[] guests` *doesn't* do. The declaration doesn't reserve ten memory locations. Indeed, a declaration like `int[] guests` doesn't really create an array. All the declaration does is set up the `guests` variable. At that point in the code, the `guests` variable still doesn't refer to a real array. (In other words, the motel has a name, but the motel hasn't been built yet.)

» `guests = new int[10]`: This second line is an assignment statement. The assignment statement reserves space in the computer's memory for ten `int` values. In terms of real estate, this line says, "I've finally built the motel. Go ahead and put guests in each room." (Again, see Figure 11-2.)

How to book hotel guests

After you've created an array, you can put values into the array's components. For instance, you want to store the fact that Room 6 contains four guests. To put the value 4 in the component with index 6, you write `guests[6] = 4`.

Now business starts to pick up. A big bus pulls up to the motel. On the side of the bus is a sign that says Noah's Ark. Out of the bus come 25 couples, each walking, stomping, flying, hopping, or slithering to the motel's small office. Only 10 of the couples can stay at the Java Motel, but that's okay because you can send the other 15 couples down the road to the old C-Side Resort and Motor Lodge.

Anyway, to register ten couples into the Java Motel, you put a couple (two guests) in each of your ten rooms. Having created an array, you can take advantage of the array's indexing and write a `for` loop, like this:

```
for (int roomNum = 0; roomNum < 10; roomNum++) {
    guests[roomNum] = 2;
}
```

This loop takes the place of ten assignment statements. Notice how the loop's counter goes from 0 to 9. Compare this with Figure 11-2 and remember that the indices of an array go from zero to one less than the number of components in the array.

However, given the way the world works, your guests don't always arrive in neat pairs, and you have to fill each room with a different number of guests. You probably store information about rooms and guests in a database. If you do, you can still loop through an array, gathering numbers of guests as you go. The code to perform such a task may look like this:

```
resultset = statement.executeQuery("select GUESTS from RoomData");
for (int roomNum = 0; roomNum < 10; roomNum++) {
    resultset.next();
    guests[roomNum] = resultset.getInt("GUESTS");
}
```

But because this book doesn't cover databases until Chapter 17, you may be better off reading numbers of guests from a plain-text file. A sample file named `GuestList.txt` is shown in Figure 11-3.

FIGURE 11-3: The GuestList.txt file.

```
1 4 2 0 2 1 4 3 0 2
```

After you've made a file, you can call on the `Scanner` class to get values from the file. The code is shown in Listing 11-1, and the resulting output is in Figure 11-4.

CROSS REFERENCE

For tips on reading from disk files, refer to Chapter 8.

```
Room    Guests
0       1
1       4
2       2
3       0
4       2
5       1
6       4
7       3
8       0
9       2
```

FIGURE 11-4: Running the code from Listing 11-1.

LISTING 11-1: **Filling an Array with Values**

```
package com.example.hotel;

import java.io.File;
import java.io.IOException;
import java.util.Scanner;

import static java.lang.System.out;

public class ShowGuests {

    public static void main(String[] args) throws IOException {
        int[] guests = new int[10];
        var diskScanner = new Scanner(new File("GuestList.txt"));

        for (int roomNum = 0; roomNum < 10; roomNum++) {
            guests[roomNum] = diskScanner.nextInt();
        }

        out.println("Room\tGuests");
```

```
        for (int roomNum = 0; roomNum < 10; roomNum++) {
            out.print(roomNum);
            out.print("\t");
            out.println(guests[roomNum]);
        }

        diskScanner.close();
    }
}
```

The code in Listing 11-1 has two `for` loops: The first loop reads numbers of guests, and the second loop writes numbers of guests.

TIP

Every array has a built-in length field. An array's *length* is the number of components in the array. So, in Listing 11-1, if you print the value of `guests.length`, you get 10.

Tab stops and other special things

In Listing 11-1, some calls to `print` and `println` use the `\t` escape sequence. It's called an *escape sequence* because you escape from displaying the letter `t` on the screen. Instead, the characters `\t` stand for a tab. The computer moves forward to the next tab stop before printing any more characters. Java has a few of these handy escape sequences. Some of them are shown in Table 11-1.

TABLE 11-1 Escape Sequences

Sequence	Meaning
\b	Backspace
\t	Horizontal tab
\n	Line feed
\f	Form feed
\r	Carriage return
\"	Double quote "
\'	Single quote '
\\	Backslash \

WARNING

In the Windows command prompt and Macintosh Terminal, tab stops are eight spaces apart. But in some environments, tab stops are only four spaces apart. With four-space stops, you don't see the sensible alignment that appears earlier, in Figure 11-4. If you want the columns to align no matter where your program runs, replace tab characters with blank spaces in Listing 11-1. (Of course, if you're fussy, you have to check for numbers with more than one digit. But that's another story.)

Make life easy for yourself

Besides what you see in Listing 11-1, you have another way to fill an array in Java: with an *array initializer.* When you use an array initializer, you don't even have to tell the computer how many components the array has. The computer figures it out for you.

Listing 11-2 shows a new version of the code to fill an array. The program's output is the same as the output from Listing 11-1. (It's the stuff shown earlier, in Figure 11-4.) The only difference between Listings 11-1 and 11-2 is the bold text in Listing 11-2. That bold doodad is an array initializer.

LISTING 11-2: Using an Array Initializer

```
package com.example.hotel;

import static java.lang.System.out;

public class ShorterShowGuests {

    public static void main(String[] args) {
        int[] guests = {1, 4, 2, 0, 2, 1, 4, 3, 0, 2};

        out.println("Room\tGuests");

        for (int roomNum = 0; roomNum < 10; roomNum++) {
            out.print(roomNum);
            out.print("\t");
            out.println(guests[roomNum]);
        }
    }
}
```

TIP

An array initializer can contain expressions as well as literals. In plain English, this means that you can put all kinds of things between the commas in the initializer. For instance, an initializer like `{1 + 3, keyboard.nextInt(), 2, 0, 2, 1, 4, 3, 0, 2}` works just fine.

UTILIZING ARRAYS

Run this code to discover some useful methods in the `java.util.Arrays` package:

```
import java.util.Arrays;

public class WriteArray {

    public static void main(String[] args) {
        int[] guests = {1, 4, 2, 0, 2, 1, 4, 3, 0, 2};
        System.out.println(guests);  // Not very useful!
        System.out.println(Arrays.toString(guests));
        Arrays.sort(guests);
        System.out.println(Arrays.toString(guests));
        Arrays.fill(guests, 0);
        System.out.println(Arrays.toString(guests));
    }
}
```

TOO MANY WORDS

In a fit of shameless narcissism, I made a rough count of the number of words in pages 1 to 7 of this book's previous edition.

```
int[] words = {0, 296, 342, 405, 363, 350, 323, 101};
```

Write a program that prompts the user for a page number and responds with the count of words on that particular page.

SPELL IT BACKWARD

Modify the code in Listing 11-1 so that it displays the rooms in reverse room-number order (starting with Room 9 and working down to Room 0).

MAKE YOUR MARKS

First the user types five numbers on the keyboard. Then the program displays five lines of characters, each line having the same number of asterisks as one of the user's numbers. A run of the program might look like this:

```
5
3
9
0
10
```

```
*****
***
*********

**********
```

ROOMS WITH A VIEW

Use my `DummiesFrame` (from Chapter 7) to create a GUI program based on the ideas in Listings 11-1 and 11-2. In your program, the frame has only one input row: a *Room number* row. If the user types 3 in the *Room number* row and then clicks the button, the program displays the number of guests in Room 3.

Stepping through an array with the enhanced for loop

Java has an enhanced `for` loop — a `for` loop that doesn't use counters or indices. Listing 11-3 shows you how to do it.

WARNING

The material in this section applies to Java 5.0 and later Java versions. But this section's material doesn't work with older versions of Java — versions such as 1.3, 1.4, and so on. For a bit more about Java's version numbers, see Chapter 2.

LISTING 11-3: Get a Load o' That for Loop!

```
package com.example.hotel;

import static java.lang.System.out;

public class EnhancedShowGuests {

    public static void main(String[] args) {
        int[] guests = {1, 4, 2, 0, 2, 1, 4, 3, 0, 2};
        int roomNum = 0;

        out.println("Room\tGuests");

        for (int numGuests : guests) {
            out.print(roomNum++);
            out.print("\t");
            out.println(numGuests);
        }
    }
}
```

Listings 11-1 and 11-3 have the same output. It's in Figure 11-4.

An *enhanced* `for` *statement* has three parts:

```
for (variableType variableName : rangeOfValues)
```

The first two parts are *variableType* and *variableName*. The loop in Listing 11-3 defines a variable named `numGuests`, and `numGuests` has type `int`. During each loop iteration, the variable `numGuests` takes on a new value. Refer to Figure 11-4 to see these values. The initial value is `1`. The next value is `4`. After that comes `2`. And so on.

Where is the loop finding all these numbers? The answer lies in the loop's *rangeOfValues*. In Listing 11-3, the loop's *rangeOfValues* is `guests`. So, during the initial loop iteration, the value of `numGuests` is `guests[0]` (which is `1`). During the next iteration, the value of `numGuests` is `guests[1]` (which is `4`). After that comes `guests[2]` (which is `2`). And so on.

WARNING

Java's enhanced `for` loop requires a word of caution: Each time through the loop, the variable that steps through the range of values stores a *copy* of the value in the original range. The variable does *not* point to the range itself.

For example, if you add an assignment statement that changes the value of `numGuests` in Listing 11-3, this statement has no effect on any of the values stored in the `guests` array. To drive this point home, imagine that business is bad and I've filled my hotel's `guests` array with zeros. Then I execute the following code:

```
for (int numGuests : guests) {
    numGuests += 1;
    out.print(numGuests + " ");
}

out.println();

for (int numGuests : guests) {
    out.print(numGuests + " ");
}
```

The `numGuests` variable takes on values stored in the `guests` array. But the `numGuests += 1` statement doesn't change the values stored in this `guests` array. The code's output looks like this:

```
1 1 1 1 1 1 1 1 1 1
0 0 0 0 0 0 0 0 0 0
```

TRACK YOUR WEIGHT

TRY IT OUT

Initialize an array with your weight for each day in a week:

```
double[] weight = {145.7, 148.3, 147.2, 146.2, 147.0, 148.5, 146.9};
```

Write code to display a list of your gains and losses during the course of the week. For example, from the first day to the next, your weight change is `2.6`. By the following day, your change is `-1.1`.

FIND THE BIGGEST

Add code to the end of Listing 11-1 so that the program finds the room(s) with the largest number of guests. The program's new output may look something like this:

```
4 guests in Room 1
4 guests in Room 6
```

To do this, have a variable named `largestSoFar`. Examine the array's elements one by one and then update the value of `largestSoFar` whenever you find a new, record-high number.

FIND THE TOTAL

Add code to the end of Listing 11-1 so that the program reports the total number of people staying at the motel. To do this, have a variable named `runningTotal`. Examine the array's values one by one, adding each such value to the big `runningTotal` value.

Do you have a room?

You're sitting behind the desk at the Java Motel. Look! Here comes a party of five. These people want a room, so you need software that checks whether a room is vacant. If one is, the software modifies the `GuestList.txt` file (refer to Figure 11-3) by replacing the number 0 with the number 5. As luck would have it, the software is on your hard drive. The software is shown in Listing 11-4.

LISTING 11-4: Do You Have a Room?

```
package com.example.hotel;

import java.io.File;
import java.io.IOException;
```

```
import java.io.PrintStream;
import java.util.Scanner;

import static java.lang.System.out;

public class FindVacancy {

    public static void main(String[] args) throws IOException {
        int[] guests = new int[10];
        int roomNum;

        var diskScanner = new Scanner(new File("GuestList.txt"));
        for (roomNum = 0; roomNum < 10; roomNum++) {
            guests[roomNum] = diskScanner.nextInt();
        }
        diskScanner.close();

        roomNum = 0;
        while (roomNum < 10 && guests[roomNum] != 0) {
            roomNum++;
        }

        if (roomNum == 10) {
            out.println("Sorry, no v cancy");
        } else {
            out.print("How many people for room ");
            out.print(roomNum);
            out.print("? ");

            var keyboard = new Scanner(System.in);
            guests[roomNum] = keyboard.nextInt();
            keyboard.close();

            var listOut = new PrintStream("GuestList.txt");
            for (roomNum = 0; roomNum < 10; roomNum++) {
                listOut.print(guests[roomNum]);
                listOut.print(" ");
            }

            listOut.close();
        }
    }
}
```

Figure 11-5 shows three runs of the code in Listing 11-4. The motel starts with two vacant rooms: Rooms 3 and 8. (Remember that the room numbers start with Room 0.) The first time you run the code, you put five people into Room 3. The second time you run the code, you put a party of ten in Room 8. (What a party!) The third time you run the code, you have no more vacant rooms. When the program discovers this, it displays the message `Sorry, no v cancy`, omitting at least one letter, in the tradition of all motel neon signs.

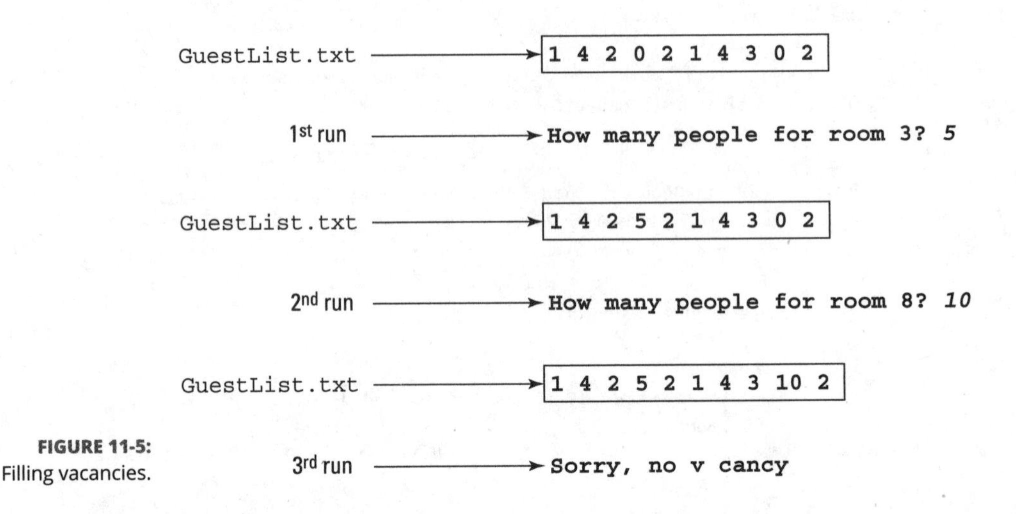

FIGURE 11-5: Filling vacancies.

In Listing 11-4, the condition `roomNum < 10 && guests[roomNum] != 0` can be tricky. If you move things around and write **guests[roomNum] != 0 && roomNum < 10**, you can get yourself into lots of trouble. For details, see this book's website (`http://javafordummies.allmycode.com`).

Writing to a file

The code in Listing 11-4 uses tricks from other chapters and sections of this book. The code's only brand-new feature is the use of `PrintStream` to write to a disk file. Think about any example in this book that calls `System.out.print` or `out.println` or their variants. What's *really* going on when you call one of these methods?

The thing called `System.out` is an object. The object is defined in the Java API. In fact, `System.out` is an instance of a class named `java.io.PrintStream` (or just `PrintStream`, to its close friends). Now each object created from the `PrintStream` class has methods named `print` and `println`. Just as each `Account` object in Listing 7-3 has a `display` method, and just as the `DecimalFormat` object in Listing 10-1 (over in Chapter 10) has a `format` method, so the `PrintStream` object named `out` has `print` and `println` methods. When you call `System.out.println`, you're calling a method that belongs to a `PrintStream` instance.

Okay, so what of it? Well, System.out always stands for some text area on your computer screen. If you create your own PrintStream object and you make that object refer to a disk file, that PrintStream object refers to the disk file. When you call that object's print method, you write text to a file on your hard drive.

In Listing 11-4, when you say

```
var listOut = new PrintStream("GuestList.txt");

listOut.print(guests[roomNum]);
listOut.print(" ");
```

you're telling Java to write text to a file on your hard drive — the GuestList.txt file.

That's how you update the count of guests staying in the hotel. When you call listOut.print for the number of guests in Room 3, you may print the number 5. So, in Figure 11-5, a number in the GuestList.txt file changes from 0 to 5. Then in Figure 11-5, you run the program a second time. When the program gets data from the newly written GuestList.txt file, Room 3 is no longer vacant. This time, the program suggests Room 8.

TIP

This is more of an observation than a tip. Say that you want to *read* data from a file named Employees.txt. To do this, you make a scanner. You call new Scanner(new File("Employees.txt")). If you accidentally call new Scanner("Employees.txt") without the new File part, the call doesn't connect to your Employees.txt file. But notice how you prepare to *write* data to a file. You make a PrintStream instance by calling new PrintStream("GuestList.txt"). You don't use new File anywhere in the call. If you goof and accidentally include new File, the Java compiler becomes angry, jumps out, and bites you.

When to close a file

Notice the placement of new Scanner calls, new PrintStream calls, and close calls in Listing 11-4. As in all the examples, each new Scanner call has a corresponding close call. And in Listing 11-4, the new PrintStream call has its own close call (the listOut.close() call). But in Listing 11-4, I'm careful to place these calls tightly around their corresponding nextInt and print calls. For example, I don't set up diskScanner at the very start of the program, and I don't wait until the very end of the program to close diskScanner. Instead, I perform all my diskScanner tasks, one after the other, in quick succession:

```
var diskScanner = new Scanner(new File("GuestList.txt")); //construct
for (roomNum = 0; roomNum < 10; roomNum++) {
```

```
        guests[roomNum] = diskScanner.nextInt(); //read
    }
    diskScanner.close(); //close
```

I do the same kind of thing with the keyboard and listOut objects.

I do this quick dance with input and output because my program uses GuestList.txt twice: once for reading numbers and a second time for writing numbers. If I'm not careful, the two uses of GuestList.txt might conflict with one another. Consider the following program:

```
// THIS IS BAD CODE
import java.io.File;
import java.io.IOException;
import java.io.PrintStream;
import java.util.Scanner;

public class BadCode {

    public static void main(String[] args) throws IOException {
        int[] guests = new int[10];

        var diskScanner = new Scanner(new File("GuestList.txt"));
        var listOut = new PrintStream("GuestList.txt");

        guests[0] = diskScanner.nextInt();
        listOut.print(5);

        diskScanner.close();
        listOut.close();
    }
}
```

Like many methods and constructors of its kind, the PrintStream constructor doesn't pussyfoot around with files. If it can't find a GuestList.txt file, the constructor creates a GuestList.txt file and prepares to write values into it. But, if a GuestList.txt file already exists, the PrintStream constructor deletes the existing file and prepares to write to a new, empty GuestList.txt file. In the BadCode class, the new PrintStream constructor call deletes whatever GuestList.txt file already exists. This deletion comes before the call to diskScanner.nextInt(). So diskScanner.nextInt() can't read whatever was originally in the GuestList.txt file. That's bad!

To avoid this disaster, I carefully separate the two uses of the GuestList.txt file in Listing 11-4. Near the top of the listing, I construct diskScanner and then read from the original GuestList.txt file and then close diskScanner. Later, toward the end of the listing, I construct `listOut` and then write to a new `GuestList.txt` file and then close `listOut`. With writing separated completely from reading, everything works correctly.

The `keyboard` variable in Listing 11-4 doesn't refer to `GuestList.txt`, so `keyboard` doesn't conflict with the other input or output variables. No harm would come from following my regular routine — putting `keyboard = new Scanner(System.in)` at the start of the program and putting `keyboard.close()` at the end of the program. But to make Listing 11-4 as readable and as uniform as possible, I place the `keyboard` constructor and the `close` call tightly around the `keyboard.nextInt` call.

TALLY HO!

Add code to the end of Listing 11-1 so that the program reports the number of empty rooms, the number of rooms with only one guest, the number of rooms with exactly two guests, and so on. To do this, create an array named `howManyRoomsWith`. Examine the motel's rooms one by one. When you encounter an empty room, add 1 to `howManyRoomsWith[0]`. When you encounter a room with one guest, add 1 to `howManyRoomsWith[1]`. And so on.

Arrays of Objects

The Java Motel is open for business, now with improved guest registration software! The people who brought you this chapter's first section are always scratching their heads, looking for the best ways to improve their services. Now, with some ideas from object-oriented programming, they've started thinking in terms of a `Room` class.

"And what," you ask, "would a `Room` instance look like?" That's easy. A `Room` instance has three properties: the number of guests in the room, the room rate, and a smoking/nonsmoking stamp. Figure 11-6 illustrates the situation.

Listing 11-5 shows the code that describes the `Room` class. As promised, each instance of the `Room` class has three fields: the `guests`, `rate`, and `smoking` fields. (A `false` value for the boolean field, `smoking`, indicates a nonsmoking room.) In addition, the entire `Room` class has a `static` field named `currency`. On my computer in the United States, this `currency` object makes room rates look like dollar amounts.

To find out what *static* means, see Chapter 10.

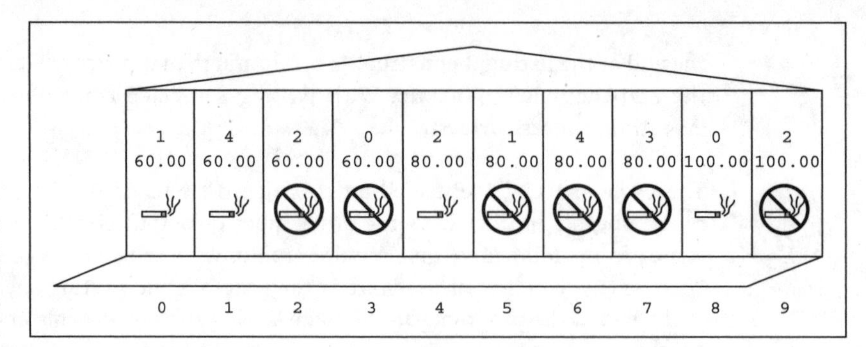

FIGURE 11-6: Another abstract snapshot of rooms in the Java Motel.

LISTING 11-5: So This Is What a Room Looks Like!

```
package com.example.hotel;

import java.text.NumberFormat;
import java.util.Scanner;

import static java.lang.System.out;

public class Room {
    private int guests;
    private double rate;
    private boolean smoking;
    private static NumberFormat currency = NumberFormat.getCurrencyInstance();

    public void readRoom(Scanner diskScanner) {
        guests = diskScanner.nextInt();
        rate = diskScanner.nextDouble();
        smoking = diskScanner.nextBoolean();
    }

    public void writeRoom() {
        out.print(guests);
        out.print("\t");
        out.print(currency.format(rate));
        out.print("\t");
        out.println(smoking ? "yes" : "no");
    }
}
```

Listing 11-5 has a few interesting quirks, but I'd rather not describe them until after you see all the code in action. That's why, at this point, I move right on to the code that calls the Listing 11-5 code. After you read about arrays of rooms (shown in Listing 11-6), check out my description of the Listing 11-5 quirks.

WARNING

This warning is a deliberate repeat of an idea from Chapter 4, Chapter 7, and from who-knows-what-other chapter: *Be careful* when you use type `double` or type `float` to store money values. Calculations with `double` or `float` can be inaccurate. For more information (and more finger wagging), see Chapters 4 and 7.

TIP

This tip has absolutely nothing to do with Java. If you're the kind of person who prefers a smoking room (with `boolean` field `smoking = true` in Listing 11-5), find someone you like — someone who can take off three consecutive days from work. Have that person sit with you and comfort you for 72 straight hours while you refrain from smoking. You might become temporarily insane while the nicotine leaves your body, but eventually you'll be okay. And your friend will feel like a real hero.

Using the Room class

Now you need an array of rooms. The code to create such a thing is in Listing 11-6. The code reads data from the `RoomList.txt` file. (Figure 11-7 shows the contents of the `RoomList.txt` file.)

Figure 11-8 shows a run of the code in Listing 11-6.

```
1
60.00
true
4
60.00
true
2
60.00
false
0
60.00
false
2
80.00
true
1
80.00
false
4
80.00
false
3
80.00
false
0
100.00
true
2
100.00
false
```

FIGURE 11-7: A file of Room data.

Room	Guests	Rate	Smoking?
0	1	$60.00	yes
1	4	$60.00	yes
2	2	$60.00	no
3	0	$60.00	no
4	2	$80.00	yes
5	1	$80.00	no
6	4	$80.00	no
7	3	$80.00	no
8	0	$100.00	yes
9	2	$100.00	no

FIGURE 11-8: A run of the code in Listing 11-6.

LISTING 11-6: Would You Like to See a Room?

```
package com.example.hotel;

import java.io.File;
import java.io.IOException;
import java.util.Scanner;

import static java.lang.System.out;

public class ShowRooms {
    public static void main(String[] args) throws IOException {

        Room[] rooms;
        rooms = new Room[10];
        var diskScanner = new Scanner(new File("RoomList.txt"));

        for (int roomNum = 0; roomNum < 10; roomNum++) {
            rooms[roomNum] = new Room();
            rooms[roomNum].readRoom(diskScanner);
        }

        out.println("Room\tGuests\tRate\tSmoking?");

        for (int roomNum = 0; roomNum < 10; roomNum++) {
            out.print(roomNum);
            out.print("\t");
            rooms[roomNum].writeRoom();
        }

        diskScanner.close();
    }
}
```

Say what you want about the code in Listing 11-6. As far as I'm concerned, only one issue in the whole listing should concern you. And what, you ask, is that issue? Well, to create an array of *objects* — as opposed to an array made up of primitive values — you have to do three things: Make the array variable, make the array itself, and then construct each individual object in the array. This is different from creating an array of `int` values or an array containing any other primitive type values. When you create an array of primitive type values, you do only the first two of these three things.

To help make sense of all this, follow along in Listing 11-6 and Figure 11-9 as you read the following points:

- `Room[] rooms;`: This declaration creates a `rooms` variable. This variable is destined to refer to an array (but doesn't yet refer to anything).
- `rooms = new Room[10];`: This statement reserves ten slots of storage in the computer's memory. The statement also makes the `rooms` variable refer to the group of storage slots. Each slot is destined to refer to an object (but doesn't yet refer to anything).
- `rooms[roomNum] = new Room();`: This statement is inside a `for` loop. The statement is executed once for each of the ten room numbers. For example, the first time through the loop, this statement says `rooms[0] = new Room()`. That first time around, the statement makes the slot `rooms[0]` refer to an actual object (an instance of the `Room` class).

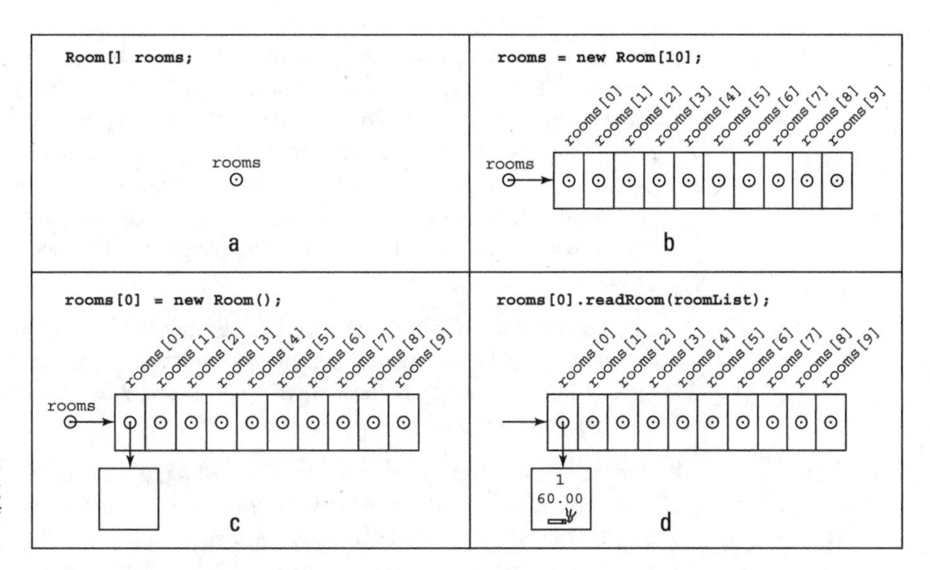

FIGURE 11-9: Steps in creating an array of objects.

Although it's technically not considered a step in array making, you still have to fill each object's fields with values. For instance, the first time through the loop, the `readRoom` call says `rooms[1].readRoom(diskScanner)`, which means, "Read data from the RoomList.txt file into the rooms[1] object's fields (the guests, rate, and smoking fields)." Each time through the loop, the program creates a new object and reads data into that new object's fields.

You can squeeze the steps together just as you do when creating arrays of primitive values. For instance, you can complete the first two steps in one fell swoop, like this:

```
Room[] rooms = new Room[10];
```

You can also use an array initializer. (For an introduction to array initializers, see the section "Make live easy for yourself," earlier in this chapter.)

Yet another way to beautify your numbers

You can make numbers look nice in plenty of ways. If you take a peek at earlier chapters, for example, you can see that Listing 7-7 uses printf and Listing 10-1 uses a DecimalFormat. But in Listing 11-5, I display a currency amount. I use the NumberFormat class with its getCurrencyInstance method.

If you compare the formatting statements in Listings 10-1 and 11-5, you don't see much difference:

- **One listing uses a constructor; the other listing calls** getCurrencyInstance. The getCurrencyInstance method is a good example of a *factory method,* which is a convenient tool for creating commonly used objects. People always need code that displays currency amounts. So the getCurrencyInstance method creates a currency format without forcing you to write a complicated DecimalFormat constructor call. In the United States, this complicated constructor call would be new DecimalFormat("$###0.00;($###0.00)").

  Like a constructor, a factory method returns a brand-new object. But unlike a constructor, a factory method has no special status. If you create your own factory method, you can name it anything you want. When you call a factory method, you don't use the keyword new.

- One listing uses DecimalFormat; the other listing uses NumberFormat. A decimal number is a certain kind of number. (In fact, a decimal number is a number written in the base-10 system.) Accordingly, the DecimalFormat class is a subclass of the NumberFormat class. The DecimalFormat methods are more specific, so for most purposes, I use DecimalFormat. But it's harder to use the `DecimalFormat` class's `getCurrencyInstance` method. For programs that involve money, I tend to use `NumberFormat`.

» Both listings use `format` methods. In the end, you just write something like `currency.format(rate)` or `decFormat.format(average)`. After that, Java does the work for you.

REMEMBER

From Chapter 4 onward, I issue gentle warnings against using types such as `double` and `float` for storing currency values. For the most accurate currency calculations, use `int`, `long`, or — best of all — `BigDecimal`.

CROSS REFERENCE

You can read more about the dangers of `double` types and currency values in Chapter 7.

The conditional operator

Listing 11-5 uses an interesting doodad called the *conditional operator*. This conditional operator takes three expressions and returns the value of just one of them. It's like a mini `if` statement. When you use the conditional operator, it looks something like this:

```
conditionToBeTested ? expression1 : expression2
```

The computer evaluates the *conditionToBeTested* condition. If the condition is true, the computer returns the value of *expression1*. But, if the condition is false, the computer returns the value of *expression2*.

So, in the code

```
smoking ? "yes" : "no"
```

the computer checks whether `smoking` has the value `true`. If so, the whole 3-part expression stands for the first string, `"yes"`. If not, the whole expression stands for the second string, `"no"`.

In Listing 11-5, the call to `out.println` causes either `"yes"` or `"no"` to display. Which string gets displayed depends on whether `smoking` has the value `true` or `false`.

TRY IT OUT

How do you learn Java? You learn it the same way you get to Carnegie Hall: Practice! Practice! Practice!

A CLASSY HOTEL

Supercharge this chapter's examples:

- Modify the code in Listing 11-4 so that it uses the Room class from Listing 11-5.
- Add to your work so that when you specify a number of incoming guests, you also enter the guests' preference for a smoking or nonsmoking room.
- Create a subclass of the Room class in Listing 11-5. In your subclass, each room has a maximum occupancy. Don't let your code put more guests in a room than the room's occupancy will allow.

DOUBLE YOUR PRESSURE

In Chapter 9, you create a `Student` class. Each student has a name and an ID number. For this programming challenge, imagine that each student has five grades — one for each of the five courses the student takes. Each grade is a `double` value from `0.0` to `4.0` (`4.0` is the best). A student's grade point average (GPA) is the average of the student's five grade values.

In this chapter's `Student` class, one of the fields is an array of five `double` values. Your program finds the student's GPA and displays it (along with the student's name and ID number) on the screen.

How to Argue with Your Code

Take a method header from Listing 11-5:

```
public void readRoom(Scanner diskScanner) {
```

This method has a parameter named `diskScanner`. The first time you call the method, you can make `diskScanner` read from the `RoomList.txt` file. The second time, you can make `diskScanner` read from the `wellNourishedKittens` file. The third time, `diskScanner` may represent the `underAppreciatedAuthors.md` file. This flexibility comes from the fact that `diskScanner` gets its value at the moment you call the method — and not a nanosecond before then.

The same kind of thing can be true of your program's `main` method. Every `main` method has a `String[] args` parameter. When you launch a program, you can send one or more words to the `args` parameter. For example, if you send

```
Java For Dummies
```

to the program, then `args[0]` is `"Java"`, `args[1]` is `"For"`, and `args[2]` is `"Dummies"`. If you send

```
"Java For Dummies" Great!
```

to the program then, because of the added quotation marks, `args[0]` is `"Java For Dummies"` and `args[1]` is `"Great!"`.

The way you send words to the `args` parameter depends on your IDE, but the general idea depends on two things:

- **There are many ways to run a program.**

  Each of these ways is called a *run configuration*. Without knowing it, you've probably been using a default, one-size-fits-most run configuration to test this book's sample programs. That default configuration served you well.

  But now, it's time for a change. A run configuration contains answers to several different questions: Which file contains the `main` method? Which version of Java will you run? Which of your drive's folders will be used? Where will the `System.out` text appear? And so on.

- **One of these answers is named *program arguments*.**

  Most IDEs have some kind of Modify Run Configuration menu option. When you select this option, you see a dialog box with places for all these answers. In that dialog box, one text field is labeled Program Arguments. Your IDE takes whatever you type in that text field and sends it to `args` when your code runs.

  You can change a configuration's program arguments before each run or create several different configurations, each with its own program arguments. For help setting up program arguments in Eclipse and IntelliJ IDEA, visit this book's website (`http://javafordummies.allmycode.com`).

TECHNICAL STUFF

You don't need an IDE to feed arguments to a program. Imagine that you have a Java class named `ThreeLittleWords`. Depending on your setup, you may be able to open a Command Prompt or Terminal window and type `java ThreeLittleWords.java Java For Dummies`. If all goes well, the `args` array will contain the words `Java For Dummies`.

Settling the argument

Listing 11-7 shows you how to use program arguments in your code.

LISTING 11-7: Generate a File of Numbers

```
package com.example.numbers;

import java.io.IOException;
import java.io.PrintStream;
import java.util.Random;

public class MakeRandomNumsFile {

    public static void main(String[] args) throws IOException {

        var generator = new Random();

        if (args.length < 2) {
            System.out.println("Usage: MakeRandomNumsFile filename number");
            System.exit(1);
        }

        var printOut = new PrintStream(args[0]);
        int numLines = Integer.parseInt(args[1]);

        for (int count = 1; count <= numLines; count++) {
            printOut.println(generator.nextInt(10) + 1);
        }

        printOut.close();
    }
}
```

Figure 11-10 shows the result of a run of Listing 11-7. Before the run, I set the program arguments to `RandomNumbers.txt 5`. The array component `args[0]` takes on the value `"RandomNumbers.txt"`, and `args[1]` becomes `"5"`. So the program's assignment statements end up having the following meaning:

```
var printOut = new PrintStream("RandomNumbers.txt");
int numLines = Integer.parseInt("5");
```

The program creates a file named `RandomNumbers.txt` and sets `numLines` to `5`. So, later in the code, the program randomly generates five values and puts those values into `RandomNumbers.txt`.

REMEMBER

To find the new `RandomNumbers.txt` file, look first in your project's root folder. For more information, refer to Chapter 6.

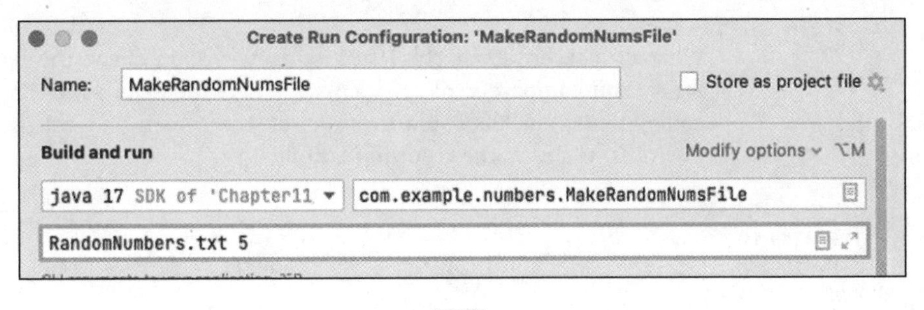

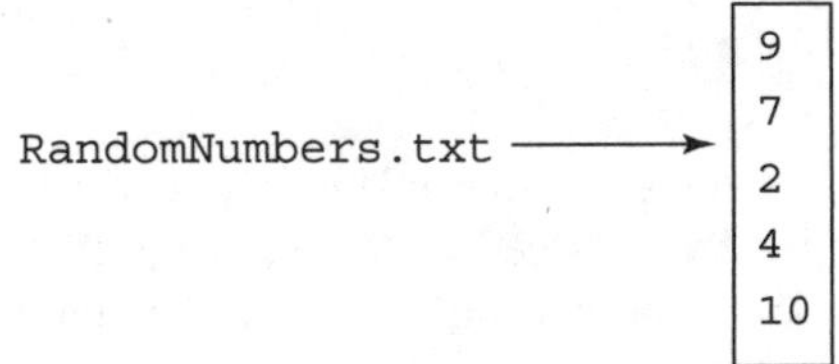

FIGURE 11-10: A run of the code in Listing 11-7.

Notice how each program argument in Listing 11-7 is a `String` value. When you look at `args[1]`, you don't see the number 5 — you see the string `"5"` with a digit character in it. Unfortunately, you can't use that `"5"` to do any counting. To get an `int` value from `"5"`, you have to apply the `parseInt` method. (Again, see Listing 11-7.)

The `parseInt` method lives inside a class named *Integer*. So, to call `parseInt`, you preface the name *parseInt* with the word `Integer`. The `Integer` class has all kinds of handy methods for doing things with `int` values.

REMEMBER

In Java, *Integer* is the name of a class, and *int* is the name of a primitive (simple) type. The two words are related, but they're not the same. The `Integer` class has methods and other tools for dealing with `int` values.

Checking for the right number of program arguments

What happens if the user makes a mistake? What if the user omits the number `5` in the text field in Figure 11-10?

Then the computer assigns `"RandomNumbers.txt"` to `args[0]` but assigns nothing to `args[1]`. This is bad. If the computer ever reaches the statement

```
int numLines = Integer.parseInt(args[1]);
```

the program crashes with an unfriendly `ArrayIndexOutOfBoundsException`.

What do you do about this? In Listing 11-7, you check the length of the `args` array. You compare `args.length` with `2`. If the `args` array has fewer than two components, you display a message on the screen and exit from the program. Figure 11-11 shows the resulting output.

FIGURE 11-11: The code in Listing 11-7 tells you how to run it.

```
Usage: MakeRandomNumsFile filename number
```

WARNING

Despite the checking of `args.length` in Listing 11-7, the code still isn't crash-proof. If the user types **five** instead of 5, the program takes a nosedive with a `NumberFormatException`. The second program argument can't be a word. The argument has to be a number (and a whole number, at that). I can add `if` statements to Listing 11-7 to make the code more bulletproof, but checking for the `NumberFormatException` is better done in Chapter 13.

When you're working with program arguments, you can enter a `String` value with a blank space in it — just enclose the value in double quote marks. For instance, you can run the code of Listing 11-7 with arguments `"My Big Fat File.txt" 7`.

The sun is about to set on this book's discussion of arrays. But before you leave the subject of arrays, think about this: An array is a row of things, and not every kind of thing fits into just one row. Take the first few examples in this chapter involving the motel. The motel rooms, numbered 0 through 9, are in one long line. But what if you move up in the world? You buy a big hotel with 50 floors and with 100 rooms on each floor. Then the data is square shaped. You have 50 rows, and each row contains 100 items. Sure, you can think of the rooms as though they're all in one long row, but why should you have to do that? How about having a 2-dimensional array? It's a square-shaped array in which each component has two indices: a row number and a column number. Alas, I have no space in this book to show you a 2-dimensional array (and I can't afford a big hotel's prices, anyway). But if you visit this book's website (`http://javafordummies.allmycode.com`), you can read all about it.

TRY IT OUT

You can never get too much practice:

MINE IS THE BIGGEST

Write a program whose program arguments include three `int` values. As its output, the program displays the largest of the three `int` values.

» **Dealing with a bunch of objects at once**

» **Using Java's cool functional programming features**

» **Developing code for multicore processors**

Chapter 12 Using Collections and Streams (When Arrays Aren't Good Enough)

Chapter 11 is about arrays. With an array, you can manage a bunch of things all at once. In a hotel management program, you can keep track of all the rooms. You can quickly find the number of people in a room or find a vacant room.

However, arrays don't always fit the bill. In this chapter, you find out where arrays fall short and how collections can save the day.

Arrays Have Limitations

Imagine that you store customer names in some predetermined order. Your code contains an array, and the array has space for 100 names:

```
String[] name = new String[100];
for (int i = 0; i < 100; i++) {
    name[i] = new String();
}
```

All is well until, one day, customer number 101 shows up. As your program runs, you enter data for customer 101, hoping desperately that the array with 100 components can expand to fit your growing needs.

No such luck. Arrays don't expand. Your program crashes with an `ArrayIndexOutOfBoundsException`.

"In my next life, I'll create arrays of length 1,000," you say to yourself. And when your next life rolls around, you do just that:

```
String[] name = new String[1000];
for (int i = 0; i < 1000; i++) {
    name[i] = new String();
}
```

But during your next life, an economic recession occurs. Rather than have 101 customers, you have only 3 customers. Now you're wasting space for 1,000 names when space for 3 names would do.

And what if no economic recession occurs? You're sailing along with your array of size 1,000, using a tidy 825 spaces in the array. The components with indices 0 through 824 are being used, and the components with indices 825 through 999 are waiting quietly to be filled.

One day, a brand-new customer shows up. Because your customers are stored in order (alphabetically by last name, numerically by identification number, whatever), you want to squeeze this customer into the correct component of your array. The trouble is that this customer belongs early on in the array, at the component with index 7. What happens then?

You take the name in component number 824 and move it to component 825. Then you take the name in component 823 and move it to component 824. Take the name in component 822 and move it to component 823. Continue doing this until you've moved the name in component 7. Then you put the new customer's name into component 7. What a pain! Sure, the computer doesn't complain. (If the computer has feelings, it probably likes this kind of busywork.) But as you move around all these names, you waste processing time, you waste power, and you waste all kinds of resources.

"In my next life, I'll leave three empty components between every two names." And of course, your business expands. Eventually you find that three aren't enough.

Collection Classes to the Rescue

The issues in the preceding section aren't new. Computer scientists have been working on these issues for a long time. They haven't discovered any magic one-size-fits-all solution, but they've discovered some clever tricks.

The Java API has a bunch of classes known as *collection* classes. Each collection class has methods for storing bunches of values, and each collection class's methods use some clever tricks. For you, the bottom line is as follows: Certain collection classes deal as efficiently as possible with the issues raised in the preceding section. If you have to deal with such issues when writing code, you can use these collection classes and call the classes' methods. Rather than fret about a customer whose name belongs in position 7, you can just call a class's `add` method. The method inserts the name at a position of your choice and deals reasonably with whatever ripple effects have to take place. In the best circumstances, the insertion is quite efficient. In the worst circumstances, you can rest assured that the code does everything the best way it can.

Using an ArrayList

One of the most versatile of Java's collection classes is the `ArrayList`. Listing 12-1 shows you how it works.

LISTING 12-1: Working with a Java Collection

```
package com.example.people;

import java.io.File;
import java.io.IOException;
import java.util.ArrayList;
import java.util.Scanner;

import static java.lang.System.out;

public class ShowNames {

    public static void main(String[] args) throws IOException {

        ArrayList<String> people = new ArrayList<>();
        var diskScanner = new Scanner(new File("names.txt"));

        while (diskScanner.hasNext()) {
            people.add(diskScanner.nextLine());
        }

        people.remove(0);
        people.add(2, "Jim Newton");
        for (String name : people) {
            out.println(name);
        }

        diskScanner.close();
    }
}
```

Figure 12-1 shows you a sample `names.txt` file. The code in Listing 12-1 reads that `names.txt` file and prints the stuff in Figure 12-2.

```
Barry Burd
Harriet Ritter
Weelie J. Katz
Harry "The Crazyman" Spoonswagler
Felicia "Fishy" Katz
Mia, Just "Mia"
Jeremy Flooflong Jones
I. M. D'Arthur
Hugh R. DaReader
```

FIGURE 12-1: Several names in a file.

```
Harriet Ritter
Weelie J. Katz
Jim Newton
Harry "The Crazyman" Spoonswagler
Felicia "Fishy" Katz
Mia, Just "Mia"
Jeremy Flooflong Jones
I. M. D'Arthur
Hugh R. DaReader
```

FIGURE 12-2: The code in Listing 12-1 changes some of the names.

All the interesting things happen when you execute the `remove` and `add` methods. The variable named `people` refers to an `ArrayList` object. When you call that object's `remove` method,

```
people.remove(0);
```

you eliminate a value from the list. In this case, you eliminate whatever value is in the list's initial position (the position numbered 0). So, in Listing 12-1, the call to `remove` takes the name `Barry Burd` out of the list.

That leaves only eight names in the list, but then the next statement,

```
people.add(2, "Jim Newton");
```

inserts a name into position number 2. (After Barry is removed, position number 2 is the position occupied by Harry Spoonswagler, so Harry moves to position 3, and Jim Newton becomes the number 2 person.)

Notice that an `ArrayList` object has two different `add` methods. The method that adds Jim Newton has two parameters: a position number and a value to be added. Another `add` method

```
people.add(diskScanner.nextLine());
```

takes only one parameter. This statement takes whatever name it finds on a line of the input file and appends that name to the end of the list. (The `add` method with only one parameter always appends its value to what's currently the end of the `ArrayList` object.)

The last few lines of Listing 12-1 contain an enhanced `for` loop. Like the loop in Listing 11-3, the enhanced loop in Listing 12-1 has the following form:

```
for (variableType variableName : rangeOfValues)
```

In Listing 12-1, the *variableType* is String, the *variableName* is name, and the *rangeOfValues* includes the things stored in the people collection. During an iteration of the loop, name refers to one of the String values stored in people. (So, if the people collection contains nine values, the for loop goes through nine iterations.) During each iteration, the statement inside the loop displays a name on the screen.

Using generics

Look again at Listing 12-1, shown earlier, and notice the funky ArrayList declaration:

```
ArrayList<String> people = new ArrayList<>();
```

Each collection class is *generified.* That ugly-sounding word means that every collection declaration should contain some angle-bracketed stuff, such as <String>. The thing that's sandwiched between < and > tells Java what kinds of values the new collection may contain. For example, in Listing 12-1, the words ArrayList<String> people tell Java that people is a bunch of strings. That is, the people list contains String objects (not Room objects, not Account objects, not Employee objects — nothing other than String objects).

WARNING

You can't use generics in any version of Java before Java 5.0, and the code in Listing 12-1 goes kablooey in any version before Java 7. For more about generics, see the later sidebar, "All about generics."

In Listing 12-1, the words ArrayList<String> people say that the people variable can refer only to a collection of String values. So, from that point on, any reference to an item from the people collection is treated exclusively as a String. If you write

```
people.add(new Room());
```

the compiler coughs up your code and spits it out because a Room (created in Chapter 11) isn't the same as a String. (This coughing and spitting happens even if the compiler has access to the Room class's code — the code in Chapter 11.) But the statement

```
people.add("George Gow");
```

is just fine. Because "George Gow" has type String, the compiler smiles happily.

ALL ABOUT GENERICS

One of the original design goals for Java was to keep the language as simple as possible. James Gosling, the language's creator, took some unnecessarily complicated features of C++ and tossed them out the window. The result was a language that was elegant and sleek. Some people said the language was *too* sleek. So, after several years of discussion and squabbling, Java became a bit more complicated. By the year 2004, Java had enum types, enhanced for loops, static import, and other interesting new features. But the most talked-about new feature was the introduction of generics:

```
ArrayList<String> people = new ArrayList<String>();
```

The use of anything like <String> was new in Java 5.0. In old-style Java, you'd write

```
ArrayList people = new ArrayList();
```

In those days, an ArrayList could store almost anything you wanted to put in it: a number, an Account, a Room, a String — anything. The ArrayList class was versatile, but with this versatility came some headaches. If you could put anything into an ArrayList, you couldn't easily predict what you would get out of an ArrayList. In particular, you couldn't easily write code that assumed you had stored certain types of values in the ArrayList. Here's an example:

```
ArrayList things = new ArrayList();
things.add(new Account());
Account myAccount = things.get(0);
//DON'T USE THIS. IT'S BAD CODE.
```

In the third line, the call to get(0) grabs the earliest value in the things collection. The call to get(0) is okay, but then the compiler chokes on the attempted assignment to myAccount. You see a message on the third line saying that whatever you get from the things list can't be stuffed into the myAccount variable. You see this message because by the time the compiler reaches the third line, it has forgotten that the item added on the second line was of type Account!

The introduction of generics fixes this problem:

```
ArrayList<Account> things = new ArrayList<Account>();
things.add(new Account());
Account myAccount = things.get(0);
//USE THIS CODE INSTEAD. IT'S GOOD CODE.
```

Adding <Account> in two places tells the compiler that things stores Account instances — nothing else. So, in the third line in the preceding code, you get a value

(continued)

from the `things` collection. Then, because `things` stores only `Account` objects, you can make `myAccount` refer to that new value.

Java 5.0 added generics to Java. But soon after the birth of Java 5.0, programmers noticed how clumsy the code for generics can be. After all, you can create generics within generics. An `ArrayList` can contain a bunch of arrays, each of which can be an `ArrayList`. So you can write

```
ArrayList<ArrayList<String>[]> mess = new ArrayList<ArrayList<String>[]>();
```

All the repetition in that `mess` declaration gives me a headache! To avoid this ugliness, Java 7 and later versions have a diamond operator: `<>`. The diamond operator tells Java to reuse whatever insanely complicated stuff you put in the previous part of the generic declaration. In this example, the `<>` tells Java to reuse `<ArrayList<String>[]>`, even though you write `<ArrayList<String>[]>` only once. Here's how the streamlined Java 7 code looks:

```
ArrayList<ArrayList<String>[]> mess = new ArrayList<>();
```

In Java 7 and later, you can write either of these `mess` declarations: the original, nasty declaration with two occurrences of `ArrayList<String>[]` or the streamlined (only mildly nasty) declaration with the diamond operator and only one `ArrayList<String>[]` occurrence.

Yes, the streamlined code is still complicated. But without all the `ArrayList<String>[]` repetition, the streamlined code is less cumbersome. The Java 7 diamond operator takes away one chance for you to copy something incorrectly and have a big error in your code.

Chapter 4 shows you how to declare a variable using the word `var`. How does `var` work in declarations with generics? That `var` word's job is to take responsibility away from the start of a declaration, so `var` doesn't play nicely with the diamond operator. In Listing 12-1, you can write

```
ArrayList<String> people = new ArrayList<>();
```

or

```
var people = new ArrayList<String>();
```

but you can't write

```
var people = new ArrayList<>(); //Not enough information!
```

Wrap it up

In Chapter 4, I point out that Java has two kinds of types: primitive types and reference types. (If you didn't read those sections, or you don't remember them, don't feel guilty. You'll be okay.) Things like int, double, char, and boolean are primitive types, and things like String, JFrame, ArrayList, and Account are reference types.

The distinction between primitive types and reference types has been a source of contention since Java's birth in 1995. Even now, Oracle's wizards are hatching plans to get around the stickier consequences of having two kinds of types. One of those consequences is the fact that collections, such as the ArrayList, can't contain values of a primitive type. For example, it's okay to write

```
ArrayList<String> people = new ArrayList<>();
```

but it's not okay to write

```
ArrayList<int> numbers = new ArrayList<>(); // BAD! BAD!
```

because int is a primitive type. So, if you want to store values like 3, 55, and 21 in an ArrayList, what do you do? Rather than store int values in the ArrayList, you store Java's Integer values:

```
ArrayList<Integer> list = new ArrayList<>();
```

In previous chapters, you see the Integer class in connection with the parseInt method:

```
int numberOfCows = Integer.parseInt("536");
```

The Integer class has many methods, such as parseInt, for dealing with int values. The class also has fields such as MAX_VALUE and MIN_VALUE, which stand for the largest and smallest values that int variables may have.

The Integer class is an example of a *wrapper class*. Each of Java's eight primitive types has a corresponding wrapper class. You can use methods and fields in Java's Double, Character, Boolean, Long, Float, Short, and Byte wrapper classes. For example, the Double class has methods named parseDouble, compareTo, toHexString, and fields named MAX_VALUE and MAX_EXPONENT.

Notice the pattern. All eight of Java's primitive types (such as int, double, char) have names that start with lowercase letters. In contrast, all of the wrapper classes (Integer, Double, Character, and so on) have names that begin with uppercase letters.

TIP

Java programmers like to begin the names of classes with uppercase letters. You can declare `public class account` or `public class employee`. But if you do, other programmers will think you're a rube.

The `Integer` class wraps the primitive `int` type with useful methods and values. In addition, you can create an `Integer` instance that wraps a single `int` value:

```
Integer myInteger = new Integer(42);
```

In this line of code, the `myInteger` variable has one `int` value inside it: the `int` value `42`. In Paul's words, wrapping the `int` value `42` into an `Integer` object `myInteger` is "something like putting lots of extra breading on okra — it makes `42` more digestible for finicky eaters like collections."

Instances of the other wrapper classes work the same way. For example, an instance of the `Double` class wraps up a single primitive `double` value.

```
Double averageNumberOfTomatoes = new Double(1.41421356237);
```

Here's a program that stores five `Integer` values in an `ArrayList`:

```
import java.util.ArrayList;

public class Main {

    public static void main(String[] args) {
        ArrayList<Integer> list = new ArrayList<>();
        fillTheList(list);
        for (Integer n : list) {
            System.out.println(n);
        }
    }

    public static void fillTheList(ArrayList<Integer> list) {
        list.add(85);
        list.add(19);
        list.add(0);
        list.add(103);
        list.add(13);
    }
}
```

In the code, notice calls like `list.add(85)` that have `int` value parameters. At this point, little Billy gets excited and says, "Look, Mom! I added the primitive `int` value `85` to my `ArrayList`!" No, Billy. That's not what's going on.

In this code, the `list` collection contains `Integer` values, not `int` values. A primitive `int` value is a lot like an instance of the `Integer` class. But a primitive `int` value isn't exactly the same as an `Integer` instance.

What's going on is called *autoboxing*. Before Java 5.0, you had to write

```
list.add(new Integer(85));
```

if you wanted to add an `Integer` to an `ArrayList`. But Java 5.0 and later Java versions can automatically wrap an `int` value inside a box. An `int` value in a parameter list becomes an `Integer` in `ArrayList`. Java's autoboxing feature makes programs easier to read and write.

Are we done yet?

Here's a pleasant surprise. When you write a program like the one shown previously in Listing 12-1, you don't have to know how many names are in the input file. Having to know the number of names may defeat the purpose of using the easily expandable `ArrayList` class. Rather than loop until you read exactly nine names, you can loop until you run out of data.

The `Scanner` class has several nice methods, such as `hasNextInt`, `hasNextDouble`, and plain old `hasNext`. Each of these methods checks for more input data. If there's more data, the method returns `true`. Otherwise, the method returns `false`.

Listing 12-1 uses the general-purpose `hasNext` method. This `hasNext` method returns `true` as long as there's anything more to read from the program's input. After the program scoops up that last `Hugh R. DaReader` line in Figure 12-1, the subsequent `hasNext` call returns `false`. This `false` condition ends execution of the `while` loop and plummets the computer toward the remainder of the Listing 12-1 code.

The `hasNext` method is quite handy. In fact, `hasNext` is so handy that it's part of a bigger concept known as an *iterator*, and iterators are baked into all of Java's collection classes.

Once and again

An iterator spits out a collection's values, one after another. To obtain a value from the collection, you call the iterator's `next` method. To find out whether the collection has any more values in it, you call the iterator's `hasNext` method. Listing 12-2 uses an iterator to display people's names.

LISTING 12-2: Iterating through a Collection

```
package com.example.people;

import java.io.File;
import java.io.IOException;
import java.util.ArrayList;
import java.util.Iterator;
import java.util.Scanner;

import static java.lang.System.out;

public class ShowNamesAgain {

    public static void main(String[] args) throws IOException {

        ArrayList<String> people = new ArrayList<>();
        var diskScanner = new Scanner(new File("names.txt"));

        while (diskScanner.hasNext()) {
            people.add(diskScanner.nextLine());
        }

        people.remove(0);
        people.add(2, "Jim Newton");

        Iterator<String> iterator = people.iterator();
        while (iterator.hasNext()) {
            out.println(iterator.next());
        }

        diskScanner.close();
    }
}
```

You can replace the enhanced `for` loop at the end of Listing 12-1 with the boldface code in Listing 12-2. When you do, you get the same output as before. (You get the output in Figure 12-2.) In Listing 12-2, the first boldface line of code creates an iterator from the `people` collection. The second and third lines call the iterator's `hasNext` and `next` methods to grab all objects stored in the `people` collection — one for each iteration of the loop. These lines display each of the `people` collection's values.

Which is better? An enhanced `for` loop or an iterator? Java programmers prefer the enhanced `for` loop because the `for` loop involves less baggage — no `iterator` object to carry from one line of code to the next. But as you see later in this chapter, the most programming-enhanced feature can be upgraded, streamlined, tweaked, and otherwise reconstituted. There's no end to the way you can improve upon your code.

So many collection classes!

The `ArrayList` class that I use in many of this chapter's examples is only the tip of the Java collections iceberg. The Java library contains many collections classes, each with its own advantages. Table 12-1 contains an abbreviated list.

TABLE 12-1 Some Collection Classes

Class Name	Characteristic
`ArrayList`	A resizable array.
`LinkedList`	A list of values, each having a field that points to the next one in the list.
`Stack`	A structure that grows from bottom to top. The structure is optimized for access to the topmost value. You can easily add a value to the top or remove the value from the top.
`Queue`	A structure that grows at one end. The structure is optimized for adding values to one end (the rear) and removing values from the other end (the front).
`PriorityQueue`	A structure, like a queue, that lets certain (higher-priority) values move toward the front.
`HashSet`	A collection containing no duplicate values.
`HashMap`	A collection of key/value pairs.

Each collection class has its own set of methods (in addition to the methods that it inherits from `AbstractCollection`, the ancestor of all collection classes).

To find out which collection classes best meet your needs, visit the Java API documentation pages at `https://docs.oracle.com/en/java/javase/17/docs/api/java.base/java/util/doc-files/coll-overview.html`.

Once again, I'd like to put you to work:

TRY IT OUT

OUR BEST PLAYERS

Use the `Player` class from Chapter 10 to make an `ArrayList` of `Player` objects. Write a program to display the names of the players whose average is .100 or higher.

OUR BIGGEST NUMBER

Create an `ArrayList` containing `Integer` values. Then step through the values in the list to find the largest value among all values in the list. For example, if the list contains the numbers 85, 19, 0, 103, and 13, display the number 103.

INSERT IN ORDER

Create an `ArrayList` containing `String` values in alphabetical order. When the user types an additional word on the keyboard, the program inserts the new word into the `ArrayList` in the proper (alphabetically ordered) place.

For example, imagine that the list starts off containing the words `"cat"`, `"dog"`, `"horse"`, and `"zebra"` (in that order). After the user types the word `fish` on the keyboard (and presses Enter), the list contains the words `"cat"`, `"dog"`, `"fish"`, `"horse"`, and `"zebra"` (in that order).

To write this program, you may find the `String` class's `compareToIgnoreCase` method and the `ArrayList` class's `size` method useful. You can find out about these methods by visiting `https://docs.oracle.com/en/java/javase/17/docs/api/java.base/java/lang/String.html#compareToIgnoreCase(java.lang.String)` and `https://docs.oracle.com/en/java/javase/17/docs/api/java.base/java/util/ArrayList.html#size()`.

Functional Programming

From 1953 to 1957, John Backus and others developed the FORTRAN programming language, which contained the basic framework for thousands of 20th century programming languages. The framework has come to be known as *imperative programming* because of its do-this-then-do-that nature.

A few years after the rise of FORTRAN, John McCarthy created another language, named *Lisp*. Unlike FORTRAN, the underlying framework for Lisp is *functional programming*. In a purely functional program, you avoid writing "do this, then do that." Instead, you write things like "Here's how you'll be transforming this into that when you get around to doing the transformation."

For one reason or another, imperative programming became the dominant mode. But in recent years, functional programming has emerged as a powerful and useful way of thinking about code.

Problem-solving the old-fashioned way

In Chapter 11, you use arrays to manage the Java Motel. But that venture is behind you now. You've given up the hotel business. (You tell people that you decided to move on. But in all honesty, the hotel was losing a lot of money. According to the United States bankruptcy court, the old Java Motel is now in Chapter 11.)

Since leaving the hotel business, you've transitioned into online sales. Nowadays, you run a website that sells books, DVDs, and other content-related items. (Barry Burd's *Java For Dummies*, 8th Edition, is currently your best seller, but that's beside the point.)

In your world, the sale of a single item looks something like the stuff in Listing 12-3. Each sale has an item and a price.

LISTING 12-3: The Sale Class

```
package com.example.sales;

public class Sale {
    private String item;
    private double price;

    public String getItem() {
        return item;
    }

    public void setItem(String item) {
        this.item = item;
    }

    public double getPrice() {
        return price;
    }

    public void setPrice(double price) {
        this.price = price;
    }

    public Sale(String item, double price) {
        this.item = item;
        this.price = price;
    }
}
```

To make use of the Sale class, you create a small program. The program totals up the sales on DVDs. The program is shown in Listing 12-4.

LISTING 12-4: Using the Sale Class

```
package com.example.sales;

import java.text.NumberFormat;
import java.util.ArrayList;

public class TallySales {

    public static void main(String[] args) {
        ArrayList<Sale> sales = new ArrayList<Sale>();
        NumberFormat currency = NumberFormat.getCurrencyInstance();

        fillTheList(sales);

        double runningTotal = 0;
        for (Sale sale : sales) {
            if (itemIsDVD(sale)) {
                runningTotal += sale.getPrice();
            }
        }

        System.out.println(currency.format(runningTotal));
    }

    static boolean itemIsDVD(Sale sale) {
        return sale.getItem().equals("DVD");
    }

    static void fillTheList(ArrayList<Sale> sales) {
        sales.add(new Sale("DVD", 15.00));
        sales.add(new Sale("Book", 12.00));
        sales.add(new Sale("DVD", 21.00));
        sales.add(new Sale("CD", 5.25));
    }
}
```

In Chapter 11, you step through an array by using an enhanced for statement. Listing 12-4 has its own enhanced for statement. But in Listing 12-4, the enhanced for statement steps through the values in a collection. Each such value is a sale. The loop repeatedly checks a sale to find out whether the item sold is a DVD. If so, the code adds the sale's price to the running total. The program's output is `$36.00` — the running total displayed as a currency amount.

The scenario in Listing 12-4 isn't unusual. You have a collection of items (a collection of sales, perhaps). You step through the items in the collection, finding the items that meet certain criteria (the sale of a DVD, for example). You grab a certain value (such as the sale price) of each item that meets your criteria. Then you do something useful with the values you've grabbed (for example, adding the values together).

Here are some other examples:

- **Step through your list of employees.** Find each employee whose performance evaluation scored 3 or higher. Give each such employee a $100 bonus and then determine the total amount of money you'll pay in bonuses.
- **Step through your list of customers.** For each customer who has shown interest in buying a smartphone, send the customer an email about this month's discount plans.
- **Step through the list of planets that have been discovered.** For each M-class planet, find the probability of finding intelligent life on that planet. Then find the average of all such probabilities.

This scenario is so common that it's worth finding better and better ways to deal with the scenario. One way to deal with it is to use some of the functional programming features in Java.

Lambda expressions

In Chapter 3 of *Java For Dummies*, the author says,

> *In Java, a method is a list of things to do. Every method has a name, and you tell the computer to do the things in the list by using the method's name in your program.*

He's a clever fellow because, in this sentence, he's careful to use the word *method*. Yes, a method is a list of things to do, but not all to-do lists are methods, and not all to-do lists have names.

A *lambda expression* is a to-do list with no name. Here's an example:

```
(sale) -> sale.getItem().equals("DVD")
```

When you put this expression in just the right place, the expression does the work of the `itemIsDVD` method in Listing 12-4. Figure 12-3 brings this point home.

```
( sale ) -> sale.getItem().equals("DVD")

boolean itemIsDVD(Sale sale ) {
    return sale.getItem().equals("DVD");
}
```

FIGURE 12-3: Does the item that's being sold happen to be a DVD?

A lambda expression is a concise way of creating a list of things to do. The list may include only one to-do item (as in Figure 12-3) or any number of items:

```
(sale) -> {
    var compareTo = "DVD";
    return sale.getItem().equals(compareTo);
}
```

When you put more than one to-do item in a lambda expression, the expression must have a complete method body with curly braces, semicolons, and the `return` keyword. Only the stuff before the arrow (->) is abbreviated.

TIP

Ugh! This section's "to-do list" terminology bothers me. I have to change to the correct wording. When you collect things to do into a method or a lambda expression, you have what programmers call a *function*. Methods are functions with names; lambda expressions are functions without names. So there!

The lambda expression in Figure 12-3 has only one parameter — the `sale` parameter. Can a lambda expression have more than one parameter? Of course, it can. Here's an example:

```
(price1, price2) -> price1 + price2
```

Figure 12-4 describes the new lambda expression's meaning.

The lambda expression in Figure 12-4 does (roughly) what the following method does:

```
double sum(double price1, double price2) {
    return price1 + price2;
}
```

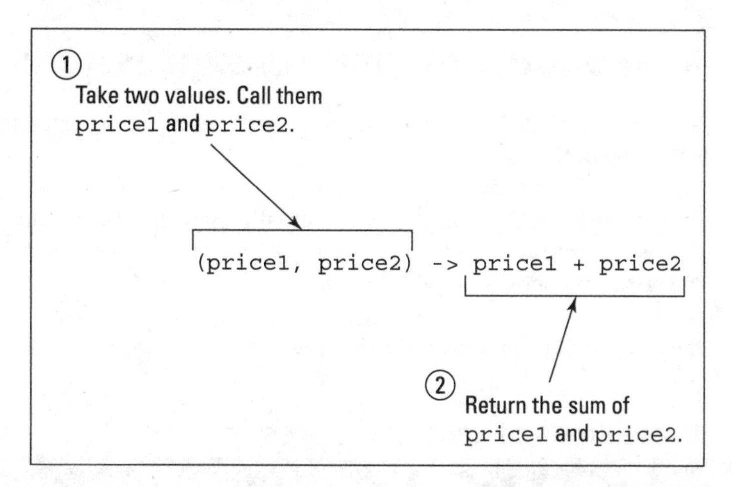
①
Take two values. Call them
price1 and price2.
(price1, price2) -> price1 + price2
②
Return the sum of
price1 and price2.

FIGURE 12-4: Add two prices.

A black sheep among the lambdas

Here's an interesting lambda expression:

```
(sale) -> System.out.println(sale.getPrice())
```

This lambda expression does (roughly) what the following method does:

```
void display(Sale sale) {
    System.out.println(sale.getPrice());
}
```

In the method's header, the word `void` indicates that the method doesn't return a value. When you call the `display` method (or you use the equivalent lambda expression), you don't expect to get back a value. Instead, you expect the code to do something in response to the call (something like displaying text on the computer's screen).

To draw a sharp distinction between returning a value and "doing something," functional programmers have a name for "doing something without returning a value" — they call that something a side effect. In functional programming, a *side effect* is considered a second-class citizen, a last resort, a tactic you use when you can't simply return a result. Unfortunately, displaying information on a screen (something that so many computer programs do) is a side effect. Any program that displays output (on a screen, on paper, or as tea leaves in a cup) isn't a purely functional program.

A taxonomy of lambda expressions

Java divides lambda expressions into about 45 different categories. Table 12-2 lists a few of the categories.

The categories in Table 12-2 aren't mutually exclusive. For example, every Predicate is a Function. (Every Predicate accepts one parameter and returns a result. The result happens to be boolean.)

The next several sections refer to some of these categories.

TABLE 12-2 A Few Kinds of Lambda Expressions

Name	Description	Example
Function	Accepts one parameter; produces a result of any type	(sale) -> sale. price
Predicate	Accepts one parameter; produces a boolean valued result	(sale) -> sale.item. equals("DVD")
BinaryOperator	Accepts two parameters of the same type; produces a result of the same type	(price1, price2) -> price1 + price2
Consumer	Accepts one parameter; produces no result	(sale) -> System.out. println(sale. price)

The interpretation of streams

The earlier section "Once and again" introduces iterators. An iterator's next method spits out a collection's values, one by one. That's great, but Java takes this concept a step further with the notion of a stream. A *stream* is like an iterator except that, with a stream, you don't have to call a next method. A stream spits out a collection's values automatically. To get values from a stream, you don't call a stream's next method. In fact, a typical stream has no next method.

With streams, you can create an assembly line that elegantly solves this chapter's sales problem. Unlike the code in Listing 12-4, the new assembly line solution uses concepts from functional programming.

The assembly line consists of several methods. Each method takes the data, transforms the data in some way or other, and hands its results to the next method in line. Figure 12-5 illustrates the assembly line for this chapter's sales problem.

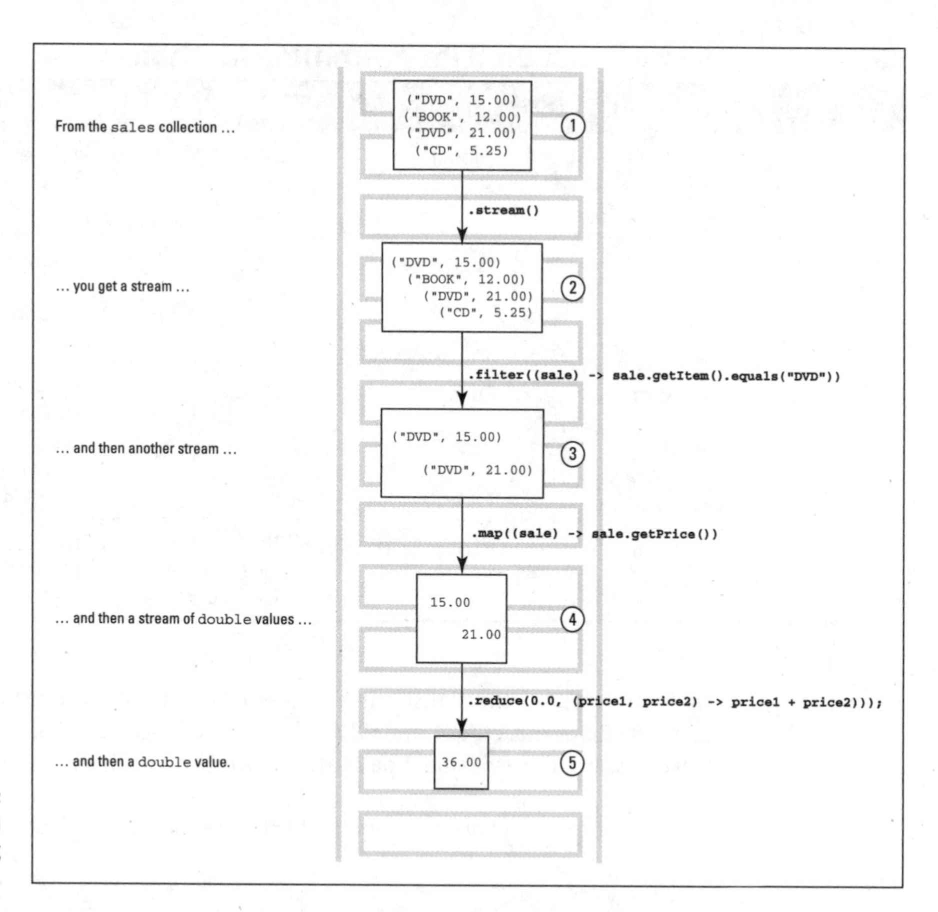

FIGURE 12-5: A functional programming assembly line.

In Figure 12-5, each box represents a bunch of raw materials as they're transformed along an assembly line. Each arrow represents a method (or, metaphorically, a worker on the assembly line).

For example, in the transition from the second box to the third box, a worker method (the `filter` method) sifts out sales of items that aren't DVDs. Imagine Lucy Ricardo standing between the second and third boxes, removing each book or CD from the assembly line and tossing it carelessly onto the floor.

The parameter to Java's `filter` method is a `Predicate` — a lambda expression whose result is `boolean`. (See Tables 12-2 and 12-3.) The `filter` method in Figure 12-5 sifts out items that don't pass the lambda expression's `true / false` test.

CROSS REFERENCE

For some help understanding the words in Column 3 of Table 12-3 (`Predicate`, `Function`, and `BinaryOperator`), see the earlier section "A taxonomy of lambda expressions."

TABLE 12-3 Some Functional Programming Methods

Method Name	Member Of	Parameter(s)	Result Type	Result Value
`stream`	`Collection` (for example, an `ArrayList` object)	(None)	`Stream`	A stream that spits out elements of the collection
`filter`	`Stream`	`Predicate`	`Stream`	A new stream containing values for which the lambda expression returns `true`
`map`	`Stream`	`Function`	`Stream`	A new stream containing the results of applying the lambda expression to the incoming stream
`reduce`	`Stream`	`BinaryOperator`	The type used by the `BinaryOperator`	The result of combining all values in the incoming stream

In Figure 12-5, in the transition from the third box to the fourth box, a worker method (the `map` method) pulls the `price` out of each `sale`. From that worker's place onward, the assembly line contains only `price` values.

To be more precise, Java's `map` method takes a `Function` such as

```
(sale) -> sale.getPrice()
```

and applies the `Function` to each value in a stream. (See Tables 12-2 and 12-3.) So the `map` method in Figure 12-5 takes an incoming stream of `sale` objects and creates an outgoing stream of `price` values.

In Figure 12-5, in the transition from the fourth box to the fifth box, a worker method (the `reduce` method) adds up the prices of DVD sales. Java's `reduce` method takes two parameters:

- The first parameter is an initial value.

  In Figure 12-5, the initial value is `0.0`.
- The second parameter is a `BinaryOperator`. (See Tables 12-2 and 12-3.)

  In Figure 12-5, the `reduce` method's `BinaryOperator` is

  ```
  (price1, price2) -> price1 + price2
  ```

The reduce method uses its BinaryOperator to combine the values from the incoming stream. The initial value serves as the starting point for all the combining. So, in Figure 12-5, the reduce method performs two additions. (See Figure 12-6.)

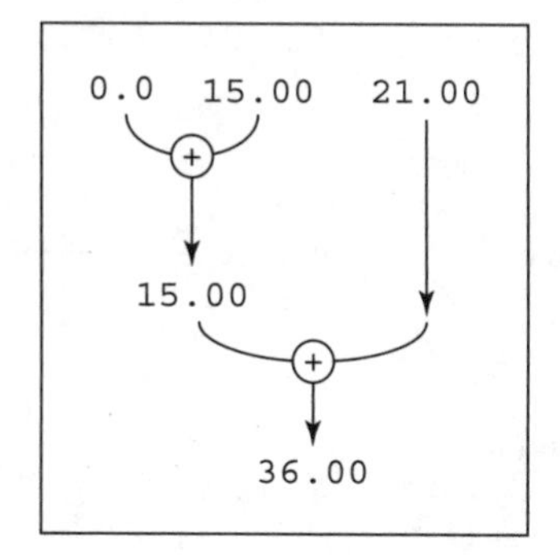

FIGURE 12-6: The reduce method adds two values at a time.

For comparison, imagine calling the method

```
reduce(10.0, (value1, value2) -> value1 * value2)
```

with the stream whose values include 3.0, 2.0, and 5.0. The resulting action is shown in Figure 12-7.

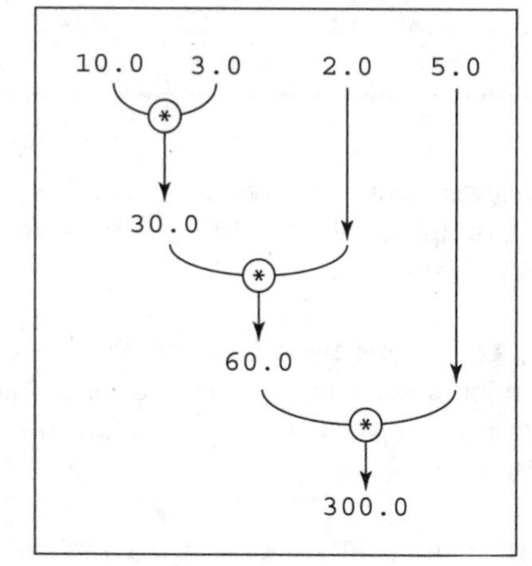

FIGURE 12-7: The reduce method multiplies values from an incoming stream.

Taken as a whole, the entire assembly line shown in Figure 12-5 adds up the prices of DVDs sold. Listing 12-5 contains a complete program using the streams and lambda expressions of Figure 12-5.

LISTING 12-5: Living the Functional Way of Life

```
package com.example.sales;

import java.text.NumberFormat;
import java.util.ArrayList;

public class TallySalesAgain {

    public static void main(String[] args) {
        ArrayList<Sale> sales = new ArrayList<>();
        NumberFormat currency = NumberFormat.getCurrencyInstance();

        fillTheList(sales);

        double total = sales.stream()
                .filter((sale) -> sale.getItem().equals("DVD"))
                .map((sale) -> sale.getPrice())
                .reduce(0.0, (price1, price2) -> price1 + price2);

        System.out.println(currency.format(total));
    }

    static void fillTheList(ArrayList<Sale> sales) {
        sales.add(new Sale("DVD", 15.00));
        sales.add(new Sale("Book", 12.00));
        sales.add(new Sale("DVD", 21.00));
        sales.add(new Sale("CD", 5.25));
    }
}
```

WARNING

The code in Listing 12-5 requires Java 8 or later. If your IDE is set for an earlier Java version, you might have to tinker with the IDE's settings. You may even have to download a newer version of Java.

The boldface code in Listing 12-5 is one big Java assignment statement. The right side of the statement contains a sequence of method calls. Each method call returns an object, and each such object is the thing before the dot in the next method call. That's how you form the assembly line.

For example, near the start of the boldface code, the name `sales` refers to an `ArrayList` object. Each `ArrayList` object has a `stream` method. In Listing 12-5, `sales.stream()` is a call to that `ArrayList` object's `stream` method.

The `stream` method returns an instance of Java's `Stream` class. (What a surprise!) So `sales.stream()` refers to a `Stream` object. (See Figure 12-8.)

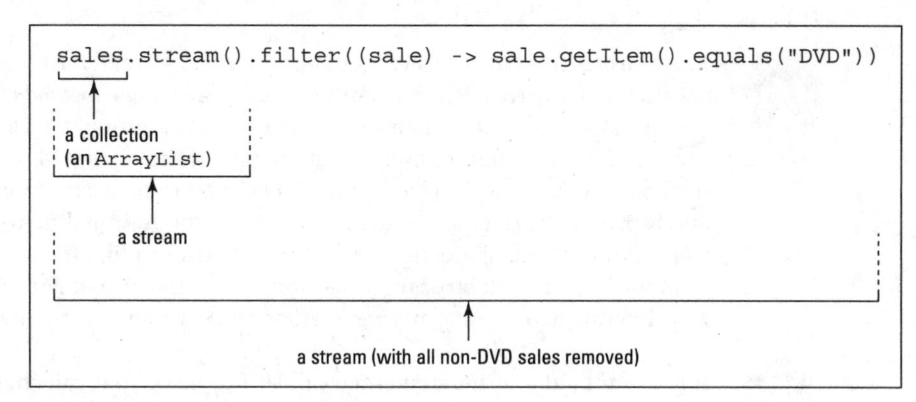

FIGURE 12-8: Getting all DVD sales.

Every Stream object has a filter method. So

```
sales.stream().filter((sale) -> sale.getItem().equals("DVD"))
```

contains a call to the Stream object's filter method. (Refer to Figure 12-8.)

The pattern continues. The Stream object's map method returns yet another Stream object — a Stream object containing prices. (See Figure 12-9.) To that Stream of prices you apply the reduce method, which yields one double value — the total of the DVD prices. (See Figure 12-10.)

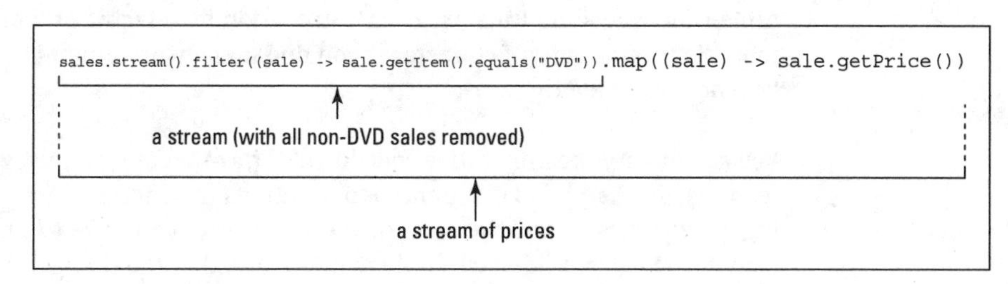

FIGURE 12-9: Getting the price from each DVD sale.

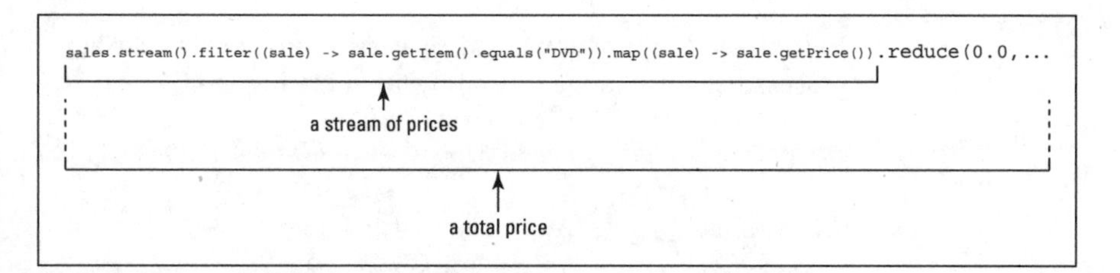

FIGURE 12-10: Getting the total price of all DVD sales.

TECHNICAL STUFF

Every stream has two kinds of methods: intermediate and terminal. An *intermediate method* is one that returns another stream. A *terminal method* is one that either returns something other than a stream or returns nothing (void). A chain of calls begins with one or more intermediate methods and ends with a single terminal method. For example, in Listing 12-5, `filter` and `map` are intermediate methods, and `reduce` is a terminal method. In a sense, the terminal method pulls values from all the intermediate methods. When Java tries to find the value of `total`, the terminal `reduce` method eagerly asks for `price1` and `price2`. At that moment, the lazy, intermediate `filter` and `map` methods wake up and start pumping out values.

REMEMBER

If you don't put a terminal method call at the end of your call chain, Java doesn't execute any of the intermediate calls. For an example, see the Try It Out paragraph at the end of this chapter.

Why bother?

The chain of method calls in Listing 12-5 accomplishes everything that the loop in Listing 12-4 accomplishes. But the code in Figure 12-10 uses concepts from functional programming. What's the big deal? Are you better off with Listing 12-5 than with Listing 12-4?

You are. For the past several years, the big trend in chip design has been *multicore* processors. With several cores, a processor can execute several statements at the same time, speeding up a program's execution by a factor of 2 or 4 or 8 or even more. Programs run much faster if you divide the work among several cores. But how do you divide the work?

You can modify the imperative code in Listing 12-4. For example, with some fancy features, you can hand different loop iterations to different cores. But the resulting code is messy. For the code to work properly, you have to micromanage the loop iterations, checking carefully to make sure that the final `runningTotal` value is correct.

In contrast, the functional code is easy to modify. To take advantage of multicore processors, you change *only one word* in Listing 12-5!

```
sales.parallelStream()
     .filter((sale) -> sale.getItem().equals("DVD"))
     .map((sale) -> sale.getPrice())
     .reduce(0.0, (price1, price2) -> price1 + price2);
```

In Listing 12-5, the `stream()` method call creates a serial stream. With a *serial stream*, Java does its processing one sale at a time. But a call to `parallelStream()` creates a slightly different kind of stream: a parallel stream. With a *parallel stream*,

Java divides the work among the number of cores in the computer's processor (or according to some other useful measure of computing power). If you have 4 million sales and four cores, each core processes 1 million of the sales.

Each core works independently of the others, and each core dumps its result into a final `reduce` method. The `reduce` method combines the cores' results into a final tally. In the best possible scenario, all the work gets done in one-fourth the time it would take with an ordinary serial stream.

TECHNICAL STUFF

When you read the preceding paragraph, don't gloss over the phrase *best possible scenario.* Parallelism isn't magic. And sometimes, parallelism isn't your friend. Consider the situation in which you have only 20 sale amounts to tally. The time it takes to divide the problem into four groups of 5 sales each far exceeds the amount of time you save in using all four cores. In addition, some problems don't lend themselves to parallel processing. Imagine that the price of an item depends on the number of similar items being sold. In that case, you can't divide the problem among four independently operating cores. If you try, each core has to know what the other cores are doing. You lose the advantage of having four threads of execution.

NO VARIABLES? NO PROBLEM!

In imperative programming, your code's pieces interact with one another. All the pieces might be updating the current price of Oracle stock shares (ticker symbol: ORCL). The simultaneous updates may conflict with one another and result in an incorrect outcome. You've experienced the same phenomenon if you've ever clicked a website's Purchase button, only to learn that the item you're trying to purchase is out of stock. Someone else completed a purchase while you were filling in your credit card information. Too many customers were grabbing for the same goods at the same time.

The source of the problem is *shared data.* How many clients share simultaneous access to Oracle's stock price? How many customers share access to a web page's Purchase button? Today's multicore processors can perform more than one instruction at a time. How many simultaneous instructions can all be trying to modify the same variable's value?

In imperative programming, a variable is a place where statements share their values with one another. Can you avoid using variables in your code?

Compare the loop in Listing 12-4 with the functional programming code in Listing 12-5. In Listing 12-4, the `runningTotal` variable is shared among all loop iterations. Because

(continued)

each iteration can potentially change the value of `runningTotal`, you can't assign each iteration to a different processor core. If you did, you'd risk having two cores updating that running total at the same time. (Chances are good that, because of the simultaneous updating, neither core would do its update correctly!) But the `filter/map/reduce` lines in Listing 12-5 make no reference to the code's `total` variable. The program's `total` variable is a final tally, not a running total. In fact, none of the variables in Listing 12-5 represents a running total. In Listing 12-5, the running total is completely anonymous.

In imperative programming, a variable is a place where statements share their values with one another. But functional programming shuns variables. So, when you do functional programming, you don't have a lot of data sharing. Many of the difficulties associated with shared data vanish into thin air.

Here's an analogy: Imagine a programming problem as a cube, and then imagine an imperative programming solution as a way of slicing the cube into manageable pieces. (See the first sidebar figure.)

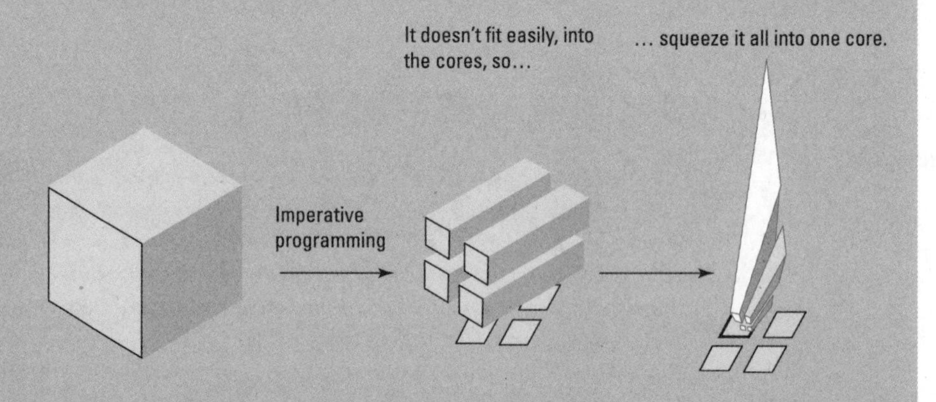

To get the most out of a four-core processor, you divide your code into four pieces — one piece for each core. But with imperative programming, your program's pieces don't fit neatly into your processor's cores. The figure shows what happens when you try to squeeze an imperative program into a multicore processor.

Functional programming also divides code into pieces, but it does so along different lines. With functional programming, the pieces of the code fit neatly into the processor's cores. (See the following sidebar figure.)

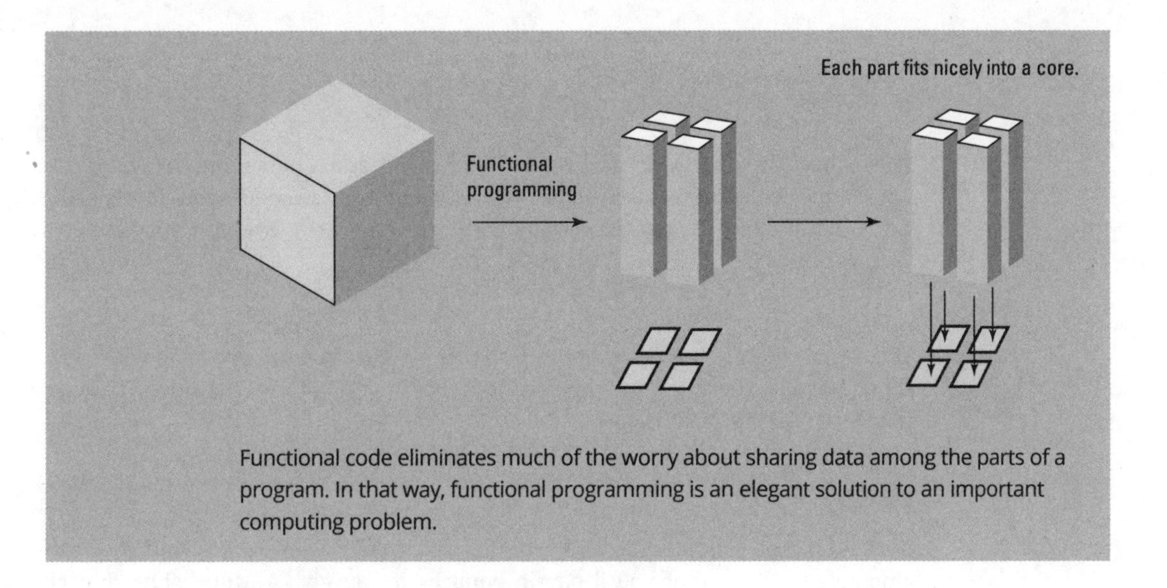
Each part fits nicely into a core.
Functional
programming

Functional code eliminates much of the worry about sharing data among the parts of a program. In that way, functional programming is an elegant solution to an important computing problem.

Method references

Take a critical look at the last lambda expression in Listing 12-5:

```
(price1, price2) -> price1 + price2)
```

This expression does roughly the same work as a `sum` method. (In fact, you can find a `sum` method's declaration in the earlier section "Lambda expressions.") If your choice is between typing a 3-line `sum` method and typing a 1-line lambda expression, you'll probably choose the lambda expression. But what if you have a third alternative? Rather than type your own `sum` method, you can refer to an existing `sum` method. Using an existing method is the quickest and safest thing to do.

As luck would have it, Java's `Double` class contains a static `sum` method. You don't have to create your own `sum` method. If you run the following code:

```
double i = 5.0, j = 7.0;
System.out.println(Double.sum(i, j));
```

the computer displays `12.0`. So, rather than type the `price1 + price2` lambda expression in Listing 12-5, you can use two colons to create a *method reference* — an expression that refers to an existing method.

```
sales.stream()
     .filter((sale) -> sale.getItem().equals("DVD"))
```

```
.map((sale) -> sale.getPrice())
.reduce(0.0, Double :: sum);
```

The expression `Double :: sum` refers to the `sum` method belonging to Java's `Double` class. When you use this `Double :: sum` method reference, you do the same thing that the last lambda expression does in Listing 12-5. Everybody is happy.

For information about static methods, see Chapter 10.

TRY IT OUT

You can always try the programming challenges that you dream up on your own. If you have no ideas to give you practice with functional programming, I have a couple of suggestions for you:

BE GENEROUS

Each employee has a name and a performance evaluation score. Find the total amount of money that you'll pay in bonuses if you give a $100 bonus to each employee whose score is 3 or higher. In your code, make optimal use of lambda expressions and method references.

SAVE TIME

Each recipe has a name, a list of ingredients (some of which involve meat products), and an estimated preparation time. Find the average time estimate for cooking one of the vegetarian recipes.

RECORD SALES

Some challenges in the Try It Out paragraphs in Chapter 9 introduce Java's `record` feature. Modify Listing 12-3 so that `Sale` is a record. Modify Listing 12-5 to use your new `Sale` record.

INTERMEDIATE METHODS ARE JUST PLAIN LAZY

In the section "The interpretation of streams," I warn that Java does nothing when you end a chain without a terminal method call. To double-check that I'm not lying about this, run Listing 12-5 with the following modified code:

```
var total = sales.stream()
        .filter((sale) -> sale.getItem().equals("DVD"))
        .map((sale) -> {
```

```
            System.out.println("Hello");
            return sale.getPrice();
        });

// Commented out: System.out.println(currency.format(total));
```

STREAMLESS VARMINT

You don't need streams to make use of Java's lambda expressions. Run this code for a glimpse at the possibilities:

```
import java.util.function.Function;

public class LambdaWithoutStream {

    public static void main(String[] args) {
        System.out.println(change("Hello", (String a) -> {return a + "!";}));
    }

    static String change(String str, Function<String, String> func) {
        return func.apply(str);
    }
}
```

» Making your code (more or less) crash-proof

» Defining your own exception class

Chapter 13

Looking Good When Things Take Unexpected Turns

September 9, 1945: A moth flies into one of the relays of the Harvard Mark II computer and gums up the works. This becomes the first recorded case of a real-life computer bug.

April 19, 1957: Herbert Bright, manager of the data processing center at Westinghouse in Pittsburgh, receives an unmarked deck of computer punch cards in the mail (which is like receiving an unlabeled CD-ROM in the mail today). Mr. Bright guesses that this deck comes from the development team for FORTRAN — the first computer programming language. He's been waiting a few years for this software. (No web downloads were available at the time.)

Armed with nothing but this good guess, Bright writes a small FORTRAN program and tries to compile it on his IBM 704. (The IBM 704 lives in its own, specially built, 2,000-square-foot room. With vacuum tubes instead of transistors, the machine has a whopping 32K of RAM. The operating system has to be loaded from tape before the running of each program, and a typical program takes between two and four hours to run.) After the usual waiting time, Bright's attempt to compile a FORTRAN program comes back with a single error: a missing comma in one of the statements. Bright corrects the error, and the program runs like a charm.

July 22, 1962: Mariner I, the first US spacecraft aimed at another planet, is destroyed when it behaves badly four minutes after launch. The bad behavior is attributed to a missing bar (like a hyphen) in the formula for the rocket's velocity.

Around the same time, orbit computation software at NASA is found to contain the incorrect statement `DO 10 I=1.10` (instead of the correct `DO 10 I=1,10`). In modern notation, this is like writing `do10i = 1.10` in place of `for (int i=1; i<=10; i++)`. The change from a comma to a period turns a loop into an assignment statement.

January 1, 2000: The Year 2000 problem wreaks havoc on the modern world.

Any historically accurate facts in these notes were borrowed from the following sources: the Computer Folklore newsgroup (`https://groups.google.com/forum/#!forum/alt.folklore.computers`), the Free On-line Dictionary of Computing (`http://foldoc.org`), *Computer* magazine (`www.computer.org/computer-magazine`), and other web pages of the IEEE (`www.computer.org`). All inaccuracies stem from this author's lunatic musings.

Garbage In

You're taking inventory. This means counting item after item, box after box, and marking the numbers of such things on log sheets, in little handheld gizmos, and into forms on computer keyboards. A particular part of the project involves entering the number of boxes you find on the shelf labeled Big Dusty Boxes That Haven't Been Opened Since Year One. Rather than break the company's decades-old habit, you decide not to open any of these boxes. You arbitrarily assign the value $3.25 to each box.

Listing 13-1 shows the software to handle this bit of inventory. The software has a flaw, which is revealed in Figure 13-1. When the user enters a whole number value, things are okay. But when the user enters something else (like the number 3.5), the program comes crashing to the ground. Surely something can be done about this. Computers are stupid, but they're not so stupid that they should fail royally whenever a user enters an improper value.

LISTING 13-1: Counting Boxes

```
package com.example.inventory;

import java.text.NumberFormat;
import java.util.Scanner;
```

```
import static java.lang.System.out;

public class InventoryA {

    public static void main(String[] args) {
        final double boxPrice = 3.25;
        var keyboard = new Scanner(System.in);
        NumberFormat currency = NumberFormat.getCurrencyInstance();

        out.print("How many boxes do we have? ");
        String numBoxesIn = keyboard.next();
        int numBoxes = Integer.parseInt(numBoxesIn);

        out.print("The value is ");
        out.println(currency.format(numBoxes * boxPrice));
        keyboard.close();
    }
}
```

```
How many boxes do we have? 3
The value is $9.75

How many boxes do we have? 3.5
Exception in thread "main" java.lang.NumberFormatException: For input string: "3.5"
        at java.lang.NumberFormatException.forInputString(Unknown Source)
        at java.lang.Integer.parseInt(Unknown Source)
        at java.lang.Integer.parseInt(Unknown Source)
        at InventoryA.main(InventoryA.java:15)

How many boxes do we have? three
Exception in thread "main" java.lang.NumberFormatException: For input string: "three"
        at java.lang.NumberFormatException.forInputString(Unknown Source)
        at java.lang.Integer.parseInt(Unknown Source)
        at java.lang.Integer.parseInt(Unknown Source)
        at InventoryA.main(InventoryA.java:15)
```

FIGURE 13-1: Three separate runs of the code in Listing 13-1.

The key to fixing a program bug is examining the message that appears when the program crashes. The inventory program's message says `java.lang.NumberFormatException`. That means a class named *NumberFormatException* is in the `java.lang` API package. Somehow, the call to `Integer.parseInt` brought this `NumberFormatException` class out of hiding.

CROSS REFERENCE

For a brief explanation of the `Integer.parseInt` method, see Chapter 11.

Well, here's what's going on. The Java programming language has a mechanism called exception handling. With *exception handling*, a program can detect that something is about to go wrong and respond by creating a brand-new object. In the official terminology, the program is said to be *throwing* an exception. That new object, an instance of the `Exception` class, is passed like a hot potato from one piece of code to another until some piece of code decides to *catch* the exception. When the exception is caught, the program executes some recovery code, buries the exception, and moves on to the next normal statement as though nothing had ever happened. The process is illustrated in Figure 13-2.

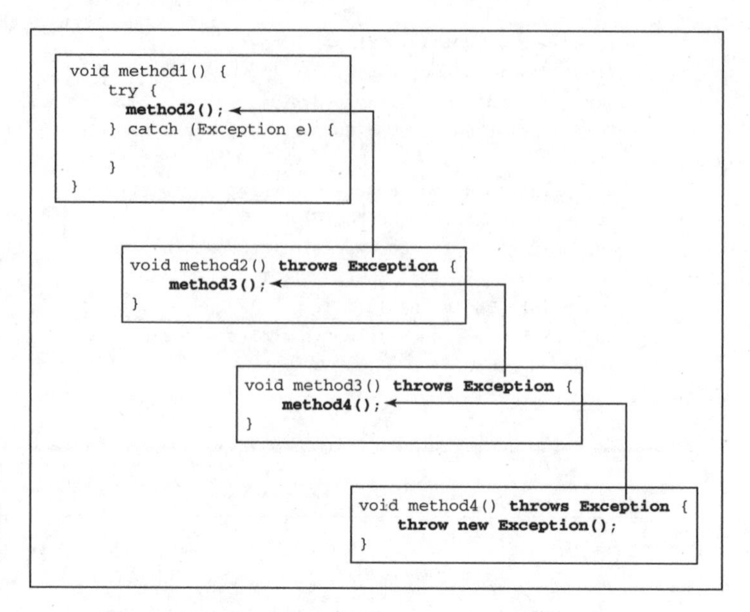

FIGURE 13-2: Throwing, passing, and catching an exception.

The whole thing is done with the aid of several Java keywords. These keywords are described in this list:

- `throw`: Creates a new exception object.
- `throws`: Indicates that a method may pass the exception up to whatever code called the method.
- `try`: Encloses code that has the potential to create a new exception object. In the usual scenario, the code inside a `try` clause contains calls to methods whose code can create one or more exceptions.
- `catch`: Deals with the exception, buries it, and then moves on.

So, the truth is out. By some chain of events like the one shown in Figure 13-2, the method `Integer.parseInt` can throw a `NumberFormatException`. When you call `Integer.parseInt`, this `NumberFormatException` is passed on to you.

TIP

The Java application programming interface (API) documentation for the `parseInt` method says, "Throws: `NumberFormatException` — if the string does not contain a parsable integer." Once in a while, reading the documentation actually pays off.

If you call yourself a hero, you'd better catch the exception so that all the other code can get on with its regular business. Listing 13-2 shows the catching of an exception.

LISTING 13-2: A Hero Counts Boxes

```
package com.example.inventory;

import java.text.NumberFormat;
import java.util.Scanner;

import static java.lang.System.out;

public class InventoryB {

    public static void main(String[] args) {
        final double boxPrice = 3.25;
        var keyboard = new Scanner(System.in);
        NumberFormat currency = NumberFormat.getCurrencyInstance();

        out.print("How many boxes do we have? ");
        String numBoxesIn = keyboard.next();

        try {
            int numBoxes = Integer.parseInt(numBoxesIn);
            out.print("The value is ");
            out.println(currency.format(numBoxes * boxPrice));
        } catch (NumberFormatException e) {
            out.print(e.getMessage());
            out.println(" ... Cannot interpret the input.");
        }

        keyboard.close();
    }
}
```

Figure 13-3 shows three runs of the code from Listing 13-2. When a misguided user types **three** instead of **3**, the program maintains its cool by displaying `For input string: "three" ... Cannot interpret the input.` The trick is to enclose the call to `Integer.parseInt` inside a `try` clause. If you do this, the computer watches for exceptions when any statement inside the `try` clause is executed. If an exception is thrown, the computer jumps from inside the `try` clause to a `catch` clause below it. In Listing 13-2, the computer jumps directly to the `catch (NumberFormatException e)` clause. The computer executes the two `print` statements inside the clause and then marches on with normal processing. In Listing 13-2, this "normal processing" means executing `keyboard.close()`.

```
How many boxes do we have? 3
The value is $9.75

How many boxes do we have? three
For input string: "three" ... Cannot interpret the input.

How many boxes do we have? -25
The value is -$81.25
```

FIGURE 13-3: Three runs of the code in Listing 13-2.

TECHNICAL STUFF

An entire `try-catch` assembly — complete with a `try` clause, `catch` clause, and what-have-you — is called a *try statement.* Sometimes, for emphasis, I call it a *try-catch statement.*

The parameter in a catch clause

Take a look at the `catch` clause in Listing 13-2 and pay particular attention to the words `(NumberFormatException e)`. This looks a lot like a method's parameter list, doesn't it? In fact, every `catch` clause is like a little mini-method with its own parameter list.

The parameter list always has an exception type name and then a parameter. The parameter is an object — an instance of the `NumberFormatException` class. When an exception is caught, Java makes the `catch` clause's parameter refer to that exception object. In other words, the name *e* in Listing 13-2 stores a bunch of information about the exception. To take advantage of this, you can call some of the exception object's methods.

In Listing 13-2, I call the exception's `getMessage` method. Lo and behold! The `getMessage` call returns the text `For input string: "three"`. That's helpful information. Java hands this information to the `e` variable whenever it can't make sense of the user's `three` input.

In addition to the `getMessage` method, each exception has a handy `printStack-Trace` method. To see how this works, rewrite the `catch` clause in Listing 13-2 as follows:

```
} catch (NumberFormatException e) {
    e.printStackTrace();
}
```

With this new `catch` clause, a run of the `inventory` program may look like the run shown in Figure 13-4. When you call `printStackTrace`, you see a list of methods that were running when the exception was thrown, along with line numbers in each of the methods. In Figure 13-4, the display includes `Integer.parseInt` and the `main` method.

```
How many boxes do we have? three
java.lang.NumberFormatException Create breakpoint : For input string: "three"
    at java.base/java.lang.NumberFormatException.forInputString(NumberFormatException.java:67)
    at java.base/java.lang.Integer.parseInt(Integer.java:668)
    at java.base/java.lang.Integer.parseInt(Integer.java:786)
    at com.example.inventory.InventoryB.main(InventoryB.java:19)
```

FIGURE 13-4: Java's stack trace.

Notice the similarity between the run in Figure 13-4 and two of the runs in Figure 13-1.

- **In Figure 13-4:** Java throws a `NumberFormatException`. The code in Listing 13-2 catches the exception and calls a method that prints a stack trace. Then the program keeps running (to call `keyboard.close()`, for example).
- **In Figure 13-1:** Java throws a `NumberFormatException`. None of the code in Listing 13-1 catches the exception, so Java does what it always does for any uncaught exceptions: It prints a stack trace and then abruptly ends the run of the program. In other words, your program crashes.

TECHNICAL STUFF

When you mix `System.out.println` calls with `printStackTrace` calls, the order in which Java displays the information isn't predictable. That's because `printStackTrace` doesn't write directly to `System.out`. The `printStackTrace` method writes to something called `System.err`, which, by default, shows up on the same screen as `System.out`. If you want, you can tweak the code so that text goes exactly where you want it to go.

TRY IT OUT

I can't think of a clever way to connect the *Try* in *TryItOut* with the `try` in `try-catch` statements. If you think of something, scribble it in the margin on this page. Then try this little challenge:

DON'T BE A QUITTER

Add `try-catch` statements to keep the following code from crashing:

```
import java.util.Scanner;

public class Main {

    public static void main(String[] args) {
        var keyboard = new Scanner(System.in);
        var words = new String[5];

        int i = 0;
        do {
```

```
            words[i] = keyboard.next();
        } while (!words[i++].equals("Quit"));

        for (int j = 0; j < 5; j++) {
            System.out.println(words[j].length());
        }

        keyboard.close();
    }
}
```

Do it yourself

In the previous sections, you confuse your inventory program by typing `three` instead of `3`. Nice work! What else can go wrong today? Are there other kinds of exceptions — ones that have nothing to do with the `NumberFormatException`? Sure, plenty of different exception types are out there. You can even create one of your own. You wanna try? If so, look at Listings 13-3 and 13-4.

LISTING 13-3: Making Your Own Kind of Exception

```
package com.example.inventory;

public class OutOfRangeException extends RuntimeException {
    public OutOfRangeException(String message) {
        super("A value is out of range.\n" + message);
    }
}
```

LISTING 13-4: Using Your Custom-Made Exception

```
package com.example.inventory;

import java.text.NumberFormat;
import java.util.Scanner;

import static java.lang.System.out;

public class InventoryC {
```

```
    public static void main(String[] args) {
        final double boxPrice = 3.25;
        var keyboard = new Scanner(System.in);
        NumberFormat currency = NumberFormat.getCurrencyInstance();

        out.print("How many boxes do we have? ");
        String numBoxesIn = keyboard.next();

        try {
            int numBoxes = Integer.parseInt(numBoxesIn);
            if (numBoxes < 0) {
                throw new OutOfRangeException("You typed " + numBoxes +
                        ". There's no such thing as a negative box.");
            }

            out.print("The value is ");
            out.println(currency.format(numBoxes * boxPrice));
        } catch (NumberFormatException e) {
            out.print(e.getMessage());
            out.println("... Cannot interpret the input.");
        } catch (OutOfRangeException e) {
            out.println(e.getMessage());
        }

        keyboard.close();
    }
}
```

Listings 13-3 and 13-4 remedy a problem that quietly cropped up earlier. Look at the last of the three runs in Figure 13-3. The user reports that the shelves have −25 boxes, and the computer takes this value without blinking an eye. The truth is that you would need a black hole (or some other exotic, space-time warping phenomenon) to have a negative number of boxes on any shelf in your warehouse. So the program should become upset if the user enters a negative number of boxes, which is what the code in Listing 13-4 does. To see what I mean, look at Figure 13-5.

The code in Listing 13-3 declares a new kind of exception class: `OutOfRangeException`. In many situations, typing a negative number would work just fine, so `OutOfRangeException` isn't built in to the Java API. However, in the inventory program, a negative number of boxes should be flagged as an anomaly.

```
How many boxes do we have? 3
The value is $9.75

How many boxes do we have? three
For input string: "three" ... Cannot interpret the input.

How many boxes do we have? -25
A value is out of range.
You typed -25. There's no such thing as a negative box.
```

FIGURE 13-5: Three runs of the code from Listings 13-3 and 13-4.

How do you know that the class in Listing 13-3 is an exception of some kind? You know because the class declaration's first line says `extends RuntimeException`. The class in Listing 13-3 is a particular kind of exception called a *runtime exception*. To learn more about runtime exceptions, see the section "The Buck Stops Here, Except When It Doesn't," later in this chapter.

TECHNICAL STUFF

When you call

```
throw new OutOfRangeException("You typed " + numBoxes +
        ". There's no such thing as a negative box.");
```

in Listing 13-4, you pass the string `"You typed " + numBoxes + ...` to the `message` parameter in Listing 13-3. The constructor in Listing 13-3 doesn't do much on its own with that `message` value. Instead, the constructor combines the message with its own `"A value is out of range.\n"` string. Then the constructor uses `super` to pass the combined string on to its superclass — the `RuntimeException` class. By the time you're done, the call `e.getMessage()` gives you a whole bunch of text: `A value is out of range ...` *blah, blah,* `... no such thing ...` *blah blah.*

CROSS REFERENCE

To read more about Java's `super` keyword, refer to Chapter 9.

Who will catch the exception?

Take one more look at Listing 13-4. Notice that more than one `catch` clause can accompany a single `try` clause. When an exception is thrown inside a `try` clause, Java starts reviewing the accompanying list of `catch` clauses. It starts at whatever `catch` clause comes immediately after the `try` clause and works its way down the program's text.

For each `catch` clause, Java asks itself, "Is the exception that was just thrown an instance of the class in this clause's parameter list?"

» If not, Java skips this `catch` clause and moves on to the next `catch` clause in line.

» If so, Java executes this `catch` clause and then skips past all other `catch` clauses that come with this `try` clause. Java goes on and executes whatever statements come after the whole `try-catch` statement.

For some concrete examples, see Listings 13-5 and 13-6.

LISTING 13-5: Yet Another Exception

```
package com.example.inventory;

class NumberTooLargeException extends OutOfRangeException {
    public NumberTooLargeException(String message) {
        super("A value is too large.\n" + message);
    }
}
```

LISTING 13-6: Catch Me If You Can

```
package com.example.inventory;

import java.text.NumberFormat;
import java.util.Scanner;

import static java.lang.System.out;

public class InventoryD {

    public static void main(String[] args) {
        final double boxPrice = 3.25;
        final int maxBoxes = 1000;
        var keyboard = new Scanner(System.in);
        NumberFormat currency = NumberFormat.getCurrencyInstance();

        out.print("How many boxes do we have? ");
        String numBoxesIn = keyboard.next();

        try {
            int numBoxes = Integer.parseInt(numBoxesIn);

            if (numBoxes < 0) {
                throw new OutOfRangeException("You typed " + numBoxes +
                        ". There's no such thing as a negative box.");
            }
```

(continued)

LISTING 13-6: (continued)

```
                if (numBoxes > maxBoxes) {
                    throw new NumberTooLargeException(numBoxes +
                            " is larger than the maximum of " + maxBoxes);
                }

                out.print("The value is ");
                out.println(currency.format(numBoxes * boxPrice));
            } catch (NumberFormatException e) {
                out.print(e.getMessage());
                out.println(" ... Cannot interpret the input.");
            } catch (OutOfRangeException e) {
                out.println(e.getMessage());
            } catch (Exception e) {
                out.println(e.getMessage());
            }

            out.println("That's that.");

            keyboard.close();
        }
    }
```

To run the code in Listings 13-5 and 13-6, you need one additional Java program file. You need the `OutOfRangeException` class in Listing 13-3.

Listing 13-6 addresses the scenario in which you have limited shelf space. You don't have room for more than 1,000 boxes, but once in a while the program asks how many boxes you have, and somebody enters the number *1001* accidentally. In cases like this, Listing 13-6 performs a quick reality check: Any number of boxes over 1,000 is tossed out as being unrealistic.

Listing 13-6 watches for a `NumberTooLargeException`, but to make life more interesting, Listing 13-6 has no `catch` clause for the `NumberTooLargeException`. In spite of this, everything still works out just fine. It's fine because `Number-TooLargeException` is declared to be a subclass of `OutOfRangeException`, and Listing 13-6 has a `catch` clause for the `OutOfRangeException`.

You see, because `NumberTooLargeException` is a subclass of `OutOfRangeException`, any instance of `NumberTooLargeException` is just a special kind of `OutOfRangeException`. So, in Listing 13-6, Java may start looking for a clause to catch a `NumberTooLargeException`. When Java stumbles upon the `OutOfRangeExceptioncatch` clause, Java says, "Okay, I've found a match. I'll execute the statements in this `catch` clause."

To keep from having to write this whole story over and over again, I introduce some new terminology. I say that the `catch` clause with parameter `OutOfRange-Exception` *matches* the `NumberTooLargeException` that's been thrown. I call this `catch` clause a *matching catch clause.*

The following list describes different things that the user may do and how the computer responds. As you read, you can follow along by looking at the runs shown in Figure 13-6:

- **The user enters an ordinary whole number, like the number *3*.** All statements in the `try` clause are executed. Then Java skips past all the `catch` clauses and executes the code that comes immediately after all the `catch` clauses. (See Figure 13-7.)

```
How many boxes do we have? 3
The value is $9.75
That's that.

How many boxes do we have? fish
For input string: "fish" ... Cannot interpret the input.
That's that.

How many boxes do we have? -25
A value is out of range.
You typed -25. There's no such thing as a negative box.
That's that.

How many boxes do we have? 1001
A value is out of range.
A value is too large.
1001 is larger than the maximum of 1000
That's that.
```

FIGURE 13-6: Four runs of the code from Listing 13-6.

- **The user enters something that's not a whole number, like the word *fish*.** The code throws a `NumberFormatException`. Java skips past the remaining statements in the `try` clause. Java executes the statements inside the first `catch` clause — the clause whose parameter is of type `NumberFormat Exception`. Then Java skips past the second and third `catch` clauses and executes the code that comes immediately after all the `catch` clauses. (See Figure 13-8.)

```
try {

    //Normal processing with no exceptions thrown

} catch (NumberFormatException e) {
    out.print(e.getMessage());
    out.println(" ... Cannot interpret the input.");
}

catch (OutOfRangeException e) {
    out.println(e.getMessage());
}

catch (Exception e) {
    out.println(e.getMessage());
}

out.println("That's that.");
```

FIGURE 13-7: No exception is thrown.

```
try {

    throw new NumberFormatException();

} catch (NumberFormatException e) {
    out.print(e.getMessage());
    out.println(" ... Cannot interpret the input.");
}

catch (OutOfRangeException e) {
    out.println(e.getMessage());
}

catch (Exception e) {
    out.println(e.getMessage());
}

out.println("That's that.");
```

FIGURE 13-8: Java throws a NumberFormat-Exception.

- **The user enters a negative number, like the number –25.** The code throws an `OutOfRangeException`. Java skips past the remaining statements in the `try` clause. Java even skips past the statements in the first `catch` clause. (After all, an `OutOfRangeException` isn't any kind of a `NumberFormatException`. The `catch` clause with parameter `NumberFormatException` isn't a match for this `OutOfRangeException`.) Java executes the statements inside the second catch clause — the clause whose parameter is of type OutOfRange Exception. Then Java skips past the third catch clause and executes the code that comes immediately after all the catch clauses. (See Figure 13-9.)

```
try {

        throw new OutOfRangeException(...);

} catch (NumberFormatException e) {
    out.print(e.getMessage());
    out.println(" ... Cannot interpret the input.");
}

catch (OutOfRangeException e) {
    out.println(e.getMessage());
}

catch (Exception e) {
    out.println(e.getMessage());
}

out.println("That's that.");
```

FIGURE 13-9: Java throws an OutOfRange-Exception.

- **The user enters an unrealistically large number, like the number 1001.** The code throws a NumberTooLargeException. Java skips past the remaining statements in the try clause. Java even skips past the statements in the first catch clause. (After all, a NumberTooLargeException isn't any kind of NumberFormatException.)

  But, according to the code in Listing 13-5, NumberTooLargeException is a subclass of OutOfRangeException. When Java reaches the second catch clause, Java says, "Hmm! A NumberTooLargeException is a kind of OutOfRangeException. I'll execute the statements in this catch clause — the clause with parameter of type OutOfRangeException." In other words, it's a match.

  Java executes the statements inside the second catch clause. Then Java skips the third catch clause and executes the code that comes immediately after all the catch clauses. (See Figure 13-10.)

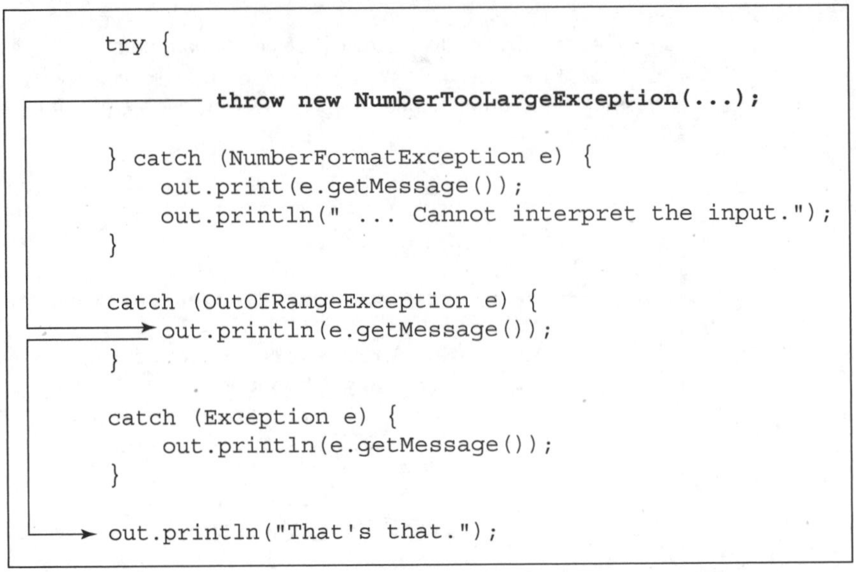

```
try {

    throw new NumberTooLargeException(...);

} catch (NumberFormatException e) {
    out.print(e.getMessage());
    out.println(" ... Cannot interpret the input.");
}

catch (OutOfRangeException e) {
    out.println(e.getMessage());
}

catch (Exception e) {
    out.println(e.getMessage());
}

out.println("That's that.");
```

FIGURE 13-10: Java throws a NumberTooLargeException.

- **Something else, something quite unpredictable, happens. (I don't know what.)** With my unending urge to experiment, I reached into the `try` clause of Listing 13-6 and added a statement that commits a serious crime. The statement divides a number by zero:

  ```
  try {
      out.println(1 / 0);
      int numBoxes = Integer.parseInt(numBoxesIn);
  ```

  When Java encounters this statement, it throws an `ArithmeticException`. Java skips past the remaining statements in the `try` clause. Then Java skips past the statements in the first and second `catch` clauses. When Java reaches the third `catch` clause, I can hear Java say, "Hmm! An `ArithmeticException` is a kind of `Exception`. I've found a matching `catch` clause — a clause with a parameter of type `Exception`. I'll execute the statements in this `catch` clause."

  So Java executes the statements inside the third `catch` clause. Then Java executes the code that comes immediately after all the `catch` clauses. (See Figure 13-11.)

When Java looks for a matching `catch` clause, Java latches on to the topmost clause that fits one of the following descriptions:

- The clause's parameter type is the same as the type of the exception that was thrown.
- The clause's parameter type is a superclass of the exception's type.

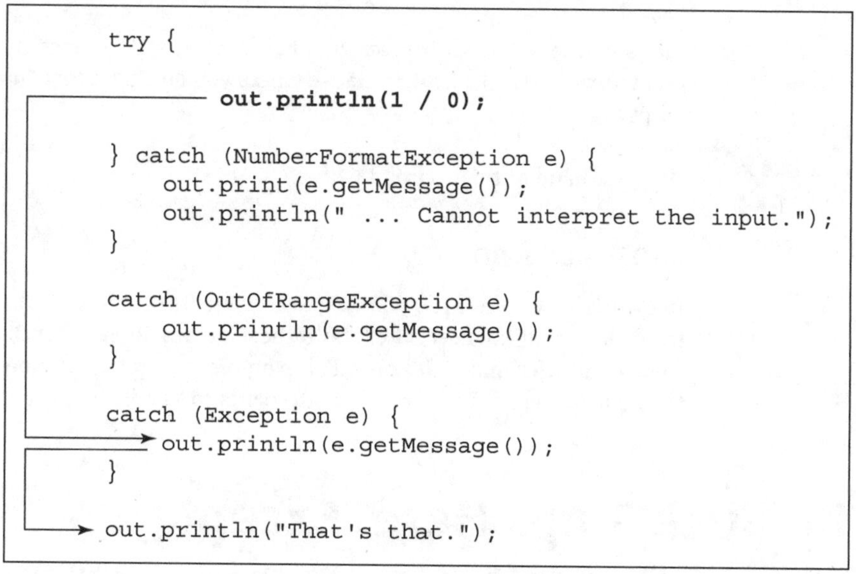

```
try {

        out.println(1 / 0);

} catch (NumberFormatException e) {
    out.print(e.getMessage());
    out.println(" ... Cannot interpret the input.");
}

catch (OutOfRangeException e) {
    out.println(e.getMessage());
}

catch (Exception e) {
    out.println(e.getMessage());
}

out.println("That's that.");
```

FIGURE 13-11: Java throws an ArithmeticException.

If a better match appears farther down the list of `catch` clauses, that's just too bad. Imagine that you added a `catch` clause with a parameter of type `NumberTooLargeException` to the code in Listing 13-6. Imagine, also, that you put this new `catch` clause *after* the `catch` clause with parameter of type `OutOfRangeException`. Then, because `NumberTooLargeException` is a subclass of the `OutOfRangeException` class, the code in the new `NumberTooLargeException` clause would never be executed. That's just the way the cookie crumbles.

Catching two or more exceptions at a time

You can catch more than one kind of exception in a single `catch` clause. For example, in a particular inventory program, you might want the same text to accompany a `NumberFormatException` and your own `OutOfRangeException`. In that case, you can rewrite part of Listing 13-6 this way:

```
} catch (NumberFormatException | OutOfRangeException e) {
    out.println(e.getMessage());
    out.println("Input a number from 0 to 1000.");
} catch (Exception e) {
    out.println(e.getMessage());
}
```

The pipe symbol, `|`, tells Java to catch either a `NumberFormatException` or an `OutOfRangeException`. If you throw an exception of either type, the program adds `Input a number from 0 to 1000` to its helpful output. If you throw an exception that's neither a NumberFormatException nor an OutOfRangeException, the program jumps to the last catch clause where you don't get that Input a number ... output.

TRY IT OUT

Try your hand at this coding task:

LOOSEN UP A BIT

In Listing 13-6, boxPrice and maxBoxes have fixed values. Make improvements to the code so that the user enters both those values. Remember that some values for these quantities make no sense. For example, a negative number of boxes is never too many boxes. Use try-catch statements to handle inappropriate user input.

The Buck Stops Here, Except When It Doesn't

Listing 13-7 has a scaled-down version of an example from Chapter 8. The code tries to read values from a disk file and then write some results to the screen, but the code doesn't work.

LISTING 13-7: Not Writing Payroll Checks

```
/*
 * This code does not compile.
 */

package com.example.payroll;

import java.io.File;
import java.util.Scanner;

import static java.lang.System.out;

public class DoNotDoPayroll {

    public static void main(String[] args) {
        out.println("Starting payroll ...");
        doPayroll();
        out.println("Payroll completed.");
    }
```

```
    public static void doPayroll() {
        var diskScanner = new Scanner(new File("EmployeeInfo.txt"));
        String name = diskScanner.nextLine();
        double amountPaid = diskScanner.nextDouble();

        out.printf("Pay to the order of %s: $%,.2f\n", name, amountPaid);

        diskScanner.close();
    }
}
```

When you construct a Scanner with System.out, nothing much can go wrong. But when you construct a Scanner with a File, you have to be careful. Maybe you misspelled the file name (EnployeeImfo.txt). Maybe your file is in the cloud, and the file is unreadable because the network is down. Maybe you lack permission to access *any* files. Who knows?

Java doesn't let you call new Scanner(new File(*anything_at_all*)) unless you acknowledge that there's some risk. The problem is that the code deep inside the Scanner constructor's body can throw an exception. This kind of exception is an instance of the FileNotFoundException class. When you try to compile the code in Listing 13-7, you see an unwelcome message such as

```
Unhandled exception: java.io.FileNotFoundException
```

The Java programming language has two kinds of exceptions. They're called *checked* and *unchecked* exceptions:

- The potential throwing of a checked exception must be acknowledged in the code.
- The potential throwing of an unchecked exception doesn't need to be acknowledged in the code.

A FileNotFoundException is one of Java's checked exception types. When you call a constructor or method that has the potential to throw a FileNotFoundException, you need to acknowledge that exception in the code.

Now, when I say that an exception is *acknowledged in the code*, what do I really mean?

```
// The author wishes to thank that FileNotFoundException,
// without which this code could not have been written.
```

No, that's not what it means to be acknowledged in the code. Acknowledging an exception in the code means one of two things:

- The statements (including any constructor or method calls) that can throw the exception are inside a `try` clause. That `try` clause has a `catch` clause with a matching exception type in its parameter list.
- The statements (including any constructor or method calls) that can throw the exception are inside a method that has a `throws` clause in its header. The `throws` clause contains a matching exception type.

If you're confused by the wording of these two bullets, don't worry. The next two listings illustrate the points made in the bullets.

Catch it soon

In Listing 13-8, the `new Scanner` call is inside a `try` clause. That `try` clause has a `catch` clause with exception type `FileNotFoundException`.

LISTING 13-8: Acknowledging with a try-catch Statement

```
package com.example.payroll;

import java.io.File;
import java.io.FileNotFoundException;
import java.util.Scanner;

import static java.lang.System.out;

public class DoPayroll {

    public static void main(String[] args) {
        out.println("Starting payroll ...");
        cutCheck();
        out.println("Payroll completed.");
    }

    public static void cutCheck() {
        try {
            var diskScanner = new Scanner(new File("EmployeeInfo.txt"));
            String name = diskScanner.nextLine();
            double amountPaid = diskScanner.nextDouble();

            out.printf("Pay to the order of %s: $%,.2f\n", name, amountPaid);
```

```
            diskScanner.close();
        } catch (FileNotFoundException e) {
            out.println(e.getMessage());
        }
    }
}
```

Figure 13-12 shows two runs of the code in Listing 13-8. For the first run, an EmployeeInfo.txt file contains the lines Barry Burd and 5000.00. For the second run, the user's hard drive contains no EmployeeInfo.txt file.

```
Starting payroll ...
Pay to the order of Barry Burd: $5,000.00
Payroll completed.

Starting payroll ...
EmployeeInfo.txt (No such file or directory)
Payroll completed.
```

FIGURE 13-12: With and without an EmployeeInfo.txt file.

When I look at Listing 13-8, my heart is filled with pride. I've acknowledged the FileNotFoundException, prevented the code from printing Pay to the order of, and avoided a nasty program crash. In fact, I handled the exception as soon as it arose. I put out that fire in the cutCheck method before it could spread and burn up the main method.

But wait! Did I do the right thing? Here's an important principle in software design: Don't be too hasty to catch an exception. Looking back at Figure 13-12, the second run ends with the words Payroll completed. Is that true? Will Barry actually receive the $5000.00 that he so rightfully deserves? No. That second run in Figure 13-12 throws a FileNotFoundException. The cutCheck method in Listing 13-8 catches the exception and doesn't tell the main method that anything went wrong. So the main method, gleeful in its ignorance of the FileNotFoundException, prints Payroll completed.

I shouldn't have shielded the FileNotFoundException from the code's main method. Instead, I should have relayed the exception from the cutCheck method up to the main method. The example in the next section shows you how.

Catch it later

In Listing 13-9, the cutCheck method says "This FileNotFoundException isn't my problem. Let main deal with it."

LISTING 13-9: A Method Admits Its Shortcomings

```
package com.example.payroll;

import java.io.File;
import java.io.FileNotFoundException;
import java.util.Scanner;

import static java.lang.System.out;

public class DoPayrollB {

    public static void main(String[] args) {
        out.println("Starting payroll ...");
        try {
            cutCheck();
            out.println("Payroll completed.");
        } catch (FileNotFoundException e) {
            out.println(e.getMessage());
        }
    }

    public static void cutCheck() throws FileNotFoundException {
        var diskScanner = new Scanner(new File("EmployeeInfo.txt"));
        String name = diskScanner.nextLine();
        double amountPaid = diskScanner.nextDouble();

        out.printf("Pay to the order of %s: $%,.2f\n", name, amountPaid);

        diskScanner.close();
    }
}
```

Figure 13-13 shows two runs of the code in Listing 13-9. For the first run, an `EmployeeInfo.txt` file contains the lines `Barry Burd` and `5000.00`, and the run ends with the words `Payroll completed`. For the second run, the user's hard drive contains no `EmployeeInfo.txt` file, and the program doesn't display those misleading words.

```
Starting payroll ...
Pay to the order of Barry Burd: $5,000.00
Payroll completed.

Starting payroll ...
EmployeeInfo.txt (No such file or directory)
```

FIGURE 13-13: Payroll completed? Maybe yes, maybe no.

The important part of Listing 13-9 is in the cutCheck method's header. That header ends with throws FileNotFoundException. By announcing that it throws a FileNotFoundException, method cutCheck passes the buck. What this throws clause really says is, "I realize that a statement inside this method has the potential to throw a FileNotFoundException, but I'm not acknowledging the exception in a try-catch statement. Java compiler, please don't bug me about this. Instead of having a try-catch statement, I'm passing the responsibility for acknowledging the exception to the main method (the method that called the cutCheck method)."

Indeed, in the main method, the call to cutCheck is inside a try clause. That try clause has a catch clause with a parameter of type FileNotFoundException. So everything is okay. Method cutCheck passes the responsibility to the main method, and the main method accepts the responsibility with an appropriate try-catch statement. Everybody's happy. Even the Java compiler is happy.

To better understand the throws clause, imagine a volleyball game in which the volleyball is an exception. When a player on the other team serves, that player is throwing the exception. The ball crosses the net and heads directly toward you. If you pound the ball back across the net, you're catching the exception. But if you pass the ball to another player, you're using the throws clause. In essence, you're saying, "Here, other player — *you* deal with this exception."

A statement in a method can throw an exception that's not matched by a catch clause. This includes situations in which the statement throwing the exception isn't even inside a try block. When this happens, execution of the program jumps out of the method that contains the offending statement. Execution jumps back to whatever code called the method in the first place.

A method can name more than one exception type in its throws clause. Just use commas to separate the names of the exception types, as in the following example:

```
throws FileNotFoundException, NumberFormatException, ArithmeticException
```

Checked or unchecked?

Sneak a peek at Listing 13-3 and notice that your newly-declared OutOfRangeException class extends Java's RuntimeException class. Why would you want to extend RuntimeException? Here's the story:

» **Any subclass of RuntimeException is unchecked.**

In fact, any descendant (sub-subclass, sub-sub-subclass, and so on) of RuntimeException is unchecked. The unchecked exceptions represent errors that can crop up almost anywhere in your program. If your code throws an unchecked exception, you should think about rewriting the code.

Some common unchecked exceptions include the `NumberFormatException` (of Listings 13-2, 13-4, and others), the `ArithmeticException`, the `IndexOutOfBoundsException`, the infamous `NullPointerException`, and many others. When you write Java code, much of your code is susceptible to these exceptions, but enclosing the code in `try` clauses (or passing the buck with `throws` clauses) is completely optional.

- **Any exception that's not a descendant of `RuntimeException` is checked.**

Unchecked exceptions represent challenges that the outside world may impose on your program. Like it or not, some files simply aren't where they should be, so you had better acknowledge this in your program and write code to make the best of it. Java's checked exceptions include the `FileNotFoundException`, (See Listings 13-8 and 13-9.) the `Printer-Exception`, the `SQLException`, and a gang of other interesting exceptions.

READER'S ACKNOWLEDGMENTS

TRY IT OUT

The following code doesn't compile because the code throws an unacknowledged `FileNotFoundException`:

```
// BAD CODE:
import java.io.File;
import java.util.Scanner;

public class Main {

    public static void main(String[] args) {
        var diskScanner = new Scanner(new File("numbers.txt"));

        int[] numerators = new int[5];
        int[] denominators = new int[5];

        int i = 0;
        while (diskScanner.hasNextInt()) {
            numerators[i] = diskScanner.nextInt();
            denominators[i] = diskScanner.nextInt();
            i++;
        }
        for (int j = 0; j < numerators.length; j++) {
            System.out.println(numerators[j] / denominators[j]);
        }
```

```
        diskScanner.close();
    }
}
```

Fix the unacknowledged `FileNotFoundException` so that the code compiles. Then notice that, depending on the values in the `numbers.txt` file, some other exceptions may be thrown during a run of the program. Add one or more `try-catch` statements to display messages about these exceptions without letting the program crash.

IF IT'S BROKEN, FIX IT

Add `try-catch` statements or `throws` clauses (or a mixture of these two things) to fix the following broken code:

```
// BAD CODE:
import java.io.DataInputStream;
import java.io.DataOutputStream;
import java.io.EOFException;
import java.io.File;
import java.io.FileInputStream;
import java.io.FileOutputStream;

public class Main {

    public static void main(String[] args) {
        var fileIn = new File("input");
        var fileInStrm = new FileInputStream(fileIn);
        var dataInStrm = new DataInputStream(fileInStrm);

        var fileOut = new File("output");
        var fileOutStrm = new FileOutputStream(fileOut);
        var dataOutStrm = new DataOutputStream(fileOutStrm);

        int numFilesCopied = 0;

        try {
            while (true) {
                dataOutStrm.writeByte(dataInStrm.readByte());
            }
        } catch (EOFException e) {
            numFilesCopied = 1;
        }
    }
}
```

When you've gotten the code to compile, create a file named `input` and run the code to see whether it creates the file named `output`.

Try, Try Again!

Once upon a time, I was a young fellow living with my parents in Philadelphia and just starting to drive a car. I was heading toward a friend's house and thinking about who-knows-what when another car came from nowhere and bashed my car's passenger door. This kind of thing is called a RunARedLightException.

Anyway, both cars were still drivable, and we were squat in the middle of a busy intersection. To avoid causing a traffic jam, we both pulled over to the nearest curb. I fumbled for my driver's license (which had a very young photo of me on it) and opened the door to get out of my car.

And that's when the second accident happened. As I was getting out of my car, a city bus was coming by. The bus hit me and rolled me against my car a few times. This kind of thing is called a DealWithLawyersException.

The truth is that everything came out just fine. I was bruised but not battered. My parents paid for the damage to the car, so I never suffered any financial consequences. (I managed to pass on the financial burden by putting the `RunARedLightException` into my `throws` clause.)

This incident helps to explain why I think the way I do about exception handling. In particular, I wonder, "What happens if, while the computer is recovering from one exception, a second exception is thrown?" After all, the statements inside a `catch` clause aren't immune to calamities.

Well, the answer to this question is anything but simple. For starters, you can put a `try` statement inside a `catch` clause. This protects you against unexpected, potentially embarrassing incidents that can crop up during the execution of the `catch` clause. But when you start worrying about cascading exceptions, you open up a slimy can of worms. The number of scenarios is large, and things can become complicated quickly. The program in Listing 13-10 helps you sort things out.

LISTING 13-10: Using Two Files

```
package com.example.payroll;

import java.io.File;
import java.io.FileNotFoundException;
```

```java
import java.util.Scanner;

import static java.lang.System.out;

public class DoPayrollC {

    public static void main(String[] args) {
        out.println("Starting payroll ...");
        try {
            cutCheck();
            out.println("Payroll completed.");
        } catch (FileNotFoundException e) {
            out.println(e.getMessage());
        }
    }

    public static void cutCheck() throws FileNotFoundException {
        var diskScanner = new Scanner(new File("EmployeeInfo.txt"));
        var diskScanner2 = new Scanner(new File("LegalInfo.txt"));
        String name = diskScanner.nextLine();
        double amountPaid = diskScanner.nextDouble();
        String disclaimer = diskScanner2.nextLine();

        out.printf("Pay to the order of %s: $%,.2f\n", name, amountPaid);
        out.println(disclaimer);

        diskScanner.close();
        diskScanner2.close();
    }
}
```

The output of a run from Listing 13-10 might look like the text in Figure 13-14.

```
Starting payroll ...
Pay to the order of Barry Burd: $5,000.00
We're not responsible for anything at all.
Payroll completed.
```

FIGURE 13-14: You'll hear from our lawyers!

But what happens if, after successfully opening the first file (`EmployeeInfo.txt`), Java can't find the second file (`LegalInfo.txt`)? Then Java throws a `FileNotFoundException` and jumps immediately out of the `cutCheck` method. Java never calls `diskScanner.close()`, and the connection to `EmployeeInfo.txt` lives on in limbo. With a small program like the one in Listing 13-10, this is no big deal. But if your code repeatedly fails to close such resources, you could be in big trouble.

That's why Java has a try-with-resources feature. A *try-with-resources statement* keeps track of all the things you've opened and closes them automatically when they're no longer needed. Listing 13-11 has the full story.

LISTING 13-11: Making Sure to Close Resources

```
package com.example.payroll;

import java.io.File;
import java.io.FileNotFoundException;
import java.util.Scanner;

import static java.lang.System.out;

public class DoPayrollD {

    public static void main(String[] args) {
        out.println("Starting payroll ...");
        try {
            cutCheck();
            out.println("Payroll completed.");
        } catch (FileNotFoundException e) {
            out.println(e.getMessage());
        }
    }

    public static void cutCheck() throws FileNotFoundException {
        try (var diskScanner = new Scanner(new File("EmployeeInfo.txt"));
             var diskScanner2 = new Scanner(new File("LegalInfo.txt"))) {
            String name = diskScanner.nextLine();
            double amountPaid = diskScanner.nextDouble();
            String disclaimer = diskScanner2.nextLine();

            out.printf("Pay to the order of %s: $%,.2f\n", name, amountPaid);
            out.println(disclaimer);
        } catch (FileNotFoundException e) {
            out.println("Abnormal return from the cutCheck method ...");
            throw e;
        }
    }
}
```

In Listing 13-11, the declarations of diskScanner1 and diskScanner2 are in parentheses after the word try. The parenthesized declarations tell Java to close diskScanner1 and diskScanner2 automatically after execution of the statements in the try clause. Java closes these connections whether you throw an exception or not.

Look at all the statements in the body of the cutCheck method. Not one of these statements contains the word close. You don't need any explicit close calls. The try-with-resources statement does all the closing for you.

You can access all kinds of resources (files, databases, connections to servers, and others) and have peace of mind knowing that Java will sever the connections automatically when necessary. Life is good!

Before we bid a fond farewell to Listing 13-11, let's take a moment to look at the cutCheck method's catch clause. As in Listing 13-9, we want the code's main method to do something about a missing LegalInfo.txt file. We could add lots of clunky if statements to make main bend to our will, but it's better to have main respond to the original FileNotFoundException. In the cutCheck method's catch cause, the variable e stands for that unsavory FileNotFoundException. So the statement throw e does the trick. That statement says "pass this despicable FileNotFoundException on to whichever method called me." And the "method that called me" is main.

TIP

In a try-with-resources statement, the catch clause is optional. In Listing 13-11, if you remove the catch clause, the program behaves almost the same way. The only thing you lose is the possibility of getting the Abnormal return from the cutCheck method output.

TRY IT OUT

BE RESOURCEFUL

Modify the code in Listing 13-8 so that it uses a try-with-resources statement.

DON'T MISS A TRICK

Modify the code in Listing 13-11 so that it checks for sensible values in the two disk files. Does the EmployeeInfo.txt file have a name and a double value? Does the LegalInfo.txt file contain any text?

» Exposing names to other classes

» Tweaking your code to find the right middle ground

Chapter 14
Sharing Names among the Parts of a Java Program

Our family's neighborhood elementary school has a yearly tradition: Each spring, the students in fifth grade take a daylong field trip to Philadelphia. When my son was in fifth grade, I went along on the trip to help supervise 4 of the 20 children. While the 4 kids and I were touring the city's historical sites, I misplaced the preplanned itinerary. I didn't know where we were supposed to go next. So, what could I do? I did what any good Philadelphia native would do — I ignored the itinerary and took the kids to the nearest cheap restaurant. I treated them all to Philadelphia's signature dish — the fat-laden, greasy, Philadelphia cheesesteak.

What could this story possibly have to do with Java? That's simple. This chapter is about access — how one class's code may gain access to another class's code. My Philadelphia anecdote is about access, too. Shortly after the school trip, my son's fifth grade teacher got wind of my shenanigans during the tour. Later, when my daughter was in fifth grade, the teacher said that I wasn't welcome to take part in that year's junket to Philadelphia. The teacher had revoked my access to those Philadelphia cheesesteaks.

Access Modifiers

Object-oriented programming is big on hiding details. Programmers who write one piece of code shouldn't tinker with the details inside another programmer's code. It's not a matter of security and secrecy — it's a matter of safety. When you hide details, you prevent the intricacies inside one piece of code from being twisted and broken by another piece of code. Your code comes in nice, discrete, manageable lumps. You keep complexity to a minimum. You make fewer mistakes. You save money. You help promote world peace.

Other chapters have plenty of examples of the use of private fields. When a field is declared private, it's hidden from all outside meddling. This hiding enhances modularity, minimizes complexity, and so on.

Elsewhere in the annals of *Java For Dummies*, 8th Edition, are examples of things that are declared public. Just like a public celebrity, a field that's declared public is left wide open. Plenty of people probably know what kind of toothpaste Elvis used, and any programmer can reference a public field, even a field that's not named *Elvis*.

In Java, the words *public* and *private* are called *access modifiers*. No doubt you've seen fields and methods without access modifiers in their declarations. A method or field of this kind is said to have *default access*. Many examples in this book use default access without making a big fuss about it. That's okay in some chapters, but not in this chapter. In this chapter, I describe the nitty-gritty details about default access.

And you can find out about yet another access modifier that isn't used in any example before this chapter. (At least, I don't remember using it in any earlier examples.) It's the `protected` access modifier. Yes, this chapter covers some of the slimy, grimy facts about protected access.

Classes and Their Members

With this topic, you can become all tangled up in terminology, so you need to get some basics out of the way. (Most of the terminology you need comes from Chapters 7 and 10, but it's worth reviewing at the start of this chapter.) Consider this chunk of Java code:

```
public class MyClass {
    int myField;            //a field (a member)
```

```
    void myMethod() {       //a method (another member)
        int myOtherField;  //a method-local variable (NOT a member)
    }
}
```

The comments on the right side of the code tell the whole story. The code has two kinds of variables: fields and method-local variables. This chapter isn't about method-local variables. It's about methods and fields.

Believe me, carrying around the phrase *methods and fields* wherever you go isn't easy. It's much better to give these things one name and be done with it. That's why both methods and fields are called *members* of a class.

At this point, you make an important distinction. Think about Java's `public` keyword. You can put `public` in front of a member. For example, you can write

```
public static void main(String[] args) {
```

or

```
public amountInAccount = 50.22;
```

You can also put the `public` keyword in front of a class. For example, you can write

```
public class Drawing {
    // Your code goes here
}
```

In Java, the `public` keyword has two slightly different meanings: one meaning for the members of a class and another meaning for the classes themselves. Most of this chapter deals with the meaning of `public` (and other such keywords) for members. The last part of this chapter (appropriately titled "Access Modifiers for Java Classes") deals with the meaning for classes.

Public and Private Access for Members

Each field is declared in a particular class, belongs to that class, and is a member of that class. The same is true of methods: Each method is declared in a particular class, belongs to that class, and is a member of that class. Can you use a certain member name in a particular place in your code? To begin answering the question, check whether that place is inside or outside the member's class:

- If the member is private, only code that's inside the member's class can refer directly to that member's name:

```
class SomeClass {
    private int myField = 10;
}

class SomeOtherClass {

    public static void main(String[] args) {
        var someObject = new SomeClass();

        System.out.println(someObject.myField);   //This doesn't work.
    }
}
```

- If the member is public, any code running in the same Java virtual machine can refer directly to that member's name.

```
class SomeClass {
    public int myField = 10;
}

class SomeOtherClass {

    public static void main(String[] args) {
        var someObject = new SomeClass();

        System.out.println(someObject.myField);   //This works.
    }
}
```

Throughout this chapter, I make claims such as the one in the previous bullet. "If a member is public, any code can refer directly to that member's name." In fact, claims of this kind aren't entirely true. In this chapter's last section, I describe Java's module feature. When two classes reside in two different modules, the ability of one class to refer to the other class's names doesn't automatically apply. For details, see the later section "From Classes Come Modules."

Figures 14-1, 14-2, and 14-3 illustrate the ideas in a slightly different way.

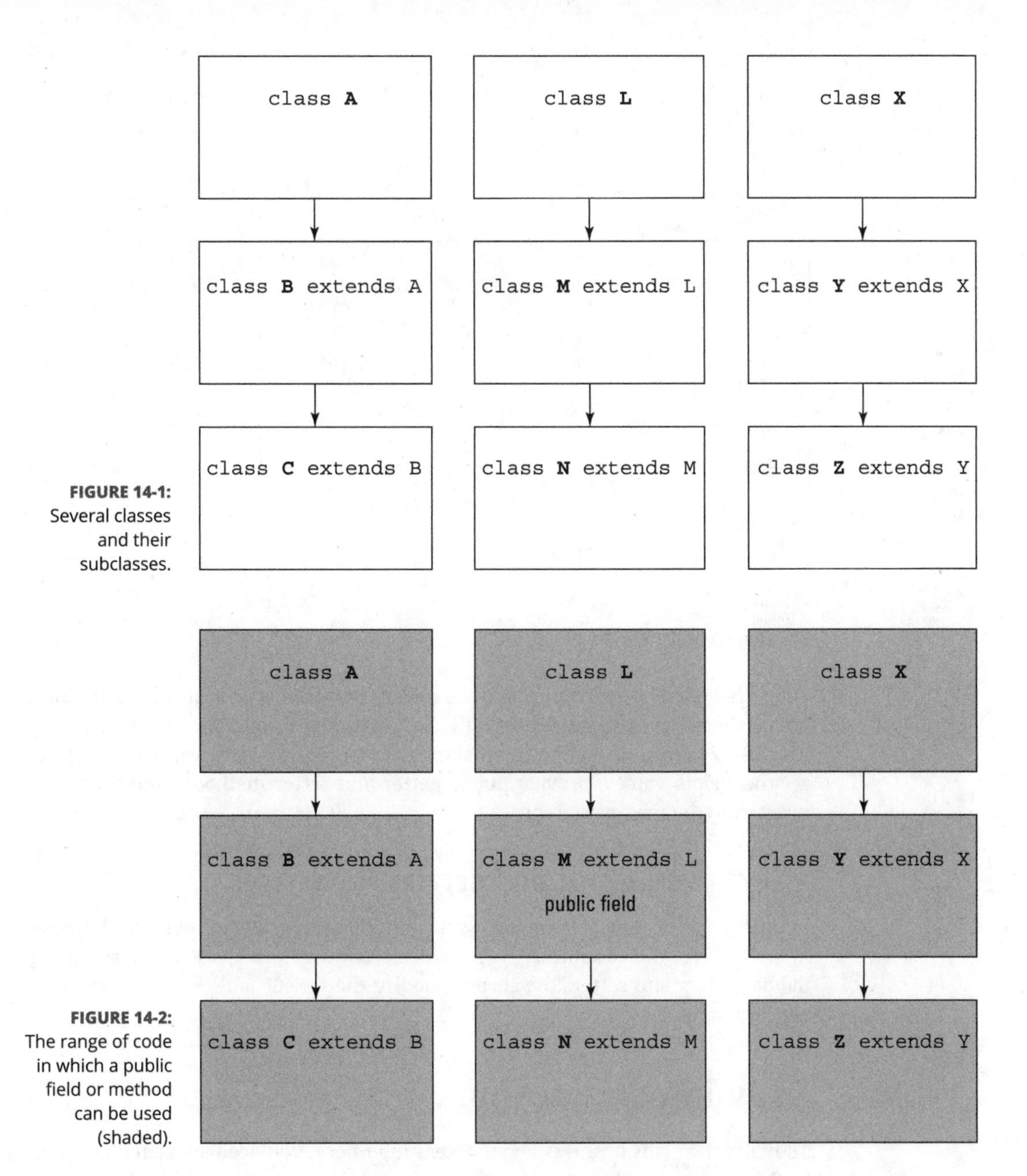

FIGURE 14-1: Several classes and their subclasses.

FIGURE 14-2: The range of code in which a public field or method can be used (shaded).

WARNING

When you see this section's examples, you may come to the wrong conclusion. You may have this little conversation with yourself: "In the example with `private int myField`, the code doesn't work. But in the example with `public int myField`, the code works. So, to have a better chance of getting my code to work, I should make my fields `public` and avoid making them `private`. Right?"

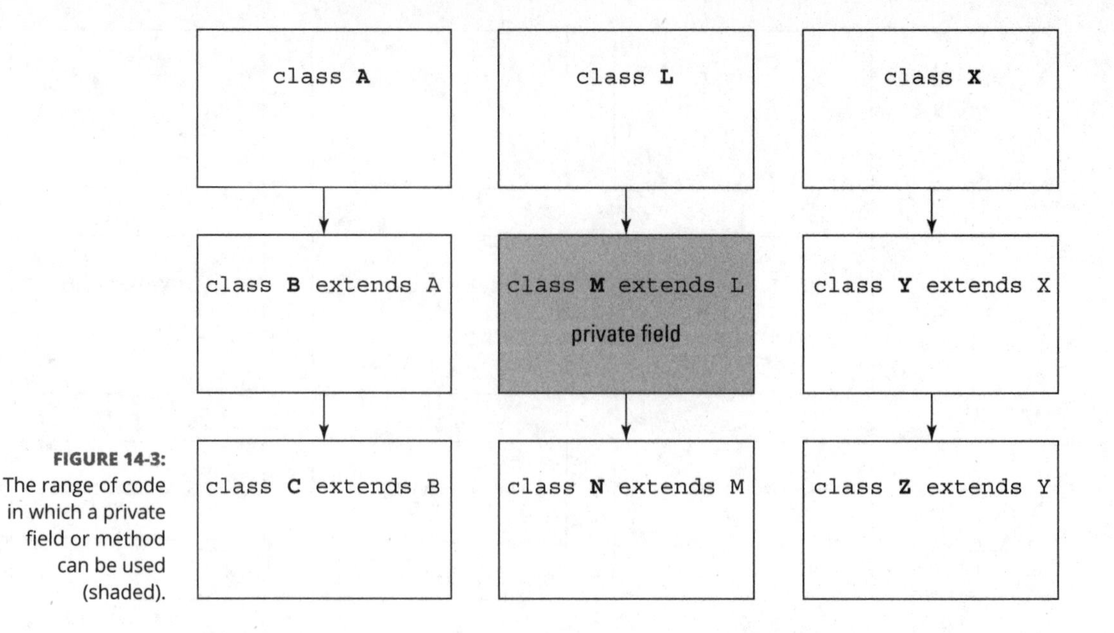

FIGURE 14-3: The range of code in which a private field or method can be used (shaded).

No, dear reader. That's not right!

Public fields are easy to use and even easier to misuse. The best way to engineer your code is to make access to each field as restrictive as possible. If a field doesn't absolutely need to be public, try making it private. If other classes have to get or set the field's values, provide public getter and setter methods. And that leads nicely into the next paragraph . . . .

EVERYTHING'S BETTER WITH GETTERS AND SETTERS

TRY IT OUT

In one of this section's examples, you can't write `someObject.myField` because, in `SomeClass`, the variable `myField` is declared to be private. Fix this by adding public getters and setters; be sure to modify the `someObject.myField` reference appropriately.

Drawing on a frame

To make clear this business about access modifiers, you need an example or two. In this chapter's first example, almost everything is public. With public access, you don't have to worry about who-can-use-what.

The code for this first example comes in several parts. The first part, which is in Listing 14-1, displays an `ArtFrame`. On the face of the `ArtFrame` is a `Drawing`. If all the right pieces are in place, running the code of Listing 14-1 displays a window like the one shown in Figure 14-4.

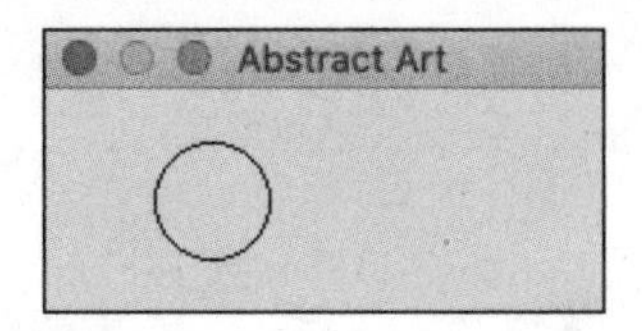

FIGURE 14-4: An ArtFrame.

LISTING 14-1: Displaying a Frame

```
package com.example.chapter14;

import com.burdbrain.drawings.Drawing;
import com.burdbrain.frames.ArtFrame;

public class ShowFrame {

    public static void main(String[] args) {
        var artFrame = new ArtFrame(new Drawing());

        artFrame.setSize(200, 100);
        artFrame.setVisible(true);
    }
}
```

The code in Listing 14-1 creates a new `ArtFrame` object, sets the object's size, and makes the object visible. Notice that Listing 14-1 starts with two `import` declarations. The first `import` declaration allows you to abbreviate the name `Drawing` from the `com.burdbrain.drawings` package. The second import declaration allows you to abbreviate the name `ArtFrame` from `com.burdbrain.frames`. You purchased the use of these packages from Burd Brain Consulting, known worldwide for its cheap, reliable software.

CROSS REFERENCE

For a review of `import` declarations, see Chapter 4.

The detective in you may be thinking, "The author must have written more code (code that I don't see here) and put that code in packages that he named *com.burdbrain.drawings* and *com.burdbrain.frames*." And, indeed, you are correct. To make Listing 14-1 work, I create something called a *Drawing*, and I'm putting all my drawings in the `com.burdbrain.drawings` package. I also need an `ArtFrame` class, and I'm putting all such classes in my `com.burdbrain.frames` package.

So, really, what's a `Drawing`? Well, if you're so eager to know, look at Listing 14-2.

LISTING 14-2: The Drawing Class

```
package com.burdbrain.drawings;

import java.awt.Graphics;

public class Drawing {
    public int x = 40, y = 40, width = 40, height = 40;

    public void paint(Graphics g) {
        g.drawOval(x, y, width, height);
    }
}
```

You don't have to know much about the code in Listing 14-2. In fact, you can get by if you notice only two things:

» **The** `Drawing` **class belongs to a package named** `com.burdbrain.drawing`**.**

» **The** `Drawing` **class and all of its members (**`x`**,** `y`**,** `width`**,** `height`**, and** `paint`**) are public.**

You can think of the listing's `paint` method as a black box because this chapter isn't really about painting things. The `paint` method in Listing 14-2 is only a vehicle to help me describe access modifiers. Even so, you may be curious about the listing's `Graphics` class and `g.drawOval` call. If so, you can read the later sidebar, "Draw on the screen (crayons not required)."

Putting a package in its place

Package names can include dots and underscores, but they can't include other punctuation characters. For example, you can name your package `good.job` or `go_away`, but you can't name it `oh, my!` or `state-of-the-art,`. A dot in a package name separates one part of the name from another.

When you put a class into a package, you have to create a directory structure that mirrors the package name's parts. For example, the code in Listing 14-2 belongs to the `com.burdbrain.drawings` package. To house that code, you must have three directories: a `com` directory, a subdirectory of `com` named `burdbrain`, and a subdirectory of `burdbrain` named `drawings`. The overall directory structure is shown in Figure 14-5.

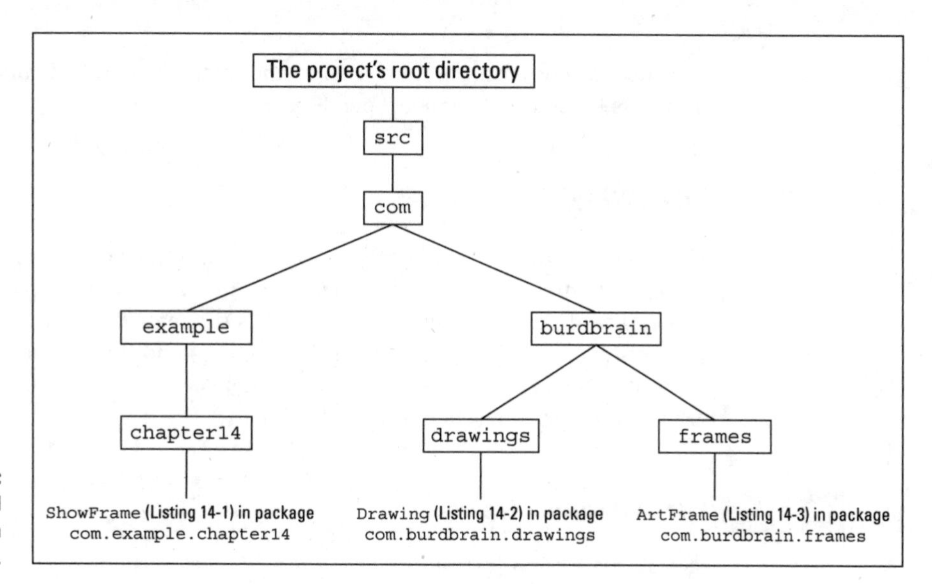

FIGURE 14-5: The files and directories in your project.

Many IDEs build this directory structure for you. Here's how it works:

- You're a programmer at the illustrious Burd Brain Consulting company. You start creating a new project. You ask for your project's root directory to be named `MyBigProject`.

  As a result, the IDE creates a directory named `MyBigProject` with a subdirectory that's typically named `src`. Java will "know" to look for code inside that `src` directory. After all, the name `src` is short for *source* code.

- The IDE asks you for a package name. You type `com.burdbrain.drawings`.

  As a result, the IDE creates the `com`, `burdbrain`, and `drawings` directories. (Refer to Figure 14-5.)

- You finish creating the project and start writing your `com.burdbrain.drawings` package's code. After a while, you select New ⇨ Package (or something like that) from the IDE's menu bar. You ask the IDE to create a second package named `com.burdbrain.frames`.

  As a result, the IDE creates the `frames` directory. (Refer to Figure 14-5.)

- You start writing your `com.burdbrain.frames` package's code.

WARNING

If you don't have your code in the appropriate directories, you get a repulsive and disgusting `NoClassDefFoundError`. Believe me, this error is never fun to see. When you see this error, you get no clues to help you figure out where the missing class is or where the compiler expects to find it. If you stay calm, you can figure out all this stuff on your own. If you panic, you'll poke around for hours. As a seasoned Java programmer, I can remember plenty of scraped knuckles that came from this heinous NoClassDefFoundError.

Making a frame

This chapter's first three listings develop one multipart example. This section has the last of three pieces in that example. This last piece isn't essential for the understanding of access modifiers, which is the main topic of this chapter. So, if you want, you can glance quickly at Listing 14-3 and then move on. On the other hand, if you want to know more about the Java Swing classes, read the sidebar "Draw on the screen (crayons not required)."

LISTING 14-3: The ArtFrame Class

```
package com.burdbrain.frames;

import com.burdbrain.drawings.Drawing;

import javax.swing.JFrame;
import java.awt.Graphics;

public class ArtFrame extends JFrame {

    Drawing;

    public ArtFrame(Drawing drawing) {
        this.drawing = drawing;
        setTitle("Abstract Art");
        setDefaultCloseOperation(EXIT_ON_CLOSE);
    }

    public void paint(Graphics g) {
        drawing.paint(g);
    }
}
```

MAKE IT AND BREAK IT

TRY IT OUT

Create your own Java project using the code from Listings 14-1 through 14-3. Run the code to make sure that it works. Then try to move the ArtFrame.java file to the com/burdbrain/drawings folder. Does your IDE automatically change the package declaration at the top of the ArtFrame.java file? If not, does your IDE offer to make that change for you? Make sure that, with a mismatch between the package declaration and the folder where the file resides, the application refuses to run.

DRAW ON THE SCREEN (CRAYONS NOT REQUIRED)

Listings 14-2 and 14-3 have all the gadgetry you need for putting a drawing on a Java frame. The code uses several names from the Java API (application programming interface). I explain most of these names in Chapters 9 and 10.

The only new name in Listings 14-2 and 14-3 is the word *paint*. The `paint` method in Listing 14-3 defers to another `paint` method — the `paint` method belonging to a `Drawing` object. (Refer to Listing 14-2.) The `ArtFrame` object creates a floating window on your computer screen. What's drawn in that floating window depends on whatever `Drawing` object was passed to the `ArtFrame` constructor.

The `paint` method in Listing 14-2 uses a standard trick for making things appear onscreen. The parameter `g` in Listing 14-2 is called a *graphics context*. In particular, `g` stores information about the setting in which Java draws pixels. What's the current ink color? What's the current text font? How do you draw an oval? The graphics context remembers all that stuff.

To make things appear, all you do is draw with this graphics context, and whatever you draw is eventually rendered on the computer screen.

Here's a little more detail: In Listing 14-2, the `paint` method takes a `g` parameter. This `g` parameter refers to an instance of the `java.awt.Graphics` class. Because a `Graphics` instance is a context, the things you draw with this context are eventually displayed on the screen. Like all instances of the `java.awt.Graphics` class, this context has several drawing methods — and one of them is `drawOval`. When you call `drawOval`, you specify a starting position (`x` pixels from the left edge of the frame and `y` pixels from the top of the frame). You also specify an oval size by putting numbers of pixels in the `width` and `height` parameters. Calling the `drawOval` method puts a little round thing into the `Graphics` context. That `Graphics` context, round thing and all, is displayed onscreen.

If you trace the flow of Listings 14-1 through 14-3, you may notice something peculiar: The `paint` methods in these listings never seem to be called. Well, for many of Java's window-making components, you just declare a `paint` method and let the method sit there quietly in the code. When the program runs, the computer calls the `paint` method automatically.

That's what happens with `javax.swing.JFrame` objects. In Listing 14-3, the frame's `paint` method is called from behind the scenes. Then the frame's `paint` method calls the `Drawing` object's `paint` method, which in turn draws an oval on the frame. That's how you get the stuff you see earlier, in Figure 14-4.

NOW IN LIVING COLOR

In Listing 14-2, I draw a circle on a frame. To fill the circle with green color, use the `Graphics` class's `setColor` and `fillOval` methods, like this:

```
g.setColor(Color.GREEN)
g.fillOval(x, y, width, height);
```

Values such as `Color.GREEN` belong to `Color` class in the `java.awt` package.

Create a frame that displays a traffic signal with its green, yellow, and red lights.

Default Access for Members

Your preferred software vendor, Burd Brain Consulting, has sold you two files: `Drawing.class` and `ArtFrame.class`. According to Chapter 2, these `.class` files are called *bytecode* files. You can use the declarations in the `Drawing.class` and `ArtFrame.class` files, but you can't modify these files. In addition, you can't see the code inside the original `Drawing.java` and `ArtFrame.java` files. Burd Brain Consulting didn't sell you the `.java` files, so you have to live with whatever happens to be inside the two `.class` files. (If only you'd purchased a copy of *Java For Dummies*, 8th Edition, which has the code for these `.java` files in Listings 14-2 and 14-3!)

Anyway, you want to tweak the way the oval looks in Figure 14-4 so that it's a bit wider. To do this, you create a subclass of the `Drawing` class — `DrawingWide` — and put it in Listing 14-4.

LISTING 14-4: A Subclass of the Drawing Class

```
package com.example.chapter14;

import com.burdbrain.drawings.Drawing;

import java.awt.Graphics;

public class DrawingWide extends Drawing {
    int width = 100, height = 30;

    public void paint(Graphics g) {
        g.drawOval(x, y, width, height);
    }
}
```

To make use of the code in Listing 14-4, you change two of the lines in Listing 14-1. (See the boldface code in Listing 14-5.)

LISTING 14-5: **How to Change the ShowFrame Code**

```
package com.example.chapter14;

// No longer needed: import com.burdbrain.drawings.Drawing;
import com.burdbrain.frames.ArtFrame;

public class ShowFrame {

    public static void main(String[] args) {
        var artFrame = new ArtFrame(new DrawingWide());

        artFrame.setSize(200, 100);
        artFrame.setVisible(true);
    }
}
```

Put Listings 14-2 through 14-5 into one project. When you run the code, you get the frame shown in Figure 14-6.

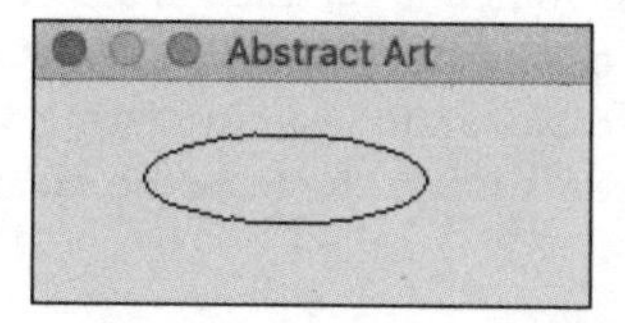

FIGURE 14-6: Another art frame.

At this point, your project has two drawing classes: the original `Drawing` class and your new `DrawingWide` class. Similar as these classes may be, they live in two separate packages. That's not surprising. The `Drawing` class, developed by your friends at Burd Brain Consulting, lives in a package whose name starts with *com.burdbrain.* But you developed `DrawingWide` on your own, so you shouldn't put it in a `com.burdbrain` package. Being a loyal employee of the Example Company, you put your code in a `com.example` package.

Your `DrawingWide` class (refer to Listing 14-4) is a subclass of the original `Drawing` class. (Refer to Listing 14-2.) In the subclass, you make reference to the original class's `x` and `y` fields. That's okay because, in the `Drawing` class, those fields are `public`. (See Figure 14-7.)

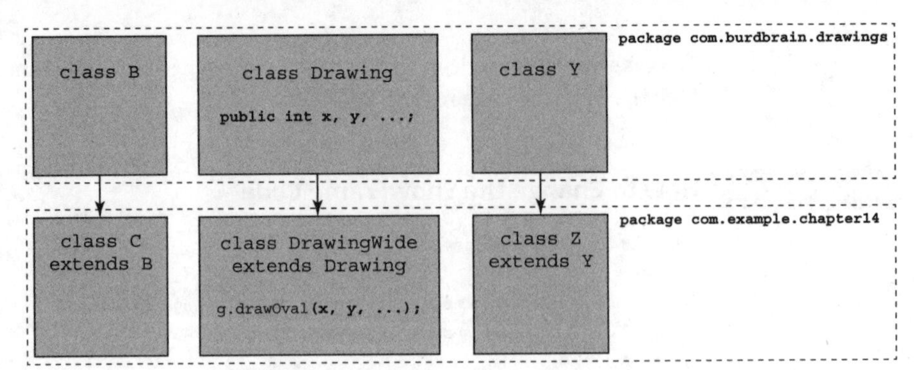

FIGURE 14-7: The range of code in which x and y can be used (shaded).

One way or another, your `DrawingWide` subclass compiles and runs as planned. You go home, beaming with the confidence of having written useful, working code.

Switching to Default access

If you're reading these paragraphs in order, you know that the last example ends happily. The code in Listing 14-4 runs like a charm. Everyone, including my wonderful editor, Paul Levesque, is happy.

But wait! Do you ever wonder what life would be like if you hadn't chosen that particular career, dated that certain someone, or read that certain *For Dummies* book? In this section, I roll back the clock a bit to show you what would have happened if one word had been omitted from the code in Listing 14-2. What if the people at Burd Brain Consulting hadn't made the `Drawing` class's fields public? (See Listing 14-6.)

LISTING 14-6: Fields with Default Access

```
package com.burdbrain.drawings;

import java.awt.Graphics;

public class Drawing {
    int x = 40, y = 40, width = 40, height = 40;

    public void paint(Graphics g) {
        g.drawOval(x, y, width, height);
    }
}
```

With this change in the Drawing class, the DrawingWide code in Listing 14-4 no longer works. Instead, you see the following error messages:

```
x is not public in com.burdbrain.drawings.Drawing;
cannot be accessed from outside package
g.drawOval(x, y, width, height);
           ^
y is not public in com.burdbrain.drawings.Drawing;
cannot be accessed from outside package
g.drawOval(x, y, width, height);
              ^
```

In Java, the default access for a member of a class is package-wide access. A member declared without the word *public, private,* or *protected* in front of it is accessible only in the package in which its class resides. Figures 14-8 and 14-9 illustrate the point.

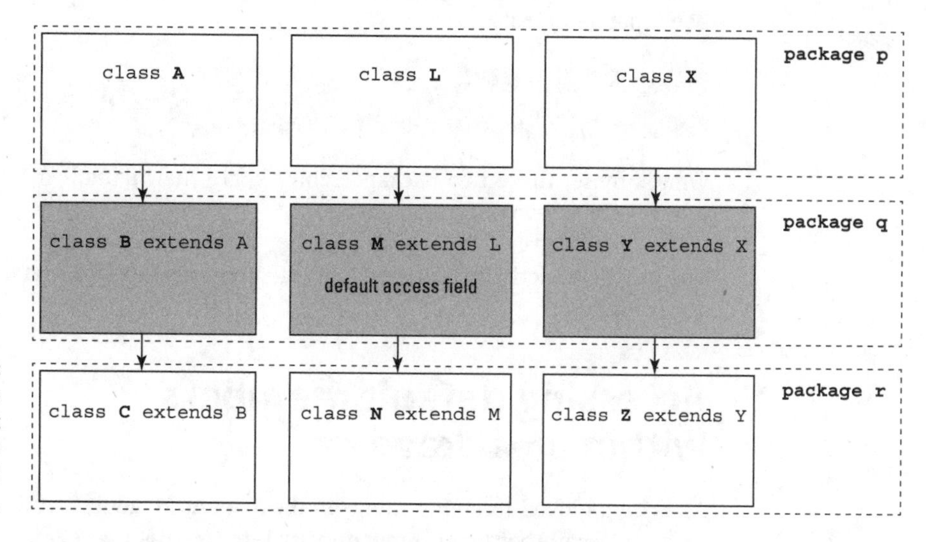

FIGURE 14-8: The range of code in which a default field or method can be used (shaded).

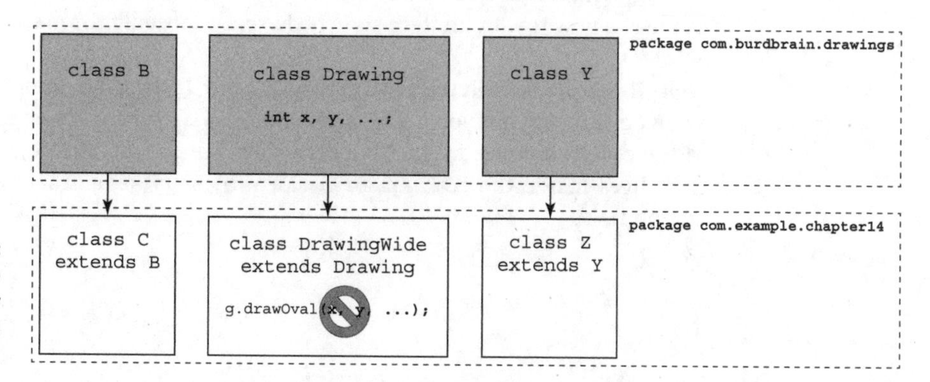

FIGURE 14-9: The range of code in which x and y can be used (shaded).

The DrawingWide class in Listing 14-4 is a subclass of the Drawing class in Listing 14-6, so you'd guess that DrawingWide can explicitly refer to the Drawing class's x and y fields. But the Drawing class's x and y fields have default access, and DrawingWide is in a different package. Sorry! It just doesn't work.

The same story about default access holds true for methods as well as fields. After all, the access rules — default and otherwise — apply to all members of classes.

REMEMBER

The access rules that I describe in this chapter don't apply to method-local variables. A method-local variable can be accessed only within its own method. For the rundown on method-local variables, see Chapter 10.

WARNING

The names of packages, with all their dots and subparts, can be slightly misleading. For example, you can create four-part package names. Imagine that you have two packages — one named com.burdbrain.drawings, and another named com.burdbrain.drawings.rectangles, The packages' names are similar, but that doesn't make one package part of the other. To import these packages, you need two separate lines:

```
import com.burdbrain.drawings;
import com.burdbrain.drawings.rectangles;
```

What's more, these two packages don't share their default members with each other. They're two different packages.

The next few sections have examples of situations in which default access yields good results.

Accessing default members within a package

As a child, I loved receiving things in the mail. As an adult, I still love it. At worst, I get junk mail that I can throw directly into the trash. At best, I get something I can use, a new toy, or an item somebody sent especially for me.

Well, if you like receiving mail, today is your lucky day. Somebody from Burd Brain Consulting sent you a subclass of the Drawing class. It's essentially the same as the code in Listing 14-4. The only difference is that this new DrawingWideBB class lives inside the com.burdbrain.drawings package. The code is shown in Listing 14-7.

LISTING 14-7: Drawing a Wide Oval

```
package com.burdbrain.drawings;

import java.awt.Graphics;

public class DrawingWideBB extends Drawing {
    int width = 100, height = 30;

    public void paint(Graphics g) {
        g.drawOval(x, y, width, height);
    }
}
```

To run the code in Listing 14-7, you have to change two lines in Listing 14-1. (See Listing 14-8.)

LISTING 14-8: Another Change to the ShowFrame Code

```
package com.example.chapter14;

import com.burdbrain.drawings.DrawingWideBB;
import com.burdbrain.frames.ArtFrame;

public class ShowFrame {

    public static void main(String[] args) {
        var artFrame = new ArtFrame(new DrawingWideBB());

        artFrame.setSize(200, 100);
        artFrame.setVisible(true);
    }
}
```

Try running Listing 14-8 along with Listings 14-3, 14-6, and 14-7. The fields `x` and `y` have default access in Listing 14-6, yet everything works just fine. The reason? It's because `Drawing` and `DrawingWideBB` are in the same package. Look back at Figure 14-8 and notice the shaded region that spans across an entire package. The code in the `DrawingWideBB` class has every right to use the `x` and `y` fields, which are defined with default access in the `Drawing` class because `Drawing` and `DrawingWideBB` are in the same package. (Compare Figure 14-10 with Figure 14-9.)

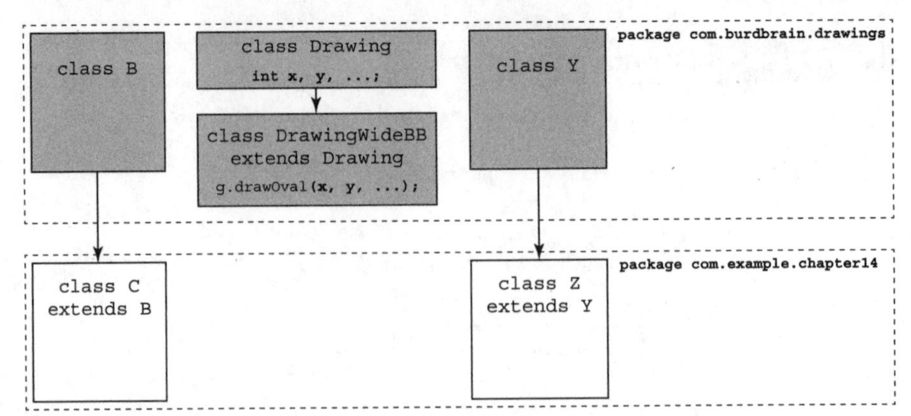

FIGURE 14-10: The range of code in which x and y can be used (shaded).

Protected Access for Members

When I was first getting to know Java, I thought the word *protected* meant "nice and secure" or something like that. "Wow, that field is protected. It must be hard to get at." Well, this notion turned out to be wrong. In Java, a member that's protected is less hidden, less secure, and available for use in more classes than one that has default access. In other words, protected access is more permissive than default access. For me, the terminology is misleading. But that's the way it is.

A class in one package and a subclass in another

Think of protected access this way. You start with a field that has default access (a field without the word `public`, `private`, or `protected` in its declaration). That field can be accessed only inside the package in which it lives. Now add the word `protected` to the front of the field's declaration. Suddenly, classes outside that field's package have some access to the field. You can now reference the field from a subclass (of the class in which the field is declared). You can also reference the field from a sub-subclass, a sub-sub-subclass, and so on. Any descendant class will do. If you want evidence, see Listing 14-9.

LISTING 14-9: Protected Fields

```
package com.burdbrain.drawings;

import java.awt.Graphics;

public class Drawing {
    protected int x = 40, y = 40, width = 40, height = 40;
```

```
    public void paint(Graphics g) {
        g.drawOval(x, y, width, height);
    }
}
```

Test the code in Listing 14-9 along with Listings 14-3 through 14-5. The `DrawingWide` class (refer to Listing 14-4) references the `x` and `y` fields that are defined in its parent `Drawing` class (refer to Listing 14-9.) That's okay, even though `DrawingWide` isn't in the same package as the `Drawing` class. It's okay because the `x` and `y` fields are protected in the `Drawing` class.

Compare Figures 14-8 and 14-11. Notice the extra bit of shading in Figure 14-11. A subclass can access a protected member of a class, even if that subclass belongs to some other package.

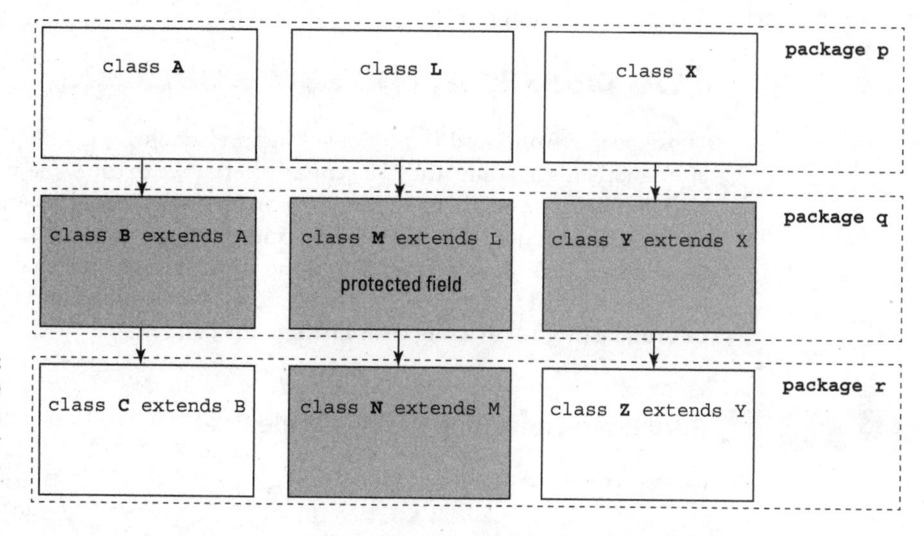

FIGURE 14-11: The range of code in which a protected field or method can be used (shaded).

Figure 14-12 shows the relationships between the `DrawingWide` class (refer to Listing 14-4), the `DrawingWideBB` class (refer to Listing 14-7), and the `Drawing` class in Listing 14-9.

TIP

Do you work with a team of programmers? Do people from outside your team use their own team's package names? If so, when they use your code, they may make subclasses of the classes you've defined. This is where protected access comes in handy. Use protected access when you want people from outside your team to make direct references to your code's fields or methods.

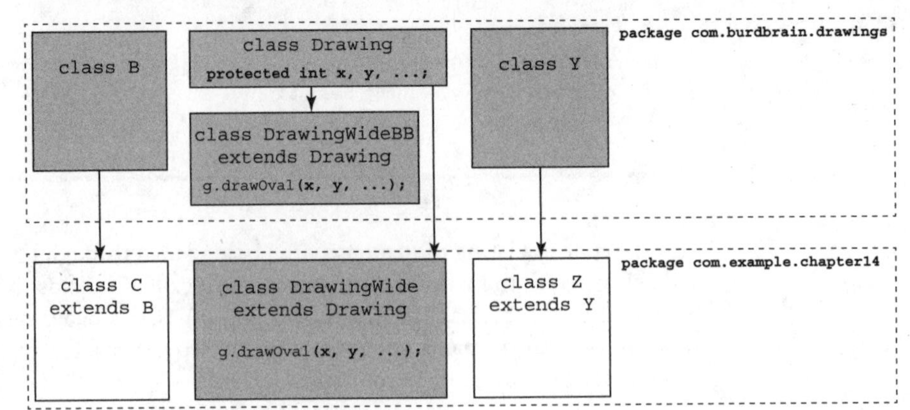

FIGURE 14-12: The range of code in which x and y can be used (shaded).

REMEMBER

For the members of a class, private access is the most restrictive. Default access is a bit less restrictive, and then comes protected access, and finally, public access, which is the least restrictive.

Two classes in the same package

Those people from Burd Brain Consulting are sending you one piece of software after another. This time, they've sent an alternative to the `ShowFrame` class — the class in Listing 14-1. This new `ShowFrameWideBB` class displays a wider oval (how exciting!), but it does this without creating a `DrawingWide` class. Instead, the new `ShowFrameWideBB` code creates a `Drawing` instance and then changes the value of the instance's `width` and `height` fields. The code is shown in Listing 14-10. You can run Listing 14-10 along with Listings 14-3 and 14-9.

LISTING 14-10: Burd Brain Consulting Draws a Wide Oval

```
package com.burdbrain.drawings;

import com.burdbrain.frames.ArtFrame;

class ShowFrameWideBB {

    public static void main(String[] args) {
        var drawing = new Drawing();
        drawing.width = 100;
        drawing.height = 30;
        var artFrame = new ArtFrame(drawing);
        artFrame.setSize(200, 100);
        artFrame.setVisible(true);
    }
}
```

Here's the story. This ShowFrameWideBB class in Listing 14-10 is in the same package as the Drawing class. (Refer to Listing 14-9.) Both classes are in the com.burdbrain.drawings package. But ShowFrameWideBB isn't a subclass of the Drawing class.

Now imagine compiling ShowFrameWideBB with the Drawing class. The ShowFrameWideBB class makes explicit reference to the Drawing class's width and height fields. So, what happens?

Well, everything goes smoothly because protected members (such as width and height, in Listing 14-9) are available in two (somewhat unrelated) places. Look again at Figure 14-11. A protected member is available to subclasses outside the package, but the member is also available to code (subclasses or not) within the member's package. Figure 14-13 shows the relationship among several classes, including the DrawingWideBB class (refer to Listing 14-10) and the Drawing class in Listing 14-9.

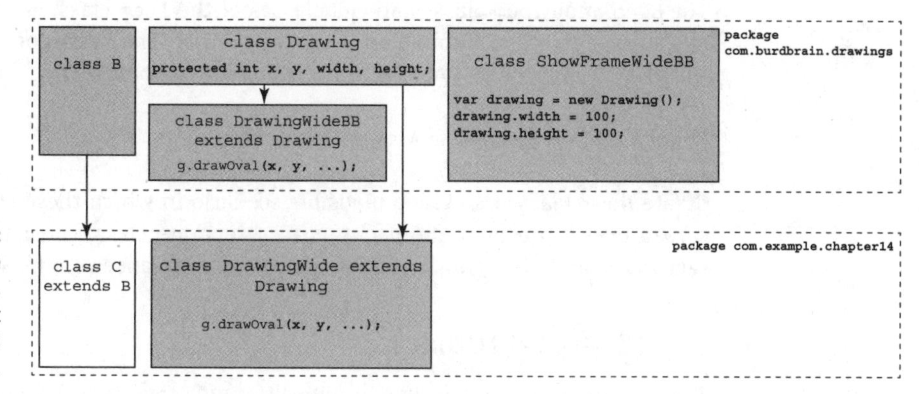

FIGURE 14-13: The range of code in which x, y, width, and height can be used (shaded).

TECHNICAL STUFF

The real story about protected access is one step more complicated than the story I describe in this section. The Java Language Specification (https://docs.oracle.com/javase/specs) mentions a hair-splitting point about code being responsible for an object's implementation. When you're first figuring out how to program in Java, don't worry about this point. Wait until you've written many Java programs. Then, when you stumble upon a variable has protected access error message, you can start worrying. Better yet, skip the worrying and take a careful look at the protected-access section in the Java Language Specification.

CROSS REFERENCE

For info about the Java Language Specification, visit Chapter 3.

TRY IT OUT

Here are some things for you to try:

BUY THIS BOOK

A Book has a title (a String) and an author (an instance of the Author class). An Author has a name (a String) and an ArrayList of Book instances. A separate class contains a main method that creates several books and several authors. The main method also displays information about the books and authors.

Put each class in its own package. Wherever possible, make your fields private, and provide public getters and setters.

NAME THAT TUNE

An Item has a name (a String) and an artist (an instance of the Artist class). Each Artist instance has a name (a String) and an ArrayList of items.

The Song and Album classes are subclasses of the Item class. Each Song instance has a genre (a value from an enum named Genre). The values of Genre are ROCK, POP, BLUES, and CLASSICAL. Each Album instance has an ArrayList of songs.

Finally, a Playlist has an ArrayList of items.

Create these classes. Devise a plausible scenario in which these classes would be spread over more than one package. In a separate class, construct instances of each class, and display information about these instances on the screen.

WHAT'S MINE IS YOURS

The following four classes live in four different .java files. Without typing these classes in an IDE's editor, decide which statements will cause the IDE to display error messages. For each such statement, decide on the least permissive access change that would eliminate the error message:

```
// THIS CODE DOES NOT COMPILE:

package com.mypackage.things;

import com.yourpackage.stuff.Stuff;
import com.yourpackage.stuff.morestuff.MoreStuff;

public class Things {
    protected int i = 0;
    private int j = 0;
    int k = 0;
```

```
    public static void main(String[] args) {
        var stuff = new Stuff();
        System.out.println(stuff.i);
        var moreStuff = new MoreStuff();
        System.out.println(moreStuff.i);
    }
}

package com.yourpackage.stuff;

import com.yourpackage.stuff.morestuff.MoreStuff;

public class Stuff {
    protected int i = 0;

    void aMethod() {
        new MoreStuff().myMethod();
    }
}

package com.yourpackage.stuff.morestuff;

import com.mypackage.things.Things;

public class MoreStuff extends Things {

  protected void myMethod() {
    System.out.println(i);
  }
}

package com.mypackage.things;

public class MoreThings extends Things {

    public void anotherMethod() {
      System.out.println(i);
      System.out.println(j);
      System.out.println(k);
    }
}
```

Access Modifiers for Java Classes

Maybe the things you read about access modifiers for members make you a tad dizzy. After all, member access in Java is a complicated subject with lots of plot twists and cliffhangers. Well, the dizziness is over. Compared with the saga for fields and methods, the access story for classes is rather simple.

A class can be either public or nonpublic. If you see something like

```
public class Drawing
```

you're looking at the declaration of a public class. But if you see plain old

```
class OnlyMyDrawing
```

the class that's being declared isn't public.

Public classes

If a class is public, you can refer to the class from anywhere in your code. Of course, some restrictions apply. You must obey all the rules in this chapter's "Putting a package in its place" section. You must account for the code's modular structure. (See this chapter's "From Classes Come Modules" section, later on.) You must also refer to a packaged class properly. For example, in Listing 14-1, you can write

```
import com.burdbrain.drawings.Drawing;
import com.burdbrain.frames.ArtFrame;
...
var artFrame = new ArtFrame(new Drawing());
```

or you can do without the import declarations and write

```
var artFrame =
    new com.burdbrain.frames.ArtFrame(new com.burdbrain.drawings.Drawing());
```

One way or another, your code must acknowledge that the `ArtFrame` and `Drawing` classes are in named packages.

Nonpublic classes

If a class isn't public, you can refer to the class only from code within the class's package.

I tried it. First, I went back to Listing 14-2 and deleted the word *public.* I turned `public class Drawing` into plain old `class Drawing`, like this:

```
package com.burdbrain.drawings;

import java.awt.Graphics;

class Drawing {
    public int x = 40, y = 40, width = 40, height = 40;

    public void paint(Graphics g) {
        g.drawOval(x, y, width, height);
    }
}
```

Then I compiled the code in Listing 14-7. Everything was peachy because Listing 14-7 contains the following lines:

```
package com.burdbrain.drawings;

public class DrawingWideBB extends Drawing
```

Because both pieces of code are in the same `com.burdbrain.drawings` package, access from `DrawingWideBB` back to the nonpublic `Drawing` class was no problem.

But then I tried to compile the code in Listing 14-3. The code in Listing 14-3 begins with

```
package com.burdbrain.frames;
```

That code isn't in the `com.burdbrain.drawings` package. So, when the computer reached the line

```
Drawing drawing;
```

from Listing 14-3, the computer went *poof!* To be more precise, the computer displayed this message:

```
com.burdbrain.drawings.Drawing is not public in com.burdbrain.drawings;
cannot be accessed from outside package
```

Well, I guess I got what was coming to me.

Things are never as simple as they seem. The rules I describe in this section apply to almost every class in this book. But Java has fancy things called *inner classes,* and inner classes follow a different set of rules. Fortunately, a typical novice programmer has little contact with inner classes. The only inner classes in this book are in Chapter 16 (and a few inner classes disguised as `enum` types). So for now, you can live quite happily with the rules that I describe in this section.

From Classes Come Modules

The cosmos is made of solar systems, and these solar systems combine to form galaxies. But the story doesn't end there. The galaxies group into galaxy clusters, which in turn group into superclusters. The same is true of Java. Fields and methods form classes, which combine to form packages, and the packages combine to form modules.

Most of the `java` and `javax` packages in this book's examples live in a module named `java.base`, but a few packages live in other modules. For example, the `java.awt` and `javax.swing` packages (imported in Listing 14-3) belong to a module named `java.desktop`.

As it is with packages, you can create your own modules. When you do, you build a wall around that module's code. Public or not, you can't access code from other peoples' modules, and other peoples' modules can't access your code.

Fortunately, the wall around your code is penetrable. Your module's code can explicitly make its packages available to the code in other modules, and the other modules can explicitly request access to your module's code.

I'm already way over the page limit for this chapter. For more information about modules, visit `https://openjdk.java.net/projects/jigsaw/quick-start`.

» Working with abstract classes

Chapter 15
Fancy Reference Types

In previous chapters, you may have read about the things that full-time and part-time employees have in common. In particular, both the FullTimeEmployee and PartTimeEmployee classes can extend the Employee class. That's nice to know if you're running a small business, but what if you're not running a business? What if you're taking care of house pets?

This chapter explores the care of house pets and other burning issues.

Java's Types

Chapter 4 explains that Java has these two kinds of types:

- **Primitive types:** Java has a total of eight primitive types: The four you use most often are int, double, boolean, and char.
- **Reference types:** Java's API has thousands of reference types, and, when you write a Java program, you define new reference types.

  Java's String type is a reference type. So are Java's Scanner, JFrame, ArrayList, and File types. My DummiesFrame is a reference type. In Chapter 8, you create your own Employee, FullTimeEmployee, and PartTimeEmployee reference types. Your first *You'll love Java!* program has a main method inside of a class, and that class is a reference type. You may not realize it, but every array belongs to a reference type.

In Java, reference types are everywhere. But until this point in the book, the only reference types you see are classes and arrays. Java has other kinds of reference types, and this chapter explores the possibilities.

The Java Interface

Think about a class (such as an `Employee` class) and a subclass (such as a `FullTimeEmployee` class). The relationship between a class and its subclass is one of inheritance. In many real-life families, a child inherits assets from a parent. And in Chapter 8, the `FullTimeEmployee` class inherits `name` and `jobTitle` fields from the `Employee` class. That's the way it works.

But consider the relationship between an editor and an author. The editor says, "By signing this contract, you agree to submit a completed manuscript by the ninth of January." Despite any excuses that the author gives before the deadline date (and, believe me, authors make plenty of excuses), the relationship between the editor and the author is one of obligation. The author agrees to take on certain responsibilities; and, in order to continue being an author, the author must fulfill those responsibilities. (By the way, there's no subtext in this paragraph — none at all.)

Now consider Barry Burd. Who? Barry Burd — that guy who writes *Java For Dummies* and certain other *For Dummies* books (all from Wiley Publishing). He's a college professor, and he's also an author. You want to mirror this situation in a Java program, but Java doesn't support multiple inheritance. You can't make Barry extend both a `Professor` class and an `Author` class at the same time.

Fortunately for Barry, Java has interfaces. An interface is a kind of reference type. In fact, the code to create an interface looks a lot like the code to create a class:

```
public interface MyInterfaceName {
  // blah, blah, blah
}
```

An interface is a lot like a class, but an interface is different. (What else is new? A cow is like a planet, but it's quite a bit different. Cows moo; planets hang in space.)

Anyway, when you read the word *interface,* you can start by thinking of a class. Then, in your head, note that

- A class can extend only one parent class, but a class can implement many interfaces.
- A parent class is a bunch of stuff that a class inherits. But an *interface* is a bunch of stuff that an implementing class is *obligated to provide*.

What about poor Barry? He can be an instance of a `Person` class with all the fields that any person has — `name`, `address`, `age`, `height`, `weight`, and so on. He can also implement more than one interface:

- Because Barry implements a `Professor` interface, he must have methods named `teachStudents`, `adviseStudents`, and `gradePapers`.
- Because he implements an `Author` interface, he must have methods named `writeChapters`, `reviewChapters`, `answerEmail`, and so on.

Declaring two interfaces

Imagine two different kinds of data. One is a column of numbers that comes from an array. Another is a table (with rows and columns) that comes from a disk file. What might these two things have in common?

I don't know about you, but I may want to display both kinds of data. So, I can write code to create a contract. The contract says, "Whoever signs this contract agrees to have a `display` method." In Listing 15-1, I declare a `Displayable` interface.

LISTING 15-1: Behold! An Interface!

```
package com.example.data;

public interface Displayable {

    void display();

}
```

Wait just a darn minute! The `display` method declaration in Listing 15-1 has a header but no body. No curly braces appear after `display()` — only a lonely-looking semicolon. What's going on here?

To answer the question, I'll let the code in Listing 15-1 speak for itself. If the code in the listing could talk, here's what the code would say:

> "As an interface, my display method has a header but no body. A class that claims to implement me (the Displayable interface) must provide (either directly or indirectly) a body for the display method. That is, a class that claims to implement Displayable must, in one way or another, provide its own code of the following kind:
>
> ```
> public void display() {
>     // Some statements go here
> }
> ```
>
> To implement me (the interface in Listing 15-1), the new code's display method must take no parameters and return nothing (also known as void)."

The Displayable interface is like a legal contract. The Displayable interface doesn't tell you what an implementing class already has. Instead, the Displayable interface tells you what an implementing class must declare in its own code.

In addition to displaying columns of numbers and tables, I may also want to summarize both kinds of data. How do you summarize a column of numbers? I don't know. Maybe you display the total of all the numbers. And how do you summarize a table? Maybe you display the table's column headings. How you summarize the data isn't my concern. All I care about is that you have some way to summarize the data.

So, I create code containing a second Java contract. The second contract says, "Whoever signs this contract agrees to have a summarize method." In Listing 15-2, I declare a Summarizable interface.

LISTING 15-2: Another Interface

```
package com.example.data;

public interface Summarizable {

    String summarize();

}
```

Any class claiming to implement the Summarizable interface must, by hook or by crook, provide an implementation of a summarize method — a method with no parameters that returns a String value.

In the declaration of an interface, a particular method might have no body of its own. A method with no body is called an *abstract method.*

An interface's abstract method is automatically public whether you use the word `public` or not. For example, in Listing 15-2, the line

```
String summarize();
```

has the same meaning as

```
public String summarize();
```

If you enjoy typing the word `public`, feel free to do so. If not, leave that word out.

Implementing interfaces

Listing 15-3 implements the `Displayable` and `Summarizable` interfaces and provides bodies for the `display` and `summarize` methods.

LISTING 15-3: Implementing Two Interfaces

```
package com.example.data;

public class ColumnOfNumbers implements Displayable, Summarizable {
    double numbers[];

    public ColumnOfNumbers(double[] numbers) {
        this.numbers = numbers;
    }

    @Override
    public void display() {
        for (double d : numbers) {
            System.out.println(d);
        }
    }

    @Override
    public String summarize() {
        double total = 0.0;
        for (double d : numbers) {
            total += d;
        }
        return Double.toString(total);
    }
}
```

REMEMBER

When you implement an interface, you provide bodies for the interface's abstract methods.

Java's compiler is serious about the use of the `implements` keyword. If you remove either of the two method declarations from Listing 15-3 without removing the `implements` clause, you see some frightening error messages in your IDE's editor. Java expects you to honor the contract that the `implements` keyword implies. If you don't honor the contract, Java refuses to compile your code. So there!

TIP

You can use Java's error messages to your advantage. Start by typing some code containing the clause `implements Displayable, Summarizable`. Because of the `implements` clause, the editor displays an error mark and lists the names of the methods that you should have declared but didn't. In this section's example, those method names are `display` and `summarize`. After a few more mouse clicks, the IDE generates simple `display` and `summarize` methods for you.

Listing 15-4 contains another class that implements the `Displayable` and `Summarizable` interfaces.

LISTING 15-4: Another Class Implements the Interfaces

```
package com.example.data;

import java.io.File;
import java.io.FileNotFoundException;
import java.util.ArrayList;
import java.util.Scanner;

public class Table implements Displayable, Summarizable {
    Scanner diskFile;
    ArrayList<String> lines = new ArrayList<>();

    public Table(String fileName) {
        try {
            diskFile = new Scanner(new File(fileName));

        } catch (FileNotFoundException e) {
            e.getMessage();
        }
        while (diskFile.hasNextLine()) {
            lines.add(diskFile.nextLine());
        }
    }
```

```
    @Override
    public void display() {
        for (String line : lines) {
            System.out.println(line);
        }
    }

    @Override
    public String summarize() {
        return lines.get(0);
    }
}
```

In Listings 15-3 and 15-4, notice several uses of the @Override annotation. Chapter 8 introduces the use of the @Override annotation. Normally, you use @Override to signal the replacement of a method that's already been declared in a superclass. But from Java 6 onward, you can also use @Override to signal an interface method's implementation. That's what I do in Listings 15-3 and 15-4.

Putting the pieces together

The code in Listing 15-5 makes use of all the stuff in Listings 15-1 to 15-4.

LISTING 15-5: Getting the Most out of Your Interfaces

```
package com.example.data;

public class Main {

    public static void main(String[] args) {
        double numbers[] = {21.7, 68.3, 5.5};
        var column = new ColumnOfNumbers(numbers);

        displayMe(column);
        summarizeMe(column);

        Table table = new Table("MyTable.txt");

        displayMe(table);
        summarizeMe(table);
    }
```

(continued)

LISTING 15-5: (continued)

```
    static void displayMe(Displayable displayable) {
        displayable.display();
        System.out.println();
    }

    static void summarizeMe(Summarizable summarizable) {
        System.out.println(summarizable.summarize());
        System.out.println();
    }
}
```

With the `MyTable.txt` file shown in Figure 15-1, the output from Listing 15-5 is shown in Figure 15-2.

```
Name  ID Balance
Barry 01 19.51
Carol 02 100.35
Myrna 03 10.07
```

FIGURE 15-1: The MyTable.txt file.

```
21.7
68.3
5.5

95.5

Name  ID Balance
Barry 01 19.51
Carol 02 100.35
Myrna 03 10.07

Name  ID Balance
```

FIGURE 15-2: Running the code in Listing 15-5.

Feast your eyes on the `displayMe` method in Listing 15-5. What kind of parameter does the `displayMe` method take? Is it a `ColumnOfNumbers`? No. Is it a `Table`? No.

The `displayMe` method knows nothing about `ColumnOfNumbers` instances or `Table` instances. All the `displayMe` method knows about is things that implement `Displayable`. That's what the `displayMe` method's parameter list says. When you hand something that implements the `Displayable` interface to the `displayMe` method, the `displayMe` method knows what it can do. The `displayMe` method can call the parameter's `display` method, because that parameter object is guaranteed to have a `display` method.

The same kind of thing is true about the summarizeMe method in Listing 15-5. How do you know that you can call summarizable.summarize() inside the body of the summarizeMe method? You can make this call because summarizable has to have a summarize() method. The rules about Java interfaces guarantee it.

That's the real power behind Java's interfaces.

CONTROL ALT DELETE

TRY IT OUT

In this section, the ColumnOfNumbers and Table classes implement the Displayable and Summarizable interfaces. What about a Deletable interface? Any class implementing the Deletable interface must have its own delete method.

Create the DeletableColumnOfNumbers class — a subclass of the ColumnOfNumbers class. In addition to all the things ColumnOfNumbers does, the DeletableColumnOfNumbers class also implements the Deletable interface. When you delete a column of numbers, you set the values of each of its entries to 0.0.

TWO KINDS OF METHODS

Inside an interface declaration, any method without a body is called an *abstract method.* If you run Java 8 or later, you can also put methods with bodies inside an interface declaration. A method with a body is called a *default method*. In an interface's code, each default method declaration starts with the default keyword:

```
public interface MyInterface {

    void method1();

    default void method2() {
        System.out.println("Hello!");
    }
}
```

In MyInterface, method1 is an abstract method, and method2 is a default method. If you create a class that implements MyInterface, like so:

```
class MyClass implements MyInterface
```

then your newly declared MyClass must declare its own method1 and provide a body for method1. Optionally, your MyClass may declare its own method2. If MyClass doesn't declare its own method2, then MyClass inherits a method2 body from MyInterface.

Create the `DeletableTable` class — a subclass of the `Table` class. In addition to all the things `Table` does, the `DeletableTable` class also implements the `Deletable` interface. When you `delete` a table, you remove all rows except the first (table heading) row. (***Hint:*** If you call the `lines` list's `remove` method starting from the `1` row and going to the `lines.size()` row, you won't be happy with the results. A call to the `remove` method modifies the list immediately, and that can mess up your loop.)

Abstract Classes

Is there anything you can say that applies to animals of every kind? If you're a biologist, maybe there is. But if you're a programmer, you can say very little. If you don't believe me, consider the wondrous variety of life on planet earth:*

- A gelada monkey spends the day on a grassy plateau. But at night the gelada goes for a snooze on the rocky, perilous edge of a mountain cliff. With any luck, the sleeping monkey doesn't toss and turn much.
- A Pompeii worm lives in an underwater tube. The temperature by the worm's head is about 72 degrees Fahrenheit (22 degrees Celsius). But at the other end of the worm, the water temperature is normally 176 degrees Fahrenheit (80 degrees Celsius). If you know one of these worms personally, don't buy any warm socks for it.
- A sea squirt lives part of its life as an animal. At a certain point in its life cycle, the sea squirt attaches itself permanently to a rock and then digests its own brain, effectively turning itself into a plant.
- A tiny water bear can survive 12 days (and maybe more) with no atmosphere in the vacuum of outer space. Even the cosmic radiation in outer space doesn't harm a water bear. In 2017, a team of scientists concluded that

---

See `www.smithsonianmag.com/science-nature/ethiopias-exotic-monkeys-147893502`

`https://serc.carleton.edu/microbelife/topics/marinesymbiosis/pompeii.html`

`www.psychologytoday.com/blog/choke/201207/how-humans-learn-lessons-the-sea-squirt`

`www.esa.int/Our_Activities/Human_Spaceflight/Research/Tiny_animals_survive_exposure_to_space`

`www.nature.com/articles/s41598-017-05796-x`

> water bears would survive most mass extinction events, including nearby supernova blasts and large asteroid impacts. So that's what I want to be in my next life — a water bear.

With so much biological diversity on our planet, the only thing I can say that applies to every animal is that every animal has a certain weight (measured in pounds or kilograms), and every animal makes (or, possibly, doesn't make) a characteristic sound. Listing 15-6 has the complete scoop.

LISTING 15-6: What a Programmer Knows about Animals

```
package com.example.species;

public class Animal {
    double weight;
    String sound;

    public Animal(double weight, String sound) {
        this.weight = weight;
        this.sound = sound;
    }
}
```

While I typed the code for the `Animal` class, I had to stop and correct several typing mistakes. The mistakes weren't really my fault: My cat was walking back and forth across my computer keyboard. And that brings me from the subject of all animals to the topic of house pets.

A house pet is an animal. But every house pet has a name — like Fluffy, Spot, or Princess. And every house pet has a recommended routine for taking care of the pet.

Of course, the care routines differ greatly from one kind of pet to another. If I had a dog, I'd have to walk the dog. But I'd never try to walk a cat. In fact, I don't even let our cat out of the house. So when I define my `HousePet` class, I want to be vague about pet care instructions. And in Java, a class that's somewhat vague is called an *abstract class*. Listing 15-7 has an example.

LISTING 15-7: What It Means to Be a House Pet

```
package com.example.species;

public abstract class HousePet extends Animal {
    String name;

    public HousePet(String name, double weight, String sound) {
        super(weight, sound);
        this.name = name;
    }

    abstract public void howToCareFor();

    public void about() {
        System.out.print(name + " weighs " + weight + " pounds");
        System.out.print(sound != null ? (" and says '" + sound + "'") : "");
        System.out.println(".");
    }
}
```

On the first line of Listing 15-7, the keyword abstract tells Java that HousePet is an abstract class. Because HousePet is an abstract class, HousePet can have an abstract method. And in Listing 15-7, howToCareFor is an abstract method. An abstract method has a header but no body. In an abstract method's declaration, there are no curly braces — only a semicolon where curly braces would normally appear.

So, when you try to execute the howToCareFor method, what happens? Well, you can't really execute the howToCareFor method in Listing 15-7. In fact, you can't even create an instance of the abstract class declared in Listing 15-7. The following lines of code are illegal:

```
// VERY BAD CODE:
HousePet myPet = new HousePet("Boop", 12.0, "Meow");
var yourPet = new HousePet("Pawz", 22.5, "Woof");
```

An abstract class has no life of its own. To use an abstract class, you have to create an ordinary (non-abstract) class that extends the abstract class. In the ordinary class, all methods have bodies. So everything works out.

CROSS REFERENCE

Before you walk away from Listing 15-7, notice the super(weight, sound) call in that listing. As in Chapter 9, the keyword super triggers a call to the superclass's constructor. In Listing 15-7, calling super(weight, sound) is like calling the Animal(double weight, String sound) constructor from Listing 15-6. The constructor assigns values to the new object's weight and sound fields.

Caring for your pet

Here's a quotation from the book *Java For Dummies*, 8th Edition:

> "To use an abstract class, you have to create an ordinary (non-abstract) class that extends the abstract class."

So, to use the HousePet class in Listing 15-7, you have to create a class that extends the HousePet class. The code in Listing 15-8 extends the abstract HousePet class and provides a body for the method named howToCareFor.

LISTING 15-8: It's a Dog's Life

```
package com.example.species;

public class Dog extends HousePet {
    int walksPerDay;

    public Dog(String name, double weight, int walksPerDay) {
        super(name, weight, "Woof");
        this.walksPerDay = walksPerDay;
    }

    @Override
    public void howToCareFor() {
        System.out.print("Walk " + name);
        System.out.println(" " + walksPerDay + " times each day.");
    }
}
```

In addition to having a name, a weight, and a sound, every dog gets walked a certain number of times per day. And now, because of the howToCareFor method's body, you know what caring for a dog means: It means walking the dog a certain number of times each day. It's a good thing the howToCareFor method is abstract in the HousePet class. You wouldn't necessarily want to walk some other kind of pet.

Take, for example, a domestic cat. "Caring" for a cat may mean not bothering it too often. And cats have other characteristics — characteristics that don't apply to dogs. For example, some cats go outdoors; others don't. You can make walksPerDay be 0 for an indoor cat, but that feels like cheating. Instead, each cat can have a boolean value representing the cat's indoor/outdoor status. Listing 15-9 has the code.

LISTING 15-9: How to Be a Cat

```
package com.example.species;

public class Cat extends HousePet {
    boolean isOutdoor;

    public Cat(String name, double weight, boolean isOutdoor) {
        super(name, weight, "Meow");
        this.isOutdoor = isOutdoor;
    }

    @Override
    public void howToCareFor() {
        System.out.println
                (isOutdoor ? "Let " : "Do not let " + name + " outdoors.");
    }
}
```

Both the Dog and Cat classes are subclasses of the HousePet class. And, because of the abstract method declaration in Listing 15-7, both the Dog and Cat classes must have howToCareFor methods. But the howToCareFor methods in the two classes are quite different. One method refers to a walksPerDay field; the other method refers to an isOutdoor field. And because the HousePet class's howToCareFor method is abstract, there's no default behavior. Either the Dog and Cat classes implement their own howToCareFor methods or the Dog and Cat classes can't claim to extend HousePet.

TECHNICAL STUFF

This paragraph describes a picky detail, and you should ignore it if you have any inclination to do so: The Dog and Cat classes must implement the howToCareFor method because the Dog and Cat classes aren't abstract. If the Dog and Cat classes were abstract (that is, if they were abstract classes extending the abstract HousePet class), then the Dog and Cat classes would not have to implement the howToCare-For method. The Dog and Cat classes could pass the implementation buck to their own subclasses. For that matter, an abstract class that implements an interface doesn't have to provide bodies for all the interfaces abstract methods. Abstract classes can take advantage of many little loopholes. But, to use these loopholes, you have to create some exotic programming examples. So, in this chapter I simplify the story and write that (a) a class that extends an abstract class must provide bodies for the abstract class's abstract methods and (b) a class that implements an interface must provide bodies for the interface's abstract methods. It's not exactly true, but it's good enough for now.

If you live in a very small apartment, you may not have room for a dog or a cat. In that case, Listing 15-10 is for you.

LISTING 15-10: You May Grow Up to Be a Fish

```
package com.example.species;

public class Fish extends HousePet {

    public Fish(String name, double weight) {
        super(name, weight, null);
    }

    @Override
    public void howToCareFor() {
        System.out.println("Feed " + name + " daily.");
    }
}
```

I could go on and on creating subclasses of the HousePet class. Many years ago, our daughter had some pet mice. Caring for the mice meant keeping the cat away from them.

In Java, subclasses multiply like rabbits.

Using all your classes

Your work isn't finished until you've tested your code. Most programs require hours, days, and even months of testing. But for this chapter's HousePet example, I do only one test. The test is in Listing 15-11.

LISTING 15-11: The Class Menagerie

```
package com.example.species;

public class Main {

    public static void main(String[] args) {
        var dog1 = new Dog("Fido", 54.7, 3);

        var dog2 = new Dog("Rover", 15.2, 2);

        var cat1 = new Cat("Felix", 10.0, false);

        var fish1 = new Fish("Bubbles", 0.1);
```

(continued)

LISTING 15-11: *(continued)*

```
        dog1.howToCareFor();
        dog2.howToCareFor();
        cat1.howToCareFor();
        fish1.howToCareFor();

        dog1.about();
        dog2.about();
        cat1.about();
        fish1.about();
    }
}
```

When you run the code in Listing 15-11, you get the output shown in Figure 15-3.

```
Walk Fido 3 times each day.
Walk Rover 2 times each day.
Do not let Felix outdoors.
Feed Bubbles daily.
Fido weighs 54.7 pounds and says 'Woof'.
Rover weighs 15.2 pounds and says 'Woof'.
Felix weighs 10.0 pounds and says 'Meow'.
Bubbles weighs 0.1 pounds.
```

FIGURE 15-3: Please don't pet the Pompeii worm.

Notice how the code in Listing 15-11 seamlessly and effortlessly calls many versions of the `howToCareFor` method. With the `dog1.howToCareFor()` and `dog2.howToCareFor()` calls, Java executes the method in Listing 15-8. With the `cat1.howToCareFor()` call, Java executes the method in Listing 15-9. And, with the `fish1.howToCareFor()` call, Java executes the method in Listing 15-10 — it's like having a big `if` statement without writing the `if` statement's code. When you add a new class for a pet mouse, you don't have to enlarge an existing `if` statement. There's no `if` statement to enlarge.

Notice also how the `about` method in the abstract `HousePet` class keeps track of the object that called it. For example, when you call `dog1.about()` in Listing 15-11, the `HousePet` class's nonspecific `about` method knows that the sound `dog1` makes is `Woof`. Everything falls into place very nicely.

TRY IT OUT

Do you like abstract art? You can use abstract classes to create abstract art!

ASCII ART

Create an abstract class named Shape. The Shape class has a size field (of type int) and an abstract show method. Extend the abstract Shape class with two other classes: a Square class and a Triangle class. In the bodies of the Square and Triangle classes' show methods, place the code that creates a text-based rendering of the shape in question. For example, a Square of size 5 looks like this:

A Triangle of size 2 looks like this:

GOOEY ART

For an extra-special challenge, create an abstract Shape class with an abstract paint method. The Shape class also has size, color, and isFilled fields. The size field has type int, the color field has type java.awt.Color, and the isFilled field has type boolean. Extend the abstract Shape class with two other classes: a Square class and a Circle class. In the bodies of the Square and Circle classes' paint methods, place the code that draws the shape in question on a Java JFrame.

Relax! You're Not Seeing Double!

If you've read this chapter's earlier sections on interfaces and abstract methods, your head might be spinning. Both interfaces and abstract classes have abstract methods. But the abstract methods play slightly different roles in these two kinds of reference types. How can you keep it all straight in your mind?

The first thing to do is to remember that no one learns about object-oriented programming concepts without getting lots of practice in writing code. If you've read this chapter and you're confused, that may be a good thing. It means you've understood enough to know how complicated this stuff is. The more code you write, the more comfortable you'll become with classes, interfaces, and all these other ideas.

The next thing to do is to sort out the differences in the way you declare abstract methods. Table 15-1 has the story.

TABLE 15-1 **Using (or Not Using) Abstract Methods**

	In an Ordinary (Non-Abstract) Class	In an Interface	In an Abstract Class
Are abstract methods allowed?	No	Yes	Yes
Can a method declaration contain the `abstract` keyword?	No	Yes	Yes
Can a method declaration contain the `default` keyword (meaning "not abstract")?	No	Yes	No
With neither the `abstract` nor the `default` keyword, a method is:	Not abstract	Abstract	Not abstract

Both interfaces and abstract classes have abstract methods. So, you may be wondering how you should choose between declaring an interface and declaring an abstract class. In fact, you might ask three professional programmers how interfaces and abstract classes differ from one another. If you do, you may get five different answers. (Yes, five answers; not three answers.)

Interfaces and abstract classes are similar beasts, and the new features in Java 8 made them even more similar than in previous Java versions. But the basic idea is about the relationships among things.

- **Extending a subclass represents an *is a* relationship.**

  Think about the relationships in this chapter's earlier section "Abstract Classes." A house pet is an animal. A dog is a house pet. A cat is a house pet. A fish is a house pet.

- **Implementing an interface represents a *can do* relationship.**

  Think about the relationships in this chapter's earlier section "The Java Interface." The first line in Listing 15-3 says `implements Displayable`. With these words, the code promises that each `ColumnOfNumbers` object can be displayed. Later in same listing, you make good on the promise by declaring a `display` method.

  Think about the relationships in this chapter's earlier section "The Java Interface." A column of numbers isn't always a summarizable thing. But in Listing 15-3, you promise that the `ColumnOfNumbers` objects will be summarizable, and you make good on the promise by declaring a `summarize` method.

If you want more tangible evidence of the difference between an interface and an abstract class, consider this: A class can implement many interfaces, but a class can extend only one other class, even if that one class is an abstract class. So, after you've declared

```
public class Dog extends HousePet
```

you can't also make `Dog` extend a `Friend` class. But you can make `Dog` implement a `Befriendable` interface. And then you can make the same `Dog` class implement a `Trainable` interface. (By the way, I've tried making my `Cat` class implement a `Trainable` interface but, for some reason, it never works.)

And, if you want an even *more* tangible difference between an interface and an abstract class, I have one for you: An interface can't contain any nonstatic, nonfinal fields. For example, if the `HousePet` class in Listing 15-7 were an interface, it couldn't have a `name` field. That simply wouldn't be allowed.

So there. Interfaces and abstract classes are different from one another. But if you're new at the game, you shouldn't worry about the difference. Just read as much code as you can, and don't get scared when you see an abstract method. That's all there is to it.

» **Responding when the user presses a key or selects a drop-down box item**

» **Putting classes inside other classes**

Chapter 16
Java's Juggling Act

Study after study shows how people perform poorly when they try to multitask. I occasionally attend two online presentations at the same time. I set up two computers, each with easily adjustable volume controls. My plan is to pivot my attention between the two computer screens, changing the volumes, moment by moment, in response to the importance of what's being presented.

This plan never works. Rather than follow both presentations, I end up following neither of them.

Most people can't concentrate on two tasks at a time, but some people can. A study at the University of Utah found that about 1 in x 40 people is a *supertasker* — someone who can drive a simulated vehicle, talk on a phone, memorize words, and do mental arithmetic all at the same time.* A series of fMRI scans has shown that supertaskers' brains don't race to keep up with all the input they receive.** Instead, their brains tune down the logical thinking and turn up the relaxing thoughts. The regions in their brains that are responsible for daydreaming and other calming activities take the front stage when these people multitask.

What about computers? How do they multitask? Since the 1960s, computers have been able to interleave many tasks in time slices lasting only fractions of a second. And nowadays, with multicore processors, each computer has several mini-brains that can all process instructions simultaneously.

---

* `https://link.springer.com/article/10.3758/PBR.17.4.479`

** `https://pubmed.ncbi.nlm.nih.gov/25223371`

Without multitasking, computer interfaces would be quite primitive. Imagine starting a download and then having to wait ten minutes before your web browser can accept mouse clicks. Better yet, start the download and wait ten minutes before any other apps accept input of any kind.

Programming for multitasking is surprisingly complex. So, in this chapter, I cover only a tiny sliver of the subject with examples from Java's Swing framework. Swing isn't the newest or slickest toolset for creating graphical interfaces in Java, but Swing's structure illustrates some important ideas in multithreaded programming.

TECHNICAL STUFF

In 2011, Oracle added a newer framework — JavaFX — to Java's bag of tricks. JavaFX provides a richer set of components than Swing, but JavaFX code requires a bit more setup than a Swing app. If you're interested in reading more about JavaFX, visit `https://openjfx.io`.

TIP

Every major Java IDE has visual tools to help you design a GUI interface. With any of these tools, you drag buttons, text fields, and other goodies from a palette onto a frame. Using the mouse, you can move and resize each component. As you design the frame visually, the tools create the frame's code automatically. For more info, check your IDE's documentation.

Juggling Two or More Calls

In previous chapters, I create windows that don't do much. A typical window displays some information but has no interactive elements. Well, the time has come to change all that. This chapter's first example is a window with a button on it. When the user clicks the button, darn it, something happens. The code is shown in Listing 16-1, and the `main` method that calls the code in Listing 16-1 is in Listing 16-2.

LISTING 16-1: A Guessing Game

```
package com.example.games;

import javax.swing.*;
import java.awt.FlowLayout;
import java.awt.event.*;
import java.util.Random;

public class GameFrame extends JFrame implements ActionListener {
    int randomNumber = new Random().nextInt(10) + 1;
    int numGuesses = 0;
```

```
    JTextField textField = new JTextField(5);
    JButton button = new JButton("Guess");
    JLabel label = new JLabel(numGuesses + " guesses");

    public GameFrame() {
        setDefaultCloseOperation(JFrame.EXIT_ON_CLOSE);
        setLayout(new FlowLayout());
        add(textField);
        add(button);
        add(label);
        button.addActionListener(this);
        pack();
        setVisible(true);
    }

    @Override
    public void actionPerformed(ActionEvent e) {
        String textFieldText = textField.getText();
        if (Integer.parseInt(textFieldText) == randomNumber) {
            button.setEnabled(false);
            textField.setText(textFieldText + " Yes!");
            textField.setEnabled(false);
        } else {
            textField.setText("");
            textField.requestFocus();
        }

        numGuesses++;
        String guessWord = (numGuesses == 1) ? " guess" : " guesses";
        label.setText(numGuesses + guessWord);
    }
}
```

LISTING 16-2: Starting the Guessing Game

```
package com.example.games;

public class ShowGameFrame {

    public static void main(String[] args) {
        new GameFrame();
    }
}
```

Some snapshots from a run of this section's code are shown in Figures 16-1 and 16-2. In a window, the user plays a guessing game. Behind the scenes, the program chooses a secret number (a number from 1 to 10). Then the program displays a text field and a button. The user types a number in the text field and clicks the button. One of two things happens next:

- **If the number the user types isn't the same as the secret number,** the computer posts the number of guesses made so far. The user gets to make another guess.
- **If the number the user types is the same as the secret number,** the text field displays `Yes!`. Meanwhile, the game is over, so both the text field and the button become disabled. Both components have that gray, washed-out look, and neither component responds to keystrokes or mouse clicks.

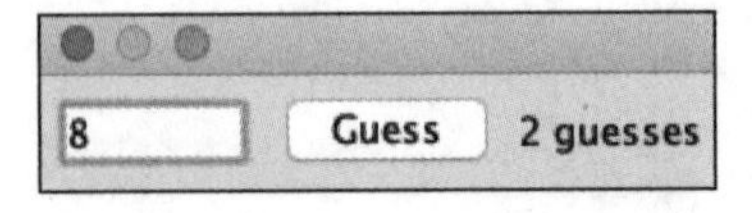

FIGURE 16-1: An incorrect guess.

FIGURE 16-2: The correct guess.

In Listing 16-1, the code to create the frame, the button, and the text field isn't earth-shattering. I do similar things in Chapters 9 and 10. The `JTextField` class is new in this chapter, but a text field isn't much different from a button or a label. Like so many other components, the `JTextField` class is defined in the `javax.swing` package. When you create a new `JTextField` instance, you can specify the number of columns. In Listing 16-1, I create a text field that's five columns wide.

CROSS REFERENCE

Listing 16-1 uses a fancy question mark and colon to decide between the singular *guess* and the plural *guesses*. If you're not familiar with this conditional operator, see Chapter 11.

TIP

Before you run this section's code, your IDE may warn you that the frame in Listing 16-2 has no `serialVersionUID` field. You can safely ignore this warning. For a bit more information about the warning, refer to Chapter 9.

Events and event handling

The big news in Listing 16-1, shown in the preceding section, is the handling of the user's button click. When you're working in a graphical user interface (GUI), anything the user does (like pressing a key, moving the mouse, clicking the mouse, or whatever) is called an *event.* The code that responds to the user's press, movement, or click is called *event-handling code.*

Listing 16-1 deals with the button-click event with three parts of its code:

- The top of the `GameFrame` class declaration says that this class implements `ActionListener`.

  By announcing that it will implement the `ActionListener` interface, the code in Listing 16-1 agrees that it will give meaning to the interface's abstract `actionPerformed` method. In this situation, *giving meaning* means declaring an `actionPerformed` method with curly braces, a body, and maybe some statements to execute.

  CROSS REFERENCE

  For the full story about Java interfaces (as opposed to graphical user interfaces), refer to Chapter 15.

- Sure enough, the code for the `GameFrame` class has an `actionPerformed` method, and that `actionPerformed` method has a body.
- Finally, the constructor for the `GameFrame` class adds `this` to the button's list of action listeners.

  Java will call the `actionPerformed` method in Listing 16-1 when the user clicks the button. Hooray!

Taken together, all three of these tricks make the `GameFrame` class handle button clicks.

CROSS REFERENCE

For more details about the use of Java's `this` keyword in Listing 16-1, see this chapter's "Don't miss this" section.

MARK FOR REMOVAL

TRY IT OUT

You can learn a lot about the code in Listing 16-1 by removing certain statements and observing the results. For each suggested removal, see whether your IDE displays any error messages. If not, try to run the program. After observing the results, put the element back and try the next suggested removal:

- Remove the entire `actionPerformed` method declaration — header and all.
- Remove the call to `setVisible(true)`.

» Remove the call to `pack()`.

» Remove the call to `button.addActionListener()`.

Follow the thread

Here's a well-kept secret: Java programs are *multithreaded*, which means that several things at a time are going on whenever you run a Java program. Sure, the computer is executing the code you've written, but it's executing other code as well (code that you didn't write and don't see). All this code is being executed at the same time. While the computer executes your `main` method's statements, one after another, the computer takes time out, sneaks away briefly, and executes statements from other, unseen methods. For most simple Java programs, these other methods are ones that are defined as part of the Java virtual machine (JVM).

For example, Java has an event-handling thread. While your code runs, the event-handling thread's code runs in the background. The event-handling thread's code listens for mouse clicks and takes appropriate action whenever a user clicks the mouse. Figure 16-3 illustrates how this business works.

Your code's thread	The event handling thread
`setLayout(new FlowLayout());`	Did the user click the mouse?
`add(textField);`	.
`add(button);`	.
`add(label);`	Did the user click the mouse?
	.
`button.addActionListener(this);`	.
`pack();`	Did the user click the mouse?
`setVisible(true);`	Yes? Okay, then. I'll call the `actionPerformed` method.

FIGURE 16-3: Two Java threads.

When the user clicks the button, the event-handling thread says, "Okay, the button was clicked. What should I do about that?" And the answer is, "Call some `actionPerformed` methods." It's as though the event-handling thread has code that looks like this:

```
if (buttonJustGotClicked()) {
    object1.actionPerformed(infoAboutTheClick);
    object2.actionPerformed(infoAboutTheClick);
    object3.actionPerformed(infoAboutTheClick);
}
```

Of course, behind every answer is yet another question. In this situation, the follow-up question is, "Where does the event-handling thread find `actionPerformed` methods to call?" And there's another question: "What if you don't want the event-handling thread to call certain `actionPerformed` methods that are lurking in your code?"

Well, that's why you call the `addActionListener` method. In Listing 16-1, the call

```
button.addActionListener(this);
```

tells the event-handling thread, "Put this code's `actionPerformed` method on your list of methods to be called. Call this code's `actionPerformed` method whenever the button is clicked."

That's how it works. To have the computer call an `actionPerformed` method, you register the method with Java's event-handling thread. You do this registration by calling `addActionListener`. The `addActionListener` method belongs to the object whose clicks (and other events) you're waiting for. In Listing 16-1, you're waiting for the button object to be clicked, and the `addActionListener` method belongs to that button object.

Don't miss this

In Chapters 9 and 10, the keyword `this` gives you access to instance variables from the code inside a method. What does the `this` keyword really mean? Well, compare it with the English phrase *state your name:*

> *I, (state your name), do solemnly swear to uphold the constitution of the Philadelphia Central High School Photography Society. . . .*

The phrase *state your name* is a placeholder. It's a space in which each person puts their own name:

- *I, Bob, do solemnly swear . . .*
- *I, Fred, do solemnly swear . . .*

Think of the pledge ("I . . . do solemnly swear . . .") as a piece of code in a Java class. In that piece of code is the placeholder phrase *state your name.* Whenever an instance of the class (a person) executes the code (that is, takes the pledge), the instance fills in its own name in place of the phrase *state your name.*

The `this` keyword works the same way. It sits inside the code that defines the `GameFrame` class. (Refer to Listing 16-1.) Whenever an instance of `GameFrame` is constructed, the instance calls `addActionListener(this)`. In that call, the `this` keyword stands for the instance itself:

```
button.addActionListener(thisGameFrameInstance);
```

By calling `button.addActionListener(this)`, the `GameFrame` instance is saying, "Add my `actionPerformed` method to the list of methods that are called whenever the button is clicked." And indeed, the `GameFrame` instance has an `actionPerformed` method. The `GameFrame` has to have an `actionPerformed` method because the `GameFrame` class implements the `ActionListener` interface. It's funny how that all fits together.

For a thought experiment, imagine that you've constructed two instances of the `GameFrame` class.

```
var frame1 = new GameFrame();
var frame2 = new GameFrame();
```

Maybe both frames (`frame1` and `frame2`) appear on the screen at the same time. Both frames contain their own copies of the `textField` variable, the `button` variable, the `label` variable, and the `actionPerfomed` method. In addition, both frames contain their own copies of `this`.

- In the `frame1` code, `this` refers to that `frame1` object.

  In the `frame1` code, calling `button.addActionListener(this)` tells Java to call the `frame1` object's `actionPerformed` method when the user clicks the `frame1` object's button.

- In the `frame2` code, `this` refers to the `frame2` object.

  In the `frame2` code, calling `button.addActionListener(this)` tells Java to call the `frame2` object's `actionPerformed` method when the user clicks the `frame2` object's button.

TRY IT OUT

THIS IS IT

In your own words, describe the uses of the keyword `this` in the following code:

```
public class Main {

    public static void main(String[] args) {
      new IntegerHolder(42).displayMyN();
      new IntegerHolder(7).displayMyN();
    }
}

    class IntegerHolder {
      private int n;

    IntegerHolder(int n) {
      this.n = n;
    }

    void displayMyN() {
      Displayer.display(this);
    }

    public int getN() {
      return n;
    }
}

class Displayer {

  public static void display(IntegerHolder holder) {
    System.out.println(holder.getN());
  }
}
```

Inside the actionPerformed method

The `actionPerformed` method in Listing 16-1 uses a bunch of tricks from the Java API. Here's a brief list of those tricks:

- Every instance of `JTextField` (and of `JLabel`) has its own getter and setter methods, including `getText` and `setText`. Calling `getText` fetches whatever string of characters is in the component. Calling `setText` changes the characters that are in the component. In Listing 16-1, judicious use of `getText` and `setText` pulls a number out of the text field and replaces the number with either nothing (the empty string `""`) or the number, followed by the word *Yes!*

- **»** Every component in the `javax.swing` package (`JTextField`, `JButton`, or whatever) has a `setEnabled` method. When you call `setEnabled(false)`, the component gets that limp, gray, washed-out look and can no longer receive button clicks or keystrokes.
- **»** Every component in the `javax.swing` package has a `requestFocus` method. When you call `requestFocus`, the component gets the privilege of receiving the user's next input. For example, in Listing 16-1, the call `textField.requestFocus()` says, "Even though the user may have just clicked the button, put a cursor in the text field. That way, the user can type another guess in the text field without clicking the text field first."

TIP

You can perform a test to make sure that the object referred to by the button variable is really the thing that was clicked. Just write `if (e.getSource() == button)`. If your code has two buttons, `button1` and `button2`, you can test to find out which button was clicked. You can write `if (e.getSource() == button1)` and `if (e.getSource() == button2)`.

COPY CAT

TRY IT OUT

Using the techniques shown in this chapter, create a program that displays a frame containing three components: a text field (`JTextField`), a button (`JButton`), and a label (`JLabel`). The user types text into the text field. Then, whenever the user clicks the button, the program copies any text that's in the text field onto the label.

Some Events Aren't Button Clicks

When you know how to respond to one kind of event, responding to other kinds of events is easy. Listings 16-3 and 16-4 display a window that converts between US and UK currencies. The code in these listings responds to many kinds of events. Figures 16-4, 16-5, 16-6, and 16-7 show some pictures of the code in action.

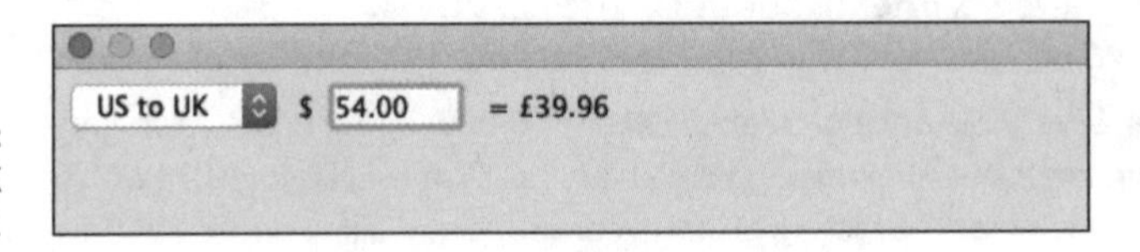

FIGURE 16-4: US-to-UK currency.

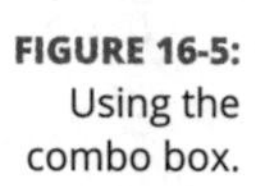

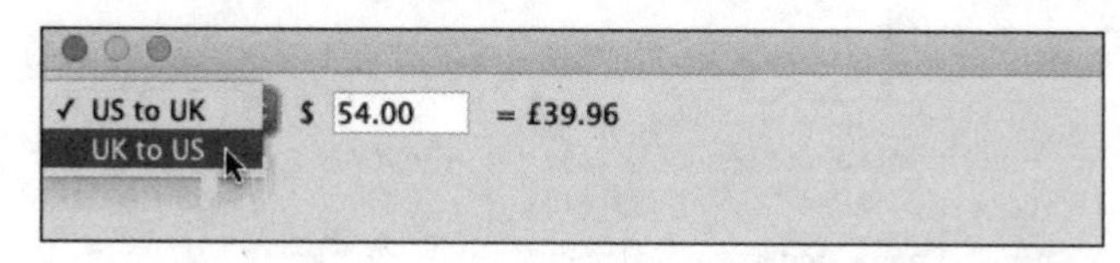

FIGURE 16-5: Using the combo box.

FIGURE 16-6: UK-to-US currency.

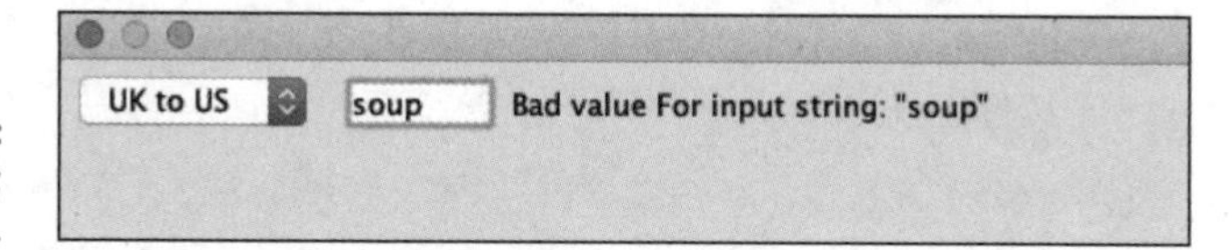

FIGURE 16-7: Junk in; junk out.

LISTING 16-3: Displaying the Local Currency

```
package com.example.money;

import javax.swing.*;
import java.awt.*;
import java.awt.event.*;
import java.text.NumberFormat;
import java.util.Locale;

public class MoneyFrame extends JFrame implements
        ItemListener, KeyListener, MouseListener {

    JComboBox<String> combo = new JComboBox<>();
    JLabel fromCurrencySymbol = new JLabel(" ");
    JTextField textField = new JTextField(5);
    JLabel resultLabel = new JLabel(" ");
    NumberFormat currencyUS = NumberFormat.getCurrencyInstance(Locale.US);
    NumberFormat currencyUK = NumberFormat.getCurrencyInstance(Locale.UK);

    public MoneyFrame() {
        setLayout(new FlowLayout(FlowLayout.LEFT));

        combo.addItem("US to UK");
        combo.addItem("UK to US");
        add(combo);
        add(fromCurrencySymbol);
        textField.setText("0.00");
        add(textField);
        add(resultLabel);
```

(continued)

LISTING 16-3: ***(continued)***

```
        combo.addItemListener(this);
        textField.addKeyListener(this);
        resultLabel.addMouseListener(this);

        setDefaultCloseOperation(JFrame.EXIT_ON_CLOSE);
        setSize(500, 100);
        setVisible(true);
    }

    void setResultText() {
        String fromCurrency = "";
        String amountString = "";
        double dollarToPound = 0.74;

        try {
            double amount = Double.parseDouble(textField.getText());

            if (combo.getSelectedItem().equals("US to UK")) {
                amountString =
                        " = " + currencyUK.format(amount * dollarToPound);
                fromCurrency = "$";
            }

            if (combo.getSelectedItem().equals("UK to US")) {
                amountString =
                        " = " + currencyUS.format(amount / dollarToPound);
                fromCurrency = "\u00A3";
            }

        } catch (NumberFormatException e) {
            amountString = "Bad value " + e.getMessage();
        }

        fromCurrencySymbol.setText(fromCurrency);
        resultLabel.setText(amountString);
    }

    @Override
    public void itemStateChanged(ItemEvent i) {
        setResultText();
    }

    @Override
    public void keyReleased(KeyEvent k) {
        setResultText();
    }
```

```
    @Override
    public void keyPressed(KeyEvent k) {
    }

    @Override
    public void keyTyped(KeyEvent k) {
    }

    @Override
    public void mouseEntered(MouseEvent m) {
        resultLabel.setForeground(Color.red);
    }

    @Override
    public void mouseExited(MouseEvent m) {
        resultLabel.setForeground(Color.black);
    }

    @Override
    public void mouseClicked(MouseEvent m) {
    }

    @Override
    public void mousePressed(MouseEvent m) {
    }

    @Override
    public void mouseReleased(MouseEvent m) {
    }
}
```

LISTING 16-4: Calling the Code in Listing 16-3

```
package com.example.money;

public class ShowMoneyFrame {

    public static void main(String[] args) {
        new MoneyFrame();
    }
}
```

Okay, so Listing 16-3 is a little long. Even so, the outline of the code in Listing 16-3 isn't too bad. Here's what the outline looks like:

```
public class MoneyFrame extends JFrame implements
        KeyListener, ItemListener, MouseListener {

    Variable declarations
    Constructor for the MoneyFrame class
    Declaration of a method named setResultText
    Methods that are required because the class implements three interfaces
}
```

The constructor in Listing 16-3 adds the following four components to the new `MoneyFrame` window:

- **A combo box:** In Figure 16-4, the combo box displays *US to UK*. In Figure 16-5, the user makes the move to select the second of the two items in the combo box: *UK to US*. In Figure 16-6, the selected item (*UK to US*) is displayed.
- **A label:** In Figure 16-4, the label displays a dollar sign.
- **A text field:** In Figure 16-4, the user types **54** in the text field.
- **Another label:** In Figure 16-4, the label displays `£39.96`.

TECHNICAL STUFF

In Java, a `JComboBox` (commonly called a *drop-down list*) can display items of any kind. In Listing 16-3, the declaration

```
JComboBox<String> combo = new JComboBox<>();
```

constructs a `JComboBox` whose entries have type `String`. That seems sensible, but if your application has a `Person` class, you can declare `JComboBox<Person> peopleBox`. In that situation, Java has to know how to display each `Person` object in the drop-down list. (It isn't a big deal. Java finds out how to display a person by looking for a `toString()` method inside the `Person` class.)

The `MoneyFrame` implements three interfaces: the `ItemListener`, `KeyListener`, and `MouseListener` interfaces. Because it implements three interfaces, the code can listen for three kinds of events. I discuss the interfaces and events in the following list:

- `ItemListener`: A class that implements the `ItemListener` interface must have an `itemStateChanged` method. When you select an item in a combo box, the event-handling thread calls `itemStateChanged`.

In Listing 16-3, when the user selects US-to-UK or UK-to-US in the combo box, the event-handling thread calls the `itemStateChanged` method. In turn, the `itemStateChanged` method calls `setResultText`. The `setResultText` method checks to see what's now selected in the combo box. If the user selects the US-to-UK option, the `setResultText` method converts dollars to pounds. If the user selects the UK-to-US option, the `setResultText` method converts pounds to dollars.

In the `setResultText` method, I use the string `"\u00A3"`. The funny-looking `\u00A3` code is Java's UK pound sign. (The `u` in `\u00A3` stands for *Unicode* — an international standard for representing characters in the world's alphabets.) If my operating system's settings defaulted to UK currency, in the runs of Java programs the pound sign would appear on its own. For information about all of this, check out the `Locale` class in 's API documentation (`https://docs.oracle.com/en/java/javase/17/docs/api/java.base/java/util/Locale.html`).

By the way, if you're thinking in terms of real currency conversion, forget about it. This program uses rates that may or may not have been accurate at one time. Sure, a program can reach out on the Internet for the most up-to-date currency rates, but at the moment you have other Javafish to fry.

- `KeyListener`: A class that implements the `KeyListener` interface must have three methods named `keyReleased`, `keyPressed`, and `keyTyped`. When you lift your finger off a key, the event-handling thread calls `keyReleased`, which calls `setResultText`, and so on.
- `MouseListener`: A class that implements the `MouseListener` interface must have `mouseEntered`, `mouseExited`, `mouseClicked`, `mousePressed`, and `mouseReleased` methods. Implementing `MouseListener` is different from implementing `ActionListener`. When you implement `ActionListener`, as in Listing 16-1, the event-handling thread responds only to mouse clicks. But with `MouseListener`, the thread responds to the user pressing the mouse, releasing the mouse, and more.

  In Listing 16-3, the `mouseEntered` and `mouseExited` methods are called whenever you move over or away from the `resultLabel`. How do you know that the `resultLabel` is involved? Just look at the code in the `MoneyFrame` constructor. The statement

  ```
  resultLabel.addMouseListener(this);
  ```

  tells Java to listen for `resultLabel` mouse events, and to call `this` class's methods when any of those mouse events occur.

Look at the mouseEntered and mouseExited methods in Listing 16-3. When mouseEntered or mouseExited is called, the computer forges ahead and calls setForeground. This setForeground method changes the color of the label's text.

Isn't modern life wonderful? The Java API even has a Color class with names like Color.red and Color.black.

Listing 16-3 has several methods that aren't really used. For example, when you implement MouseListener, your code has to have its own mouseReleased method. You need the mouseReleased method not because you're going to do anything special when the user releases the mouse button but rather because you made a promise to the Java compiler and you have to keep it.

TRY IT OUT

In an earlier section, you create a program that copies text from a text field to a label whenever the user clicks a button. Modify the program so that the user doesn't have to click a button. The program automatically updates the label's text whenever the user modifies the text field's content.

The Inner Sanctum

Here's big news! You can define a class inside of another class! For the user, Listing 16-5 behaves the same way as Listing 16-1. But in Listing 16-5, the GameFrame class contains a class named MyActionListener.

LISTING 16-5: A Class within a Class

```
package com.example.games;

import javax.swing.*;
import java.awt.FlowLayout;
import java.awt.event.*;
import java.util.Random;

public class GameFrame extends JFrame {

    int randomNumber = new Random().nextInt(10) + 1;
    int numGuesses = 0;

    JTextField textField = new JTextField(5);
    JButton button = new JButton("Guess");
    JLabel label = new JLabel(numGuesses + " guesses");
```

```
    public GameFrame() {
        setDefaultCloseOperation(JFrame.EXIT_ON_CLOSE);
        setLayout(new FlowLayout());
        add(textField);
        add(button);
        add(label);
        button.addActionListener(new MyActionListener());
        pack();
        setVisible(true);
    }

    class MyActionListener implements ActionListener {

        @Override
        public void actionPerformed(ActionEvent e) {
            String textFieldText = textField.getText();

            if (Integer.parseInt(textFieldText) == randomNumber) {
                button.setEnabled(false);
                textField.setText(textFieldText + " Yes!");
                textField.setEnabled(false);
            } else {
                textField.setText("");
                textField.requestFocus();
            }
            numGuesses++;
            String guessWord = (numGuesses == 1) ? " guess" : " guesses";
            label.setText(numGuesses + guessWord);
        }
    }
}
```

The MyActionListener class in Listing 16-5 is an inner class. An *inner class* is a lot like any other class, but within an inner class's code, you can refer to the enclosing class's fields. For example, several statements inside MyActionListener use the name textField, and textField is declared in the enclosing GameFrame class — not inside the MyActionListener class.

Notice that the code in Listing 16-5 uses the MyActionListener class only once. (The only use is in a call to button.addActionListener.) So, I ask, do you really need a name for something that's used only once? No, you don't. You can substitute the entire definition of the inner class inside the call to button.addActionListener. When you do this, you have an *anonymous inner class.* Listing 16-6 shows you how it works.

LISTING 16-6: A Class with No Name (Inside a Class with a Name)

```
package com.example.games;

import javax.swing.*;
import java.awt.FlowLayout;
import java.awt.event.*;
import java.util.Random;

public class GameFrame extends JFrame {

    int randomNumber = new Random().nextInt(10) + 1;
    int numGuesses = 0;

    JTextField textField = new JTextField(5);
    JButton button = new JButton("Guess");
    JLabel label = new JLabel(numGuesses + " guesses");

    public GameFrame() {
        setDefaultCloseOperation(JFrame.EXIT_ON_CLOSE);
        setLayout(new FlowLayout());
        add(textField);
        add(button);
        add(label);
        button.addActionListener(new ActionListener() {

            @Override
            public void actionPerformed(ActionEvent e) {
                String textFieldText = textField.getText();

                if (Integer.parseInt(textFieldText) == randomNumber) {
                    button.setEnabled(false);
                    textField.setText(textFieldText + " Yes!");
                    textField.setEnabled(false);
                } else {
                    textField.setText("");
                    textField.requestFocus();
                }
                numGuesses++;
                String guessWord = (numGuesses == 1) ? " guess" : " guesses";
                label.setText(numGuesses + guessWord);
            }
        });
        pack();
        setVisible(true);
    }
}
```

Inner classes are good for things like event handlers, such as the `actionPerformed` method in this chapter's examples. The most difficult thing about an *anonymous* inner class is keeping track of the parentheses, the curly braces, and the indentation. My humble advice is, start by writing code without any inner classes, as in the code from Listing 16-1. Later, when you become bored with ordinary Java classes, experiment by changing some of your ordinary classes into inner classes.

MAINTAIN ANONYMITY

TRY IT OUT

In a previous section, you create a program that copies text from a text field to a label whenever the user clicks a button. Modify the code so that it has an inner class. Then if you're ambitious, modify the code so that it has an anonymous inner class.

MARY HAD A LITTLE LAMBDA

In Listing 16-6, replace the `addActionListener` call with the following code:

```
button.addActionListener(e -> {
    String textFieldText = textField.getText();

    if (Integer.parseInt(textFieldText) == randomNumber) {
        button.setEnabled(false);
        textField.setText(textFieldText + " Yes!");
        textField.setEnabled(false);
    } else {
        textField.setText("");
        textField.requestFocus();
    }
    numGuesses++;
    String guessWord = (numGuesses == 1) ? " guess" : " guesses";
    label.setText(numGuesses + guessWord);
});
```

Notice the similarities (and differences) between an anonymous inner class and a lambda expression.

» Inserting values into a database

» Making queries to a database

Chapter 17

Using Java Database Connectivity

The year is 1998. I'm scheduled to introduce Java Database Connectivity to a group of computer science professors at the ITiCSE conference in Ireland. Every day, for six days, I spend all afternoon driving from one town to another. Every evening, my family members visit the town's sites while I sit in our hotel room, pounding away at my laptop. The demo I have planned for the conference presentation isn't working, and nothing I do makes the error messages go away.

On the day of the conference, my demo still isn't working. I do a scaled-down presentation — one that should be called "Simulating Java Database Connectivity." Like any experienced instructor, I make up an excuse for the presentation's wimpy results. "JDBC is still in beta," I say. Fortunately, the conference attendees seem to believe me.

A month later, back home in New Jersey, I discover the flaw in the demo I had planned. Rather than type `database customers`, I should have typed `database-customers`. The thing that kept me from touring towns in Ireland and reduced my conference presentation from substance to fluff was a single blank space. It wasn't even a character I could write on a piece of paper.

So much for my confession. This chapter covers *Java Database Connectivity** (JDBC), and your experience with JDBC will be joyous and fruitful. The JDBC classes provide common access to most database management systems. Just get a driver for your favorite vendor's system, and you're ready to go.

Creating a Database and a Table

The crux of JDBC is contained in two packages: `java.sql` and `javax.sql`, both of which are in the Java API. This chapter's examples use the classes in `java.sql`. The first example is shown in Listing 17-1.

LISTING 17-1: Creating a Database and a Table

```
package com.example.accounts;

import java.sql.Connection;
import java.sql.DriverManager;
import java.sql.SQLException;
import java.sql.Statement;

public class CreateTable {

    public static void main(String[] args) {

        final String CONNECTION = "jdbc:sqlite:AccountDatabase.db";

        try (Connection conn = DriverManager.getConnection(CONNECTION);
             Statement statement = conn.createStatement()) {

            statement.executeUpdate("""
                    create table ACCOUNTS
                    (NAME VARCHAR(32) NOT NULL PRIMARY KEY,
                    ADDRESS VARCHAR(32),
                    BALANCE FLOAT )""");
            System.out.println("ACCOUNTS table created.");
```

* Apparently, there's no evidence in any of Oracle's literature that the acronym JDBC actually stands for Java Database Connectivity. But that's okay. If Java Database Connectivity isn't the correct terminology, it's close enough. In the Java world, JDBC certainly doesn't stand for John Digests Bleu Cheese.

```
        } catch (SQLException e) {
            System.out.println(e.getMessage());
        }
    }
}
```

Running the examples in this chapter is a bit trickier than running other chapters' examples. To talk to a database, you need an intermediary piece of software known as a *database driver.* Database drivers come in all shapes and sizes, and many of them are quite expensive. But Listing 17-1 points to a freebie driver: the Sqlite JDBC driver.

When you install Java, you don't get this Sqlite driver. You need a separate file named `sqlite-jdbc.jar`, which you can download from `https://github.com/xerial/sqlite-jdbc`.

Even after you've downloaded a copy of `sqlite-jdbc.jar`, your IDE might not know where you've put the file on your computer's hard drive. It's usually not enough to put `sqlite-jdbc.jar` in a well-known directory. Instead, you have to tell your IDE exactly where to find your `sqlite-jdbc.jar` file. Here's what you do in two commonly used IDEs:

- **Eclipse:** Choose Project ⇨ Properties. In the resulting dialog box, select Java Build Path and then select the Libraries tab. On the Libraries tab, select Modulepath. Click the Add External JARs button, and then navigate to the `sqlite-jdbc.jar` file on your computer's hard drive.
- **IntelliJ IDEA:** Choose File ⇨ Project Structure. In the resulting dialog box, select Libraries. Click the Plus Sign (+) icon and, in the resulting drop-down box, select Java. Navigate to the `sqlite-jdbc.jar` file on your computer's hard drive.

TIP

For more detailed instructions the use of Sqlite with Java, visit `www.sqlitetutorial.net/sqlite-java`.

Seeing what happens when you run the code

During a successful run of the code in Listing 17-1, you see an `ACCOUNTS table created` message. That's about it. The code has no other visible output because most of the output goes to a database.

If you poke around a bit, you can find direct evidence of the new database's existence. Using your computer's File Explorer or Finder, you can navigate to the project folder containing the code in Listing 17-1 (the project's root folder). Just look for a file named `AccountDatabase.db`.

Unfortunately, you can't see what's inside the database unless you run a couple more programs. Read on!

If you don't want to use Sqlite, you have to replace the `CONNECTION` string in this chapter's examples. Which other string you use depends on the kind of database software you have, and on other factors. Check your database vendor's documentation.

Using SQL commands

These days, people work with two different kinds of databases:

- **Relational:** A relational database has any number of *tables*. In a particular table, each column represents a property. Figure 17-1 has a visual representation of the table described in Listing 17-1.
- **NoSQL (Not Only SQL):** In a *NoSQL database*, data isn't organized into tables. Instead, each chunk of data may be a document, a graph, a set of key-value pairs, or some other structure.

NAME	ADDRESS	BALANCE
Barry Burd	222 Cyber Lane	24.02
Joe Dow	111 Luddite Street	55.63

FIGURE 17-1: Two deadbeat customers.

The code in Listing 17-1 talks to a relational database. The heart of the code lies in the call to `executeUpdate`. The `executeUpdate` call contains a string — a text block of the kind you see in Chapter 5.

If you're familiar with Structured Query Language, or SQL, the command strings in the calls to `executeUpdate` make sense to you. If not, pick up a copy of *SQL For Dummies*, 9th Edition, by Allen G. Taylor (Wiley). One way or another, don't go fishing around this chapter for an explanation of the `create table` command. You won't find an explanation, because the big `create table` string in Listing 17-1 isn't part of Java. This command is just a string of characters you feed to Java's `executeUpdate` method. This string, which is written in SQL, creates a new database table with three columns (columns for a customer's `NAME`, the customer's `ADDRESS`, and the account's `BALANCE`). When you write a Java database program, that's what you do. You write ordinary SQL commands and surround those commands with calls to Java methods.

Connecting and disconnecting

Aside from the call to the `executeUpdate` method, the code in Listing 17-1 is copy-and-paste stuff. Here's a rundown on what each part of the code means:

- `DriverManager.getConnection`: Establish a session with a particular database.

  The `getConnection` method lives in a Java class named `DriverManager`. In Listing 17-1, the call to `getConnection` creates an `AccountDatabase.db` file and opens a connection to that file. Of course, you may already have an `AccountDatabase.db` file before you start running the code in Listing 17-1. If you do, the call to `getConnection` uses your existing `AccountDatabase.db` file.

  In the `CONNECTION` string, notice the colons. The code doesn't simply name the `AccountDatabase.db` file — it tells the `DriverManager` class which protocols to use to connect with the file. The code `jdbc:sqlite:` — which is a lot like the `http:` in a web address — tells the computer to use the `jdbc` protocol to talk to the `sqlite` protocol, which in turn talks directly to your `AccountDatabase.db` file.

- `conn.createStatement`: Make a statement.

  It seems strange, but in Java Database Connectivity, you create a single `statement` object. After you've created a `statement` object, you can use that object many times, with many different SQL strings, to issue many different commands to the database. So, before you start calling the `statement.executeUpdate` method, you have to create an actual `statement` object. The call to `conn.createStatement` creates that `statement` object for you.

- `try`-with-resources: Release resources, come what may!

  As Ritter always says, you're not being considerate of others if you don't clean up your own messes. Every connection and every database statement lock up some system resources. When you're finished using these resources, you release them.

  In Listing 17-1, Java's `try`-with-resources block automatically closes and releases your resources at the end of the block's execution. In addition, `try`-with-resources takes care of all the messy details associated with failed attempts to catch exceptions gracefully.

CROSS REFERENCE

For the scoop about `try`-with-resources, see Chapter 13.

Putting Data in the Table

Like any other tabular configuration, a database table has columns and rows. When you run the code in Listing 17-1, you get an empty table. The table has three columns (NAME, ADDRESS, and BALANCE) but no rows. To add rows to the table, run the code in Listing 17-2.

LISTING 17-2: Inserting Data

```
package com.example.accounts;

import java.sql.Connection;
import java.sql.DriverManager;
import java.sql.SQLException;
import java.sql.Statement;

public class AddData {

    public static void main(String[] args) {

        final String CONNECTION = "jdbc:sqlite:AccountDatabase.db";

        try (Connection conn = DriverManager.getConnection(CONNECTION);
             Statement statement = conn.createStatement()) {

            statement.executeUpdate("""
                    insert into ACCOUNTS values
                    ('Barry Burd', '222 Cyber Lane', 24.02)""");
            statement.executeUpdate("""
                    insert into ACCOUNTS values
                    ('Joe Dow', '111 Luddite Street', 55.63)""");

            System.out.println("Rows added.");

        } catch (SQLException e) {
            System.out.println(e.getMessage());
        }
    }
}
```

Listing 17-2 uses the same strategy as the code in Listing 17-1: Create Java strings containing SQL commands and make those strings be arguments to Java's executeUpdate method. In Listing 17-2, I put two rows in the ACCOUNTS table — one for me and another for Joe Dow. (Joe, I hope you appreciate this.)

TIP

For the best results, put all this chapter's listings in the same project. That way, you don't have to add the `sqlite-jdbc.jar` file to more than one project. You can also count on the `AccountDatabase.db` file being readily available to all four of this chapter's code listings.

Retrieving Data

What good is a database if you can't get data from it? In this section, you query the database you created in previous sections. The code to issue the query is shown in Listing 17-3.

LISTING 17-3: Making a Query

```
package com.example.accounts;

import java.sql.Connection;
import java.sql.DriverManager;
import java.sql.ResultSet;
import java.sql.SQLException;
import java.sql.Statement;
import java.text.NumberFormat;

import static java.lang.System.out;

public class GetData {

    public static void main(String[] args) {
        NumberFormat currency = NumberFormat.getCurrencyInstance();
        final String CONNECTION = "jdbc:sqlite:AccountDatabase.db";

        try (Connection conn = DriverManager.getConnection(CONNECTION);
             Statement statement = conn.createStatement();
             ResultSet resultset = statement.executeQuery
                     ("select * from " + "ACCOUNTS")) {

            while (resultset.next()) {
                out.print(resultset.getString("NAME"));
                out.print(", ");
                out.print(resultset.getString("ADDRESS"));
                out.print(" ");
                out.println(currency.format
                        (resultset.getFloat("BALANCE")));
            }
```

(continued)

LISTING 17-3: *(continued)*

```
        } catch (SQLException e) {
            out.println(e.getMessage());
        }
    }
}
```

REMEMBER

To use a database other than Sqlite, change the value of CONNECTION in each of this chapter's examples.

A run of the code from Listing 17-3 is shown in Figure 17-2. The code queries the database and then steps through the rows of the database, printing the data from each of the rows.

FIGURE 17-2: Retrieving data from the database.

```
Barry Burd, 222 Cyber Lane $24.02
Joe Dow, 111 Luddite Street $55.63
```

Listing 17-3 calls executeQuery and supplies the call with an SQL command. For those who know SQL commands, this particular command gets all data from the ACCOUNTS table (the table you create in Listing 17-1).

The thing returned from calling executeQuery is of type java.sql.ResultSet. (That's one of the differences between the executeUpdate and executeQuery methods: executeQuery returns a result set, and executeUpdate doesn't.) A *result set* is much like a database table. Like the original table, the result set has rows and columns. Each row contains the data for one account. In this example, each row has a name, an address, and a balance amount.

After you call executeQuery and get your result set, you can step through the result set one row at a time. To do this, you go into a little loop and test the condition resultset.next() at the top of each loop iteration. Each time around, the call to resultset.next() does two things:

- It moves you to the next row of the result set (the next account) if another row exists.
- It tells you whether another row exists by returning a boolean value — true or false.

If the condition `resultset.next()` is `true`, the result set has another row. The computer moves to that other row, so you can march into the body of the loop and scoop data from that row. On the other hand, if `resultset.next()` is `false`, the result set has no more rows. You jump out of the loop and start closing everything.

Now imagine that Java is pointing to a row of the result set and you're inside the loop in Listing 17-3. You're retrieving data from the result set's row by calling the result set's `getString` and `getFloat` methods. Back in Listing 17-1, you set up the `ACCOUNTS` table with the columns `NAME`, `ADDRESS`, and `BALANCE`. Here in Listing 17-3, you're getting data from these columns by calling your `get`*`SomeTypeOrOther`* methods and feeding the original column names to these methods. After you have the data, you display the data on the computer screen.

TIP

Each Java `ResultSet` instance has several nice `get`*`SomeTypeOrOther`* methods. Depending on the type of data you put into a column, you can call methods `getArray`, `getBigDecimal`, `getBlob`, `getInt`, `getObject`, `getTimestamp`, and several others.

Destroying Data

It's true. All good things must come to an end. By writing this, I'm referring both to this book's content and to the information in this chapter's `AccountDatabase.db` file.

To get rid of the database table you create in Listing 17-1, run the code in Listing 17-4.

LISTING 17-4: Arrivederci, Database Table

```
package com.example.accounts;

import java.sql.Connection;
import java.sql.DriverManager;
import java.sql.SQLException;
import java.sql.Statement;

public class DropTable {

    public static void main(String[] args) {
        final String CONNECTION = "jdbc:sqlite:AccountDatabase.db";
```

(continued)

LISTING 17-4: *(continued)*

```
        try (Connection conn = DriverManager.getConnection(CONNECTION);
             Statement statement = conn.createStatement()) {

            statement.executeUpdate("drop table ACCOUNTS");
            System.out.println("ACCOUNTS table dropped.");
        } catch (SQLException e) {
            System.out.println(e.getMessage());
        }
    }
}
```

When you run this code, you wipe the slate clean. Your AccountDatabase.db file no longer contains an ACCOUNTS table. So, if you want to run Listing 17-1 again (perhaps with a change or two), you can.

Who knows? You may even create a table to store your favorite *Java For Dummies* jokes.

TRY IT OUT

Naturally, I have some things for you to try:

HIGH ROLLERS ONLY

Rerun the code in Listing 17-3. This time, use the following string in the executeQuery call:

```
"select * from ACCOUNTS where BALANCE > 30"
```

WATCH YOUR PRIMARY KEY

Run the AddData program (from Listing 17-2) two times in a row without modifying any of the program's code. What error messages do you see? Why?

NOT TOO TAXING

Create a table containing three columns: an item name, a price, and a tax rate. Store data in several rows of the table.

Retrieve the data from the table and display a row of output for each row in the table. Each row of output contains the item name followed by the price with tax added. For example, if an item's price is $10 and the item's tax rate is 0.05 (meaning 5 percent), the item's output row contains the number $10.50.

On the last line of the program's output, display the total of all items' tax-added prices.

One Step Beyond

If you want to work with databases, JDBC is a good place to start. But for industrial-strength projects, JDBC may not be enough. For big databases, you need high-level tools.

Think about a Java class named Account. The code looks something like this:

```
public class Account {
    String name;
    String address;
    double balance;
}
```

An instance of the Account class is like a row of this chapter's ACCOUNTS table. So, you can run software that automatically creates a row for each instance in your code. With this software, you don't have to worry about updating a database. Instead, you create classes and objects the way you do in any other Java program. The software inserts data and performs queries on your behalf.

Products like Hibernate, TopLink, JPA (Java Persistence API), and MyBatis are called *object-relational mapping* (ORM) frameworks because they synchronize data between Java objects and database relations. For many professional programmers, ORM is the way to go.

5 The Part of Tens

IN THIS PART . . .

Explore the best resources for Java on the web.

Read the Dear Barry advice column.

IN THIS CHAPTER

» Checking out this book's website

» Finding resources from Oracle

» Reading more about Java

Chapter 18

Ten Packs of Java Websites

Before starting this chapter, I paused to count the number of tabs I had open in my web browser. I counted 64 tabs. What about you? Are you a taba-holic too?

This Book's Website

For all matters related to the technical content of this book, visit `http://javafordummies.allmycode.com`. And don't forget: If you have questions about anything you read in this book, send email to me at `JavaForDummies@allmycode.com`, post a question on `www.facebook.com/allmycode`, or tweet to the Burd with `@allmycode`.

For Business Issues Related to This Book

For example, to ask, "How can I purchase 100 more copies of *Java For Dummies*?" visit `www.dummies.com`.

Download the Java Development Kit

Get the open source version of Java at https://adoptium.net. For Oracle's official version of Java, visit www.oracle.com/java/technologies.

Your Grandparents' Java Download Site

People who want to run Java programs but don't *have* to write Java programs should visit www.java.com.

The Horse's Mouth

Check the official Java API documentation at https://docs.oracle.com/en/java/javase/17/docs/api. And, to settle any argument about the way the language behaves, read the rigorous Java language specification at https://docs.oracle.com/javase/specs.

Join Java User Groups

I happen to share leadership roles in the New York JavaSIG (www.javasig.com) and the Garden State Java User Group (www.gsjug.org). Both groups have regular online meetings and in-person meetings. Come join us!

Find the Latest News about Java

Bookmark https://foojay.io/, https://dev.java and https://community.oracle.com/community/java.

Find News, Reviews, and Sample Code

For articles by the experts, visit InfoQ at www.infoq.com and TheServerSide at www.theserverside.com. You can always find good reading at these two sites.

Got a Technical Question about Anything?

If you're stuck and need help, search for answers and post questions at Stack Overflow (`http://stackoverflow.com`).

You can also post questions at Coderanch — "A friendly place for programming greenhorns" (`https://coderanch.com`). For questions specific to Java, visit `https://coderanch.com/f/33/java`.

Become Involved in the Future of Java

The *Java Community Process* (JCP) ". . . is the mechanism for developing standard technical specifications for Java technology. Anyone can register . . . and participate in reviewing and providing feedback for the Java Specification Requests (JSRs)." For more info about JCP membership, visit `www.jcp.org/en/participation/membership_ind`.

» How to learn more

» What to do when you're looking for a job

» Other free advice

Chapter 19
Ten Bits of Advice for New Software Developers

I enjoy hearing from the people who read my books. "Nice job!" one reader says. Another reader asks, "Can I run your book's examples on older versions of Java?" Yet another posts this comment: "You're Barry Burd. Does that mean you're related to Larry Bird?"

In all the questions I receive from readers, one popular theme is "What to do next?" More specifically, people ask me what else to learn, what else to read, how to get practice writing software, how to find work, and other questions of that kind. I'm flattered to be asked, but I'm reluctant to think of myself as an authority on such matters. No two people give you the same answers to these questions, and if you ask enough people, you're sure to find disagreement.

This chapter contains ten pieces of advice based on questions I've received from readers. But remember that, in addition to these ten hints for living and learning, I have one additional, overriding piece of advice:

> Think critically about the advice you receive. When in doubt, trust your intuition.

Collect opinions. Talk to people about the issues. Try things and, if they work (or even if they don't work but they show some promise), keep doing them. If they show no promise, try other things. Sharing is important, too. Don't forget to share.

How Long Does It Take to Learn Java?

The answer depends on you — on your goals, on your existing knowledge, on your capacity to think logically, on the amount of spare time you have, and on your interest in the subject.

The more excited you are about computer programming, the quicker you learn. The more ambitious your goals, the longer it takes to achieve them.

REMEMBER

There's no such thing as "knowing all about Java." No matter how much you know, you always have more to learn. I've written several Java books and, as far as I'm concerned, I've barely scratched the surface.

Which of Your Books Should I Read?

Funny you should ask! I've written several books, including these three (all from Wiley):

- *Beginning Programming with Java For Dummies*
- *Java For Dummies*
- *Java Programming for Android Developers For Dummies*

Each book starts from scratch, so you don't need to know anything about app development to read any of these books. But each book covers (roughly) twice as much material as the previous book in the list. For example, *Java For Dummies* goes twice as fast and covers twice as much material as *Beginning Programming with Java For Dummies*. Which book you read depends on your level of comfort with technical subjects. If you're in doubt about where to start, find some sample pages from any of these books to help you determine which book is best for you.

Are Books Other than Yours Good for Learning Java and Android Development?

Yes. I'd love to recommend some, but I'm not conscientious enough to carefully read and review other peoples' books.

Which Computer Programming Language(s) Should I Learn?

The answer depends on your goals. If you plan to write code for a living, the answer depends on the job opportunities where you live. The TIOBE Programming Community Index (`www.tiobe.com/index.php/content/paperinfo/tpci`) provides monthly ratings for popular programming languages. But the TIOBE Index might not apply specifically to your situation. In June 2021, the Haskell language ranked only 47th among the languages used around the world. But maybe there's a hotbed of Haskell programming in the town where you live.

Do you want to write applications for large enterprises? Then Java is a must-have language. Do you want to write code for the iPhone? You probably want to learn Swift. Do you want to create web pages? Learn HTML, CSS, and JavaScript.

Which Skills Other than Computer Coding Should I Learn?

Sorry to disappoint you, but you're asking someone who has an axe to grind. I'm a college professor. I believe that no learning, no matter how impractical it might seem to be, is ever wasted.

If you insist on a more definitive answer, go learn a little about databases. Database work isn't necessarily coding, but it's important stuff. Also, read as much as you can about *software engineering* — the study of techniques for the effective design and maintenance of computer code. Don't be afraid of math, either (because learning math stretches your logical-thinking muscles). And, whenever you can, hone your communication skills. The better you communicate, the more valuable your work is to other people.

How Should I Continue My Learning as a Software Developer?

Practice, practice, practice. Take the examples you find in my book (or anywhere else) and think of ways to change the code. Add an option here or a button there.

Find out what happens when you try to improve the code. If it works, think of another way to make a change. If it doesn't work, search the documentation for a solution to your problem. If the documentation doesn't help (and often, documentation doesn't help), search the web for answers to your problem. Post questions at an online forum. If you don't find an answer, put aside the problem for a while and let it incubate in your mind.

REMEMBER

You don't learn programming by only reading about it — you have to scrape some knuckles while writing code and seeking solutions. Only after trying, failing, and trying again can you appreciate the work involved in developing computer software.

How Else Should I Continue My Learning as a Developer?

How did you know that I have a second suggestion? I recommend finding like-minded people where you live and getting together with them regularly. These days, you can find tech user groups in almost every corner of the globe. Find a Java user group that meets in your area and attend the group's meetings frequently. If you're a novice, you might not understand much of the discussion, but you'll be exposed to the issues that concern today's Java developers.

Look for more tech groups and attend their meetings. Find meetings about other programming languages, other technologies, and other topics that aren't solely about technology. Meet people face-to-face and find out which topics will be in next year's books.

To complement those face-to-face meetings, search the web for screen-to-screen meetings. You can find free online technical sessions almost any day of the year.

How Can I Land a Job Developing Software?

Do all the things you'd normally do when you look for a job, but don't forget my answer to the previous question. User groups are fantastic places for networking.

Go to meetings and be a good listener. Don't think about selling yourself. Be patient and enjoy the ride. I landed a great consulting opportunity only after several years of attending one group's meetings. In the meantime, I learned a lot about software (and quite a bit about dealing with other people).

I Still Don't Know What to Do with My Life

That's not a question. But it's okay anyway.

Everyone has to make ends meet. If you manage to put food on your table, the next step is to find out what you love to do. I've spent a lifetime teaching college students, writing books, and developing computer code — and I love doing all of it. (Well, I love most of it. I detest grading papers, and I dislike proofreading my own work.)

Fortunately, I can make money teaching, writing, and developing. I could make more money working 9-to-5 for a big company, managing a software team, or creating the next big start-up, but I don't like doing those things. My life has been enriched because I do what I like doing, whether I'm working or not.

My advice is, find the best match of the things you like to do and the things that help you earn a living. Compromise, if you must, but be honest with yourself about the things that make you happy. (Of course, these things shouldn't make other people unhappy.)

Finally, be specific about your likes and dislikes. For example, saying, "I'd like to be rich" isn't specific at all. Saying, "I'd like to create a great game" is more specific, but you can do better. Saying, "I like to design game software, but I need a partner who can do the marketing for me" is quite specific, and makes quite a tidy set of goals.

If I Have Other Questions, How Can I Contact You?

Send email to `javafordummies@allmycode.com`. Follow me on Facebook (`/allmycode`) or Twitter (`@allmycode`). Visit my ugly-but-informative website: `www.allmycode.com`. Attach two tin cans to a very long string. Put a note in an old pneumatic tube. Train a carrier pigeon to fly to my office. Hire a chimpanzee to . . .

Index

Symbols

A

B

C

D

E

F

G

H

I

J

K

L

M

N

O

P

Q

R

S

T

U

V

W

About the Author

Barry Burd received a master's degree in computer science at Rutgers University and a PhD in mathematics at the University of Illinois. As a teaching assistant in Champaign–Urbana, Illinois, he was elected five times to the university-wide List of Teachers Ranked as Excellent by Their Students.

Since 1980, Dr. Burd has been a professor in the department of mathematics and computer science at Drew University in Madison, New Jersey. He has spoken at conferences in the United States, Europe, Australia, and Asia, and in 2020, he was honored to be named a Java Champion. He is the author of many articles and books, including *Beginning Programming with Java For Dummies*, *Java Programming For Android Developers For Dummies*, and *Flutter For Dummies*, all from Wiley.

Dr. Burd lives in Madison, New Jersey, with his wife of n years, where n > 40. In his spare time, he enjoys being a workaholic.

Dedication

For

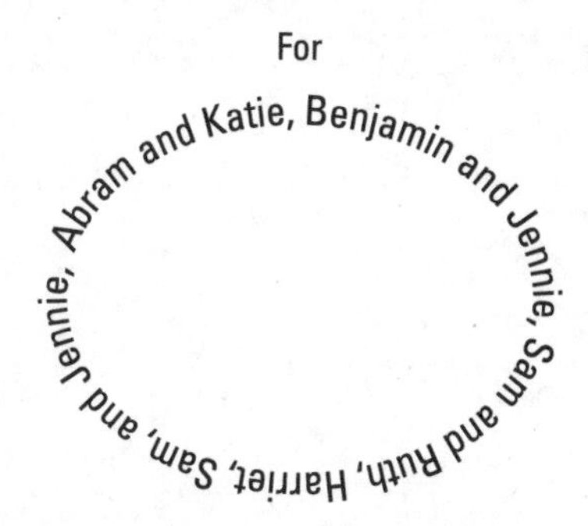
Abram and Katie, Benjamin and Jennie, Sam and Ruth, Harriet, Sam, and Jennie,

Author's Acknowledgments

I heartily and sincerely thank Paul Levesque, for his work on so many of my books in this series.

Thanks also to Kelsey Baird, for her hard work and support in so many ways.

Thanks to Chad Darby and Becky Whitney, for their efforts in editing this book.

Thanks to the staff at Wiley Publishing for helping to bring this book to bookshelves.

Thanks to Frank Greco and the leaders of the New York JavaSIG: Jeanne Boyarsky, Sai Sharan Donthi, Rodrigo Graciano, Chandra Guntur, Vinay Karle, Justin Lee, Lily Luo, and Neha Sardana. Thanks to Michael Redlich and the leaders of the "Garden State Java User Group" Chandra (again), Caitlin Mahoney, Scott Selikoff, Neha (again), and Paul Syers. Thanks to my colleagues, the faculty members in the mathematics and computer science department at Drew University: Sarah Abramowitz, Chris Apelian, Seth Harris, Emily Hill, Steve Kass, Diane Liporace, Yi Lu, Ziyuan Meng, Ellie Small, and even that maniac Steve Surace. Finally, a special thanks to Richard Bonacci, Cameron McKenzie, Scott Stoll, and Gaisi Takeuti for their long-term help and support.

Publisher's Acknowledgments

Acquisitions Editor: Kelsey Baird

Senior Project Editor: Paul Levesque

Copy Editor: Becky Whitney

Tech Editor: Chad Darby

Production Editor: Saikarthick Kumarasamy

Cover Image: © kowalskichal/Shutterstock